Foundation Engineering

By Karl Terzaghi and Ralph B. Peck

Soil Mechanics in Engineering Practice

By Ralph B. Peck, Walter E. Hanson, and Thomas H. Thornburn

Foundation Engineering

FRONTISPIECE

Karl Terzaghi (1883-)

Founder and guiding spirit of soil mechanics, outstanding engineering geologist, and pre-eminent foundation engineer. He was the first to make a comprehensive investigation of the engineering properties of soils; he created or adapted most of the theoretical concepts needed for understanding and predicting the behavior of masses of soil; and he devised the principal techniques for applying scientific methods to the design and construction of foundations and earth structures.

Foundation Engineering

RALPH B. PECK
Research Professor
of Soil Mechanics
University of Illinois

WALTER E. HANSON
Engineer of Bridge and
Traffic Structures
Illinois Division of Highways

THOMAS H. THORNBURN
Research Associate Professor
of Civil Engineering
University of Illinois

New York · JOHN WILEY & SONS, Inc.
London · CHAPMAN & HALL, Limited

sixth printing, february, 1959

Library of Congress Catalog Card Number: 52–13138

Printed in the United States of America

Foreword

In a broad sense, foundation engineering is the art of selecting, designing, and constructing the elements that transfer the weight of a structure to the underlying soil or rock. In practice, however, the actual construction is usually not carried out by the organization responsible for the design; the role of the engineer is generally considered to consist only of the selection of the type of foundation, the design of the substructure, and the supervision of construction. It is the purpose of this book to provide the basic information necessary to carry out this role.

The art of foundation engineering had its origins in antiquity. It developed with the accumulation of experience but without the aid of science until, in about 1920, it had reached a considerable degree of refinement. Yet, occasional inexplicable failures indicated that the limitations of the empirical procedures were not properly understood.

In the early 1920's there began a concerted scientific effort to determine the physical laws responsible for the behavior of the subsurface materials from which foundations derive their support. The new field of endeavor, known as soil mechanics, attracted and still holds the attention of many workers. It has provided new techniques for selecting the appropriate types of foundation under a given set of conditions and for predicting the performance of the completed substructure. It has by no means decreased the importance of the accumulated experience of the ages, but it has defined the limits within which the traditional techniques are applicable and has provided new techniques suitable under the circumstances in which the traditional procedures are not valid.

In recent years the power of science has become increasingly apparent, and there has been a tendency to discount the importance of the vast store of knowledge acquired during past generations by trial and error. This attitude has been reflected in many engineering schools by the replacement of courses in foundation engineering by others in soil mechanics, and by the prevalent opinion that detailed training in soil mechanics must precede and may even eliminate the need for training in foundation engineering.

In reality, soil mechanics is only one of the bodies of knowledge upon which the foundation engineer may draw. If studied to the exclusion of the other aspects of the art, it leads to the erroneous and dangerous impression that all problems in foundation engineering are susceptible of direct scientific solution. Unfortunately, the vagaries of nature and the demands of economy combine to eliminate this possibility.

The purpose of this book is to introduce the undergraduate student to the field of foundation engineering and to provide him with the ability to investigate and evaluate subsurface conditions, select the most suitable types of foundation for a given site, judge the performance of each type in service, and design the structural elements of the type finally selected. Where soil mechanics contributes to these purposes it is introduced, but it is nowhere presented merely for its own sake. Experience in the classroom has indicated that this approach develops in the student an excellent comprehension of the successful practice of foundation engineering.

Because much of the information concerning the characteristics of soils is common to the fields of foundation, highway, railway, and airport engineering, the curricula of several engineering schools contain a basic course in the properties of soils. To meet the needs of such a course, the first part of this book has been expanded beyond the limits adequate for a discussion of the foundations of structures. Appropriate sections for omission by students concerned only with the latter subject will be obvious to the experienced instructor.

Ralph B. Peck
Walter E. Hanson
Thomas H. Thornburn

December 1952

Acknowledgments

The authors are grateful for many helpful comments by several people who have read all or parts of early versions of the manuscript, and especially to Professor E. J. Daily and Mr. H. O. Ireland, whose criticisms were based upon use of the material for several years in the classroom.

Much information has been drawn from *Soil Mechanics in Engineering Practice,* by K. Terzaghi and the senior author. The pen portraits are the work of Gene Wilder. The successive manuscripts were capably and painstakingly typed by Mrs. Ruth B. Howard.

The authors also acknowledge with thanks the use of the original copies of illustrative matter for the following plates at the beginning of the various chapters: 5 and 9, Raymond Concrete Pile Company; 10, Wm. H. Mueser; 11, DeLeuw, Cather and Company; 12, Robert E. White; 14, Louisiana Highway Commission; 16, U. S. Bureau of Reclamation; 17 and 18, Dr. Leonardo Zeevaert; 19, Mrs. Warren F. Goodell; 20, Roberts and Schaefer Company; 21, Holabird and Root and Burgee.

Contents

Page

PART C. SELECTION OF FOUNDATION TYPE AND BASIS FOR DESIGN

Symbols

The symbols in this text conform generally to those suggested in 1941 by the American Society of Civil Engineers (*Soil Mechanics Nomenclature,* Manual of Engineering Practice No. 22), although numerous exceptions have been made to avoid confusion, especially in relation to well-established symbols in structural design.

The dimensions of the quantities in the following list are those most commonly used in the text, and are also generally used in practice. The foundation engineer deals, on one hand, with soil mechanics, in which both metric and English units are common. On the other hand, he deals with several branches of structural engineering, in which various English units are used. This situation leads to an assortment of dimensions that is undesirable from an academic point of view. Yet, because of the prevailing habits in engineering practice, any attempt to reduce the variety of units would not be realistic.

Conversion from one system to another is facilitated by the following relation

$$1 \text{ ton/sq ft} = 2 \text{ kips/sq ft} \approx 1 \text{ kg/sq cm} \approx 34 \text{ ft of water} = 15 \text{ lb/sq in}$$

in which the ton is the short ton of 2000 lb. Other useful conversion factors are 1 lb = 454 g, and 1 ft = 30.5 cm.

If no dimension is added to a symbol, the symbol indicates a pure number.

A (cm^2 or ft^2) = area
A_r(%) = area ratio of sampling spoon
A_s (in.^{2}) = area of tensile steel (concrete design)
a (in.) = width of column (concrete design)
B (ft) = width of footing or foundation
B' (ft) = adjusted width of rectangular footing, for use in curves of Fig. 19.2
b (in.) = width of beam (concrete design)
C (any dimension) = constant; coefficient
C_c = compression index
C.g. = center of gravity (design plates)
c (tons/ft^2) = cohesion
c (ft) = distance from neutral axis to extreme fiber (concrete design)
c (in.) = constant in Engineering News pile-driving formula

c_v (cm²/sec) = coefficient of consolidation
D (ft) = depth
D_e (in.) = external diameter of sampling spoon
D_f (ft) = depth of foundation (Fig. 14.1)
D_i (in.) = internal diameter of sampling spoon
D_P (mm) = diameter of grain corresponding to percentage P on grain-size curve
D_r = relative density of cohesionless soil
D_{10} (mm) = effective grain size (expressed in centimeters in equations)
DL (tons) = dead load (design plates)
d (ft) = distance; distance from given pile to center of gravity of pile group; depth from top of slab or beam to center of tension reinforcement (concrete design)
E (tons/ft²) = modulus of elasticity
E_i (tons/ft²) = initial tangent modulus
e = void ratio
e_{max} = void ratio in loosest state
e_{min} = void ratio in densest state
Δe = change in void ratio
e (ft) = eccentricity
F.S. = factor of safety (design plates)
f (lb/in²) = maximum fiber stress
f_c (lb/in.²) = allowable compressive stress in concrete
f'_c(lb/in.²) = ultimate compressive strength of concrete
f_s (lb/in.²) = allowable stress in reinforcing steel
g (cm/sec²) = acceleration of gravity
H (lb) = total horizontal load
ΣH (lb) = resultant horizontal load
H (ft or cm) = distance; thickness of stratum or fill, except when used in connection with theory of consolidation. In this event, H = half-thickness of layer drained top and bottom.
H (ft) = height of fall of ram of pile driver
H_c (ft) = maximum height of fill not associated with creep of subsoil
ΔH (ft or cm) = position head (hydraulics)
h (ft or cm) = hydraulic head; height
h_c (cm) = height of capillary rise
h_{cc} (cm) = height of complete saturation by capillary rise
h_s (ft) = height of uniform surcharge equivalent to wheel loads behind abutment
I (ft⁴) = moment of inertia
I_l = liquidity index
I_w(%) = plasticity index
i = hydraulic gradient
i_c = critical hydraulic gradient
i_p (g/cm³ or lb/ft³) = pressure gradient (hydraulics)
j = ratio of moment arm of resisting couple to depth d (concrete design)

k = coefficient; ratio of depth of neutral axis to depth d (concrete design); ratio of least width of column or pedestal to width of footing
k (cm/sec) = coefficient of permeability
k_h (lb/ft²/ft) = coefficient of horizontal component of earth pressure
k_p (lb/ft²/ft) = coefficient of passive earth pressure
k_v (lb/ft²/ft) = coefficient of vertical component of earth pressure
L (ft or cm) = length; length of footing
L_w(%) = liquid limit
LL (tons) = live load (design plates)
l (ft or cm) = length
M (ft–lb) = total moment
ΣM (ft–lb) = resultant total moment
m (in./ft) = batter of pile (horizontal/vertical)
m_v (cm²/g) = coefficient of volume compressibility
N = number of blows per foot in standard penetration test
N' = uncorrected number of blows per foot in standard penetration test
N_q = bearing capacity factor
N_γ = bearing capacity factor
N_ϕ = $\tan^2 (45° + \phi/2)$
n = porosity; number of piles in cluster; ratio of modulus of elasticity of steel to that of concrete
n_1 = number of piles in row
n_b = number of bars in central width equal to short dimension of footing
n_t = total number of bars required by moment at critical section of footing
P (%) = per cent of grains smaller than given size
P (lb) = total load; pile reaction
P (ft) = perimeter of group of piles
P_A (lb/ft) = total active earth pressure per lineal foot of wall
P_h (lb/ft) = horizontal component of total active earth pressure per lineal foot of wall
P_P (lb/ft) = total passive earth pressure per lineal foot of wall
P_{hs} (lb/ft) = horizontal component of earth pressure per lineal foot of wall due to surcharge on backfill of abutment
P_v (lb/ft) = vertical component of total active earth pressure per lineal foot of wall
P_w(%) = plastic limit
p (tons/ft²) = normal stress; total pressure
$\bar{p}$ (tons/ft²) = intergranular pressure
Δp (tons/ft²) = change in pressure
p_c (tons/ft²) = confining pressure; all-around pressure
p_o (tons/ft²) = original overburden pressure
p_0' (tons/ft²) = maximum pressure under which soil has previously been consolidated
p_1 (tons/ft²) = major principal stress
p_3 (tons/ft²) = minor principal stress
Q (cm³) = total discharge (hydraulics)

Q_a (lb) = safe load on pile
q (cm^3/sec) = discharge per unit of time (hydraulics)
q (tons/ft^2) = load per unit of area; soil pressure beneath footing
q_a (tons/ft^2) = allowable soil pressure
q_d (tons/ft^2) = net ultimate bearing capacity
q_d' (tons/ft^2) = maximum intensity of loading that can be supported by the soil at the base of the footing
q_n (lb/ft^2) = net soil pressure for structural design of footing (column load divided by area of footing)
q_t (tons/ft^2) = toe pressure
q_u (tons/ft^2) = unconfined compressive strength
R (lb) = resultant force
R (lb/in.^{2}) = design constant M/bd^2 (concrete design)
r (cm or ft) = radius
S = ratio of long to short side of footing
S_r(%) = degree of saturation
S_t = sensitivity
S_w(%) = shrinkage limit
S (ft) = settlement
s (ft) = spacing of piles in row
s (in.) = penetration of pile under last blow of hammer; spacing of stirrups (concrete design)
s (tons/ft^2) = shearing resistance per unit of area
s_s = specific gravity of solid constituents
T_s (g/cm) = surface tension of water
T_v = time factor (theory of consolidation)
t (sec) = time
t (tons/ft^2) = shearing stress
U(%) = degree of consolidation
U = uniformity coefficient = D_{60}/D_{10}
u (lb/in.^{2}) = bond stress per unit of area of surface of bar (concrete design)
u_w (tons/ft^2) = porewater pressure
V(cm^3 or ft^3) = total volume
V (lb) = total vertical load
ΣV (lb) = resultant total vertical load
V (lb/ft) = total shear on section of beam or retaining wall
V_g (cm^3 or ft^3) = total volume of gas
V_s (cm^3 or ft^3) = total volume of solids
V_v (cm^3 or ft^3) = total volume of voids
V_w (cm^3 or ft^3) = total volume of water
v (cm/sec or ft/sec) = discharge velocity (hydraulics)
v (lb/in.^{2}) = allowable stress in diagonal tension (concrete design)
W (g or lb) = total weight
W_c (lb/ft) = weight of concrete in retaining wall
W_H (lb) = weight of ram of pile driver
W_s (g or lb) = total weight of solids
W_s (lb/ft) = weight of soil above heel of retaining wall
W_w (g or lb) = total weight of water
W.T. = water table (design plates)

$w(\%)$ = water content in per cent of dry weight
x (ft) = distance
$\bar{x}$ (ft) = distance from reference line to center of gravity
z (cm or ft) = depth
α (deg) = angle; contact angle
β (deg) = slope angle of surface of backfill or retaining wall
γ (g/cm^3 or lb/ft^3) = unit weight (soil, water, and air)
γ' (g/cm^3 or lb/ft^3) = submerged unit weight
γ_d (g/cm^3 or lb/ft^3) = unit weight of soil if water is entirely replaced by air
γ_s (g/cm^3 or lb/ft^3) = unit weight of solid constituents
γ_{sat} (g/cm^3 or lb/ft^3) = unit weight of saturated soil
γ_w (g/cm^3 or lb/ft^3) = unit weight of water
ϵ = strain
μ = Poisson's ratio
Σ = summation
Σ_0 (in.^{2}/in.) = sum of perimeters of reinforcing bars in one set (concrete design)
ϕ (deg) = angle of internal friction; angle of shearing resistance
ϕ_a (deg) = angle of shearing resistance for partly saturated clay
ϕ_{cq}(deg) = consolidated-quick angle of shearing resistance

Part A

PROPERTIES OF SUBSURFACE MATERIALS

The behavior of every foundation depends primarily on the engineering characteristics of the underlying deposits of soil or rock. Therefore, the foundation engineer must be able to distinguish among the various deposits of different character, to identify their principal constituents, and to determine their physical properties. The purpose of Part A is to provide the information necessary to accomplish these objectives.

Arthur Casagrande (1902–   )

Professor of Soil Mechanics and Foundation Engineering, Harvard University. He has been responsible for many advances in soil mechanics, particularly in the development of procedures and apparatus for testing soils, and in the techniques for identifying and classifying soil materials. In 1936 he organized the First International Conference on Soil Mechanics and Foundation Engineering. Through the stimulus of this conference, through his extensive consulting practice, and through his outstanding ability as a teacher, he has exerted a powerful influence for the acceptance of soil mechanics in foundation engineering.

PLATE 1.

CHAPTER 1

Identification of Soils and Rocks

1.1. Significant Characteristics of Soil and Rock

Definitions. All foundations constructed by the civil engineer ultimately derive their support from the underlying soil or rock. Hence, soil and rock constitute the principal structural materials with which the foundation engineer deals, and a knowledge of the origin, manner of occurrence, and characteristics of these materials is essential for the successful practice of foundation design and construction.

The terms rock and soil, as used by the civil engineer, imply a clear distinction between two kinds of foundation materials. *Rock* is considered to be a natural aggregate of mineral grains connected by strong and permanent cohesive forces. *Soil,* on the other hand, is regarded as a natural aggregate of mineral grains, with or without organic constituents, that can be separated by gentle mechanical means such as agitation in water. These definitions are convenient and are generally understood, and they are used in this book. Nevertheless, in reality there is no sharp distinction between rock and soil. Even the strongest and most rigid rocks may be weakened by the processes of weathering, and some intact rocks are as weak and compressible as soils. Since the engineer must deal with whatever foundation materials he encounters, he must be prepared to recognize and determine the significant properties of all materials, whether they be sound unaltered rock, weathered partly decomposed rock, or soil.

Significant Properties of Foundation Materials. For different foundation materials the properties of significance to the foundation engineer are likely to be different. Many sound rocks, for example, are at least as strong and as rigid as the concrete that would be used in the substructure of a bridge or building. Hence, the intrinsic properties of the rocks themselves are of little importance and receive little attention in this book. Yet formations of sound rock almost always contain zones or surfaces of weakness, such as joints or bedding planes. These discontinuities represent structural defects that may influence the supporting capacity of the rock. It is, therefore, of great importance that the engineer observe the character, spacing, and general pattern of these defects wherever they are found.

Rocks that have become altered by the processes of physical and chemical weathering require particular attention. The weathering is likely to be most extensive adjacent to joints and other discontinuities; therefore, the pattern of such discontinuities remains one of the most significant characteristics of the rock formation. In addition, the strength and compressibility of the altered rock also require investigation because many weathered rocks, especially in humid or tropical regions, are locally much weaker and more compressible than concrete.

If the subsurface materials are soils, the intrinsic properties of the materials themselves become of outstanding importance although discontinuities and variations in the soil mass still deserve careful consideration. The strength and compressibility of the soils usually control the design of the foundation and therefore may require extensive investigation. Hence, the properties of soils deserve and receive particular attention in this book.

Purpose of Identification and Classification. In all branches of civil engineering and especially in foundation engineering, experience is a priceless asset. Indeed, the accumulated experience of generations of foundation engineers including those of our own times constitutes the essence of the art. Yet, unless the individual experiences of all engineers can be summarized into a body of knowledge that can be assimilated readily, they lose their value to the engineering profession.

In a general way, it has been found that rocks and soils can be classified into groups within each of which the significant engineering properties are somewhat similar. Consequently, proper classification of subsurface materials is an important step in connection with any foundation job, because it provides the first clues to the experiences that may be anticipated during and after construction. The effort required to learn how to identify and classify rocks and soils properly is, therefore, a good investment.

It should be recognized that the identification and classification of the products of nature constitute an artificial procedure, because the products of nature are infinitely varied and do not lend themselves to separation into distinct categories. As a result, various arbitrary systems of classification have been developed, each with certain advantages and disadvantages for a particular purpose. Furthermore, as attempts are made to refine any one system, the system inevitably becomes more complicated and ultimately becomes so cumbersome that it defeats its own purpose.

To avoid this difficulty, it is preferable to make use of relatively simple systems of classification with small numbers of categories, to one of which a given rock or soil can be assigned. More detailed information concerning the given rock or soil can best be summarized by stating the numerical results, known as *index properties*, of certain physical tests known as *classification tests*. If the classification tests are properly chosen, soils or rock materials having similar index properties are likely to exhibit similar engineering behavior.

The usefulness of index properties is so great that they will be discussed in detail in the following pages. In addition to their value in the correlation of construction experience, they provide a means for checking the correctness of the field identification of a given material. If the material has been improperly identified, the index properties will indicate the error and will lead to correct classification. Thus, even the beginner with no experience of his own can be assured that his classifications are appropriate, and he can then take advantage of the experiences of other engineers. The techniques for acquiring this knowledge are described in Arts. 1.2 to 1.9, inclusive.

1.2. Description and Field Identification of Rocks

Classification of Rocks. The rock materials encountered at the surface of the earth or beneath the surface soil are commonly classified into three groups according to their mode of origin. *Igneous* rocks are considered to be the primary rocks formed by the cooling of molten magmas, or by the recrystallization of older rocks under heat and pressure great enough to render them fluid. *Sedimentary* rocks are the products of deposition of plant and animal remains, and of materials formed by the chemical decomposition and physical disintegration of igneous, sedimentary, or metamorphic rocks. *Metamorphic* rocks are those produced by internal processes acting on pre-existing rocks of any kind, with the limitation that the rocks concerned remained essentially solid during their transformations.

It is usually possible to assign a given rock to one of these three groups after visual inspection, provided one is familiar with the typical characteristics of each. The most important properties utilized in identification are texture, structure, and mineralogical composition.

The *texture* of a rock refers to the size, shape, and manner of aggregation of the fragments, particles, or crystals of which the rock is composed. The *structure* refers to the gross features usually seen in large rock masses such as systems of cracks and joints, or planes of stratification. Since a relatively small number of rock minerals makes up a very large proportion of the common rocks, it requires but a little practice to learn to distinguish and classify most rocks.

Igneous Rocks. Igneous rocks are classified primarily on the basis of texture and color. Since they were formed by solidification from the liquid state, they are very hard and have textures that vary from coarsely *crystalline* to *glassy* depending upon the rate of cool-

Table 1.1. A Field Classification of the Igneous Rocks

Color	Light		Intermediate	Dark	
Principal Minerals / Texture	Quartz and Feldspar, Other Minerals Minor	Feldspar, Little or No Quartz	Feldspar and Hornblende	Augite and Feldspar	Augite, Hornblende, Olivine
Very Coarse, Irregular, Crystalline	Pegmatite	Syenite pegmatite	Diorite pegmatite	Gabbro pegmatite	
Coarse and Medium Crystalline	Granite	Syenite	Diorite Dolerite	Gabbro Dolerite	Peridotite
Fine Crystalline	Aplite			Diabase	
Dense	Felsite			Basalt	
Glassy	Volcanic glass				
Porous (Gas Openings)	Pumice		Scoria or vesicular basalt		
Fragmental	Tuff (fine), breccia (coarse), cinders (variable)				

Modified after Trefethen.

ing. The principal mineral constituents are light-colored *quartz* and *feldspar* and dark-colored *hornblende, biotite, augite,* and *olivine.* Table 1.1 is a simple guide to field classification of the more common types of igneous rocks.

The structure of igneous rock is usually described as *massive.* This term indicates a lack of structural features, but in reality joints and cracks may be found in all igneous rocks. If the rock masses were deformed at the same time that crystallization was taking place, they may have acquired a banded flow structure commonly called *gneissic.* Most of the igneous materials that cooled very rapidly either contain gas bubbles or else have a fragmental structure.

Sedimentary Rocks. The sedimentary rocks may be divided into three general groups in accordance with the origin of the sediment. Those consisting of rock or mineral fragments derived from preexisting material belong to the *clastic* group; those deposited from organisms living in water are called *organic;* and those precipitated by chemical activity or left by evaporation are *chemical.* The texture of the clastic and organic sediments may vary from coarse-grained to microscopic, but the chemical sediments are usually microscopic. Only in classifying the clastic sediments is grain size of any importance. This is illustrated in Table 1.2, by means of which the common sedimentary rocks can be classified.

The predominant minerals found in most sediments are *quartz, calcite,* and *clay minerals.* Consequently, about 99 per cent of all sedimentary rocks may be classified as sandstone, limestone, or shale. Sandstone and shale can usually be differentiated by visual inspection on the basis of particle size. An estimate of the amount of lime in a given rock can be made by treating the rock with dilute hydrochloric acid and noting the vigor of the effervescence. The most important minor constituents are *iron oxides, feldspar,* and *carbonaceous material.* With only the slight knowledge of mineralogy required to use Table 1.2 most sedimentary rocks can be identified in the field.

The most important structural characteristic of sedimentary rocks from the geological point of view is their *bedding* or *stratification.* A knowledge of the thickness, mineralogical character, and position of the beds is also of great importance to the engineer. Of equal or greater significance from the engineering standpoint may be information regarding structural defects such as joints, cracks, fault zones, and solution channels or cavities.

Metamorphic Rocks. The metamorphic rocks are differentiated from igneous and sedimentary rocks primarily on the basis of structure and mineralogical composition. The agents of metamorphism such as *heat, pressure,* and *hydrothermal solutions* lead to the crystallization and orientation of minerals not found in the parent rock. The most obvious characteristic of metamorphic rocks is the *foliated* structure which refers to the nearly parallel arrangement of *platy* or *needle-like* minerals. Some of the typical metamorphic minerals found in foliated rocks are *chlorite, sericite, talc, hornblende,* and *biotite.* Depending upon the grain size of the crystals the structure

Table 1.2. A Simplified Classification of Common Sedimentary Rocks

Group	Grain Size	Composition		Name
Clastic	Appreciable quantity of grains more than 2 mm diameter	Rounded pebbles in medium-grained matrix.*		Conglomerate
		Angular coarse rock fragments, often quite variable		Breccia
	More than 50% of grains fall in range 0.06–2.00 mm diameter	Medium quartz grains	Less than 10% of other minerals	Siliceous sandstone
			Appreciable quantity of clay minerals	Argillaceous sandstone
			Appreciable quantity of calcite	Calcareous sandstone
			Appreciable amount of iron oxide cement	Ferruginous sandstone
			Over 25% feldspar	Arkose
			25–50% feldspar and darker minerals	Graywacke
	More than 50% of grains fall in range 0.002–0.06 mm diameter	Fine to very fine quartz grains with clay minerals		Siltstone (if laminated, shale)
	Predominantly grains less than 0.002 mm diameter	Microscopic clay minerals	Less than 10% other minerals	Shale
			Appreciable calcite	Calcareous shale
			Appreciable carbonaceous material	Carbonaceous shale
			Appreciable iron oxide cement	Ferruginous shale
Organic	Variable	Calcite and fossils		Fossiliferous limestone
	Medium to microscopic	Calcite and appreciable dolomite		Dolomitic limestone or dolomite
	Variable	Carbonaceous material		Bituminous coal
Chemical	Microscopic	Calcite		Limestone
		Dolomite		Dolomite
		Quartz		Chert, flint, etc.
		Iron compounds with quartz		Iron formation
		Halite		Rock salt
		Gypsum		Rock gypsum

* The matrix of a conglomerate or of a breccia is the material filling the spaces between the larger rock fragments of which the sediment is composed.

may be described as *gneissic* (coarse), *schistose* (medium), or *slaty* (microscopic).

A large group of metamorphic rocks, most of which are derived from sediments, possesses a *massive* rather than a foliated structure. These may be distinguished from the parent sediment by their greater density, hardness, and crystallinity. Most rocks in this group are composed primarily of *calcite, dolomite,* or *quartz.* Table 1.3 is a

Table 1.3. A Classification of Common Metamorphic Rocks

Texture	Structure: Foliated		Structure: Massive
Coarse crystalline	Gneiss		Conglomerite Granite gneiss
Medium crystalline	Schist	Sericite Mica Talc Chlorite Hematite, etc.	Marble Quartzite Serpentine Soapstone
Fine to microscopic crystalline	Phyllite Slate		Hornfels Anthracite coal

highly simplified classification of metamorphic rocks showing the types most frequently encountered. Although metamorphic rocks show such structural defects as cracks and joints, the most important engineering characteristics are the softness of schists and the high susceptibility of all foliated rocks to weathering.

1.3. Description and Field Identification of Soils

Principal Types of Soil. The principal terms used by civil engineers to describe soils are gravel, sand, silt, and clay. Most natural soils consist of a mixture of two or more of these constituents, and many contain an admixture of organic material in a partly or fully decomposed state. The mixture is then given the name of the constituent that appears to have the most influence on its behavior, and the other constituents are indicated by adjectives. Thus a silty clay has most of the properties of a clay but contains a significant amount of silt, and an organic silt is composed primarily of silt-sized mineral matter but contains a significant amount of organic material.

Gravels and sands are known as coarse-grained soils, and silts and clays as fine-grained. The distinction is based on whether the individual particles can be differentiated by the naked eye. The methods of describing coarse-grained soils differ from those appropriate for fine-grained soils; therefore, the procedures are discussed under separate headings.

Coarse-grained Soil Materials. The coarse-grained soil materials are mineral fragments that may be identified primarily on the basis of grain size.

Grains having a diameter greater than about 2.0 mm are classified as *gravel.* However, if the diameter exceeds about 200 mm (8 in.) the term *boulder* is usually applied.

If the grains are visible to the naked eye, but are less than about 2.0 mm in size, the soil is described as a *sand.* This name is usually further modified as *coarse, medium,* or *fine.* The boundaries between these various size fractions must be chosen arbitrarily. In the United States the MIT classification, given in Table 1.4, is commonly used.

Table 1.4. Grain Size of Coarse-Grained Soil Constituents

MIT Classification

Gravel	Larger than 2 mm
Coarse sand	2 mm to 0.6 mm
Medium sand	0.6 mm to 0.2 mm
Fine sand	0.2 mm to 0.06 mm

The description of sand and gravel should include an estimate of the quantity of material in the different size ranges, as well as a statement of the *shape* and *mineralogical composition* of the grains. The particle shape often has an important influence on the density and stability of a deposit. It may be described as *rounded, subrounded, subangular,* or *angular.* The presence of weak rock materials such as shale, or minerals such as mica, may influence the durability or compressibility of the deposit. Fragments of these materials may be identified with the aid of a small magnifying glass.

If all the particles of a coarse-grained material are approximately the same size, the material is called *uniform* or *poorly graded.* If the soil contains a good representation of all particle sizes ranging from coarse to fine, it is designated as *well-graded.* Some natural soil deposits are composed of a mixture of particles of uniformly coarse size and uniformly fine size but are lacking in particles of the intermediate sizes. Such deposits are said to be *gap-graded.*

Fine-grained Soil Materials. Inorganic silt, which constitutes the coarser portion of the microscopic soil fraction, possesses little or no plasticity or cohesion. The least plastic varieties consisting primarily of very fine rounded quartz grains are called *rock flour*. The more plastic varieties containing an appreciable quantity of flake-shaped particles are called *plastic silt*.

Clay is predominantly an aggregate of microscopic and submicroscopic flake-shaped crystalline minerals. It is characterized by the typical colloidal properties of plasticity, cohesion, and the ability to adsorb ions. These properties are exhibited over a wide range of water content.

The distinction between silt and clay cannot be based on grain size because the significant physical properties of the two materials are related only indirectly to the size of the particles. Furthermore, since both are microscopic, physical properties other than grain size must be used as criteria for field identification.

The *dry strength* provides one basis for distinction. A small briquet of the soil is molded and allowed to dry in the air. It is then broken and a fragment about ⅛ inch in size is pressed between thumb and forefinger. A clay fragment can be broken only with great effort, whereas a silt fragment crushes readily.

Since silts are considerably more permeable than clays, the *shaking test* may also be used to distinguish between the two materials. In this test a small amount of soil is mixed with water to a very soft consistency in the palm of the hand. The back of the hand is then lightly tapped. If the material is silt, the water will rise quickly to the surface of the soil and give it a shiny or glistening appearance. If the pat is then deformed either by squeezing or by stretching, the water will flow back into it and leave the surface with a dull appearance. Since clay soils contain much smaller voids than silts and are much less permeable, their appearance does not change during the shaking test. An estimate of the relative proportions of silt and clay in an unknown soil mixture can be made by noting whether the reaction is rapid, slow or nonexistent.

The property of *plasticity* is characteristic of clays and may also be used as the basis for a simple field test. At certain moisture contents a soil that contains appreciable quantities of clay can be deformed and remolded in the hand without disintegration. Thus, if a sample of moist soil can be manipulated between the palms of the hands and fingers and rolled out into a long thin thread, it unquestionably contains a significant amount of clay. As moisture is lost during continued manipulation, the soil approaches a nonplastic condition and becomes crumbly. Just before the crumbly state is reached, a highly plastic clay can be rolled into a thread with a diameter of approximately ⅛ in. which is able to support its own weight when a foot or more long. A silt, on the other hand, can seldom be rolled into a thread with a diameter as small as 1/8 in. without severe cracking, and is completely lacking in tensile strength unless small amounts of clay are present.

A fourth procedure, known as the *dispersion test,* is also useful for distinguishing between silt and clay, and for making a rough estimate of the relative amounts of sand, silt, and clay in a material. A small quantity of the soil is dispersed with water in a glass cylinder or test tube and then allowed to settle. The coarser particles fall out first and the finest particles remain in suspension the longest. Ordinarily, sands settle in 30 to 60 seconds. Material of silt size settles in 15 to 60 minutes whereas that of clay size remains in suspension for at least several hours and usually for several days unless the particles of clay combine in groups or floccules (see Art. 1.5).

Table 1.5 is a summary of the field identification tests that may be used to identify fine-grained soils.

Table 1.5. Summary of Identification Tests for Fine-grained Soils

Test	Silts	Clays
Dry strength	Very low. Even oven-dry strength is low. Powder easily rubs off surface of the samples.	High to very high. Exceptionally high if oven-dry. Powder will not rub off the surface.
Shaking test	Rapid reaction. Typical silts have a livery appearance when shaken.	Sluggish or no reaction. Surface of the sample remains lustrous.
Plasticity	Plastic thread has little strength. Dries quickly. Crumbles easily as it dries below plastic range.	Plastic thread has high to very high strength. Dries slowly. Usually stiff and tough as it dries below plastic range.
Dispersion	Settle in 15 to 60 minutes. (Sands settle in 30 to 60 seconds.)	Settle in several hours or days unless flocculation occurs.

Organic Soil Materials. Very small quantities of organic matter often have a significant influence on the physical properties of soils. Most organic soils are weaker and more compressible than soils having the same mineral composition but lacking in organic matter. The presence of an appreciable quantity of organic material can usually be recognized by the dark-gray to black color and the odor of decaying vegetation which it lends to the soil.

Organic silt is a fine-grained more or less plastic soil containing mineral particles of silt size and finely divided particles of organic matter. Shells and visible fragments of partly decayed vegetative matter may also be present.

Organic clay is a clay soil which owes some of its significant physical properties to the presence of finely divided organic matter.

Highly organic soil deposits such as *peat* or *muck* may be distinguished by a dark-brown to black color, by the presence of fibrous particles of vegetable matter in varying states of decay, and by the characteristic organic odor.

Combinations of organic and mineral soil materials are not always easily recognized, particularly if the organic content is small. Nevertheless, the engineer should always be suspicious of the presence of organic matter if the soil has a dark-brown, dark-gray, or black color. If the organic odor cannot be distinguished, it can sometimes be brought out by a slight amount of heat.

1.4. Index Properties of Soils

In the preceding article, simple methods have been described for identifying the major soil components. These methods, however, represent only the first step in the adequate description of soil materials. They must be supplemented by other procedures leading to quantitative results that may be related to the physical properties with which the engineer is directly concerned. As stated in Art. 1.1, the tests required for this purpose are known as classification tests, and the results as the index properties of the soils.

By performing suitable classification tests and determining the corresponding index properties, the engineer acquires the means for describing a given soil accurately without the use of verbal descriptions that are subject to misunderstanding on account of vague terminology. The development of the ability to think of soils in terms of numerical values of their index properties should be one of the foremost aims of every engineer who deals with foundations.

Index properties may be divided into two general types, soil *grain properties* and soil *aggregate properties*. The soil grain properties are the properties of the individual grains of which the soil is composed, without reference to the manner in which these grains are arranged in a soil deposit. Thus it is possible to determine the grain properties of any soil sample whether disturbed or undisturbed. Soil aggregate properties, on the other hand, depend upon the structure and arrangement of the particles in the soil mass. Although soil grain properties are commonly used for identification purposes, the engineer should realize that the soil aggregate properties have a greater influence on the engineering behavior of a soil.

1.5. Soil Grain Properties

Size and Shape of Grains. The most important grain property of coarse-grained soils is the grain-size distribution. This is determined by performing a *mechanical analysis.* The sizes of coarse-grained constituents can be determined by means of a set of sieves. The finest sieve commonly used in the field or in the laboratory is the no. 200 U.S. Standard sieve in which the width of the openings is 0.074 mm. Since this size is very close to the boundary between silt and sand, the no. 200 sieve is frequently used to separate the coarse-grained material from the fine-grained when only a rough mechanical analysis is desired.

To determine the grain-size distribution of any soil containing fine-grained material, a wet method of mechanical analysis must be used. All wet methods of analysis are based on Stokes's law which expresses the velocity at which a spherical particle falls through a fluid medium as a function of the diameter and specific gravity of the particle. A suspension of the soil is first agitated and then allowed to stand at rest. After a given time has elapsed, all particles coarser than a certain size have settled below some plane at an arbitrary depth in the suspension. This size can be computed by means of Stokes's law. The corresponding density of the suspension at the arbitrary depth is a measure of the quantity of soil smaller than the computed size. By making density measurements at the same depth at various times, the grain-size distribution can be determined.

For engineering purposes, the density is usually measured by means of a hydrometer. Hence, the method, first suggested by G. J. Bouyoucos and later modified by A. Casagrande, is called the *hydrometer method.* A sample of about 50 g of soil is dispersed in one liter of distilled water and poured into a standard sedimentation cylinder. The suspension in the cylinder is then shaken for approximately one minute and the cylinder set upright on a plane horizontal surface. A special hydrometer of streamlined design is inserted in the suspension, and readings of the density are taken after various intervals of time. It is customary to take readings at 30 seconds, 1 minute, 2 minutes, 4 minutes, 8 minutes, 15 minutes, etc. The computations based on these readings lead to the grain-size distribution on the assumption that all the particles are spherical. Actually, the finest soil particles are not spherical, but are known to be flake-shaped or needle-shaped. The grain size that is actually determined, therefore, is the diameter of a sphere that would settle out of suspension at the same rate as the soil particle.

One of the most common sources of error involved in wet mechanical analysis is inadequate dispersion of the fine-grained soil particles. The laboratory technician may believe that he is determining the sizes of the individual grains, whereas in reality he may be measuring the sizes of floccules composed of several grains. To avoid flocculation, a small amount of an electrolyte, known as a *dispersing agent*, is added to the suspension. There is no way to determine by ordinary laboratory tests when ultimate dispersion has been obtained. It is, therefore, necessary to exercise critical judgment and to resort to several different methods of dispersion if there is reason to doubt the validity of the data obtained.

The use of the electron microscope enables the research investigator to determine the actual size and shape of fine-grained soil particles, but such refinement is neither economical nor practicable in the routine classification of soils.

The results of a mechanical analysis are usually presented in the form of a grain-size distribution curve. The percentage P of material finer than a given size is plotted as the ordinate to a natural scale, and the corresponding grain size D_P, in millimeters, as the abscissa to a logarithmic scale. A plot of this type has the advantage that materials of equal uniformity are represented by curves of identical shape whether the soil is coarse-grained or fine-grained. Moreover, the shape of the curve is indicative of the grading. Uniform soils are represented by nearly vertical lines, and well-graded soils by S-shaped curves that extend across several cycles of the logarithmic scale. Figure 1.1 shows grain-size curves for soils of several types.

The grain-size characteristics of soils can be compared most conveniently by a study of certain significant numerical values derived from the distribution curves. The most common procedure used by engineers was first suggested by Allen Hazen on the basis of his studies of filter sands. Hazen found that the permeability of clean filter sands in the loose state could be correlated with numerical values designated as D_{10}, the *effective grain size;* and $U = D_{60}/D_{10}$, the *uniformity coefficient.* The effective size is the diameter of the particle corresponding to $P = 10$ per cent on the grain-size plot. Hence, 10 per cent of the particles are finer and 90 per cent are coarser than the effective size. The values of effective grain size and uniformity coefficient as defined by Hazen have been utilized to express the grain-size characteristics of mixed-grained natural soils. The relationship between the shape of the grain-size curve and the uniformity coefficient is illustrated in Fig. 1.1.

Mineralogical Composition. The most important grain property of fine-grained soil materials is the mineralogical composition. If the soil particles are smaller than about 0.002 mm, the influence of the force of gravity on each particle is insignificant compared with

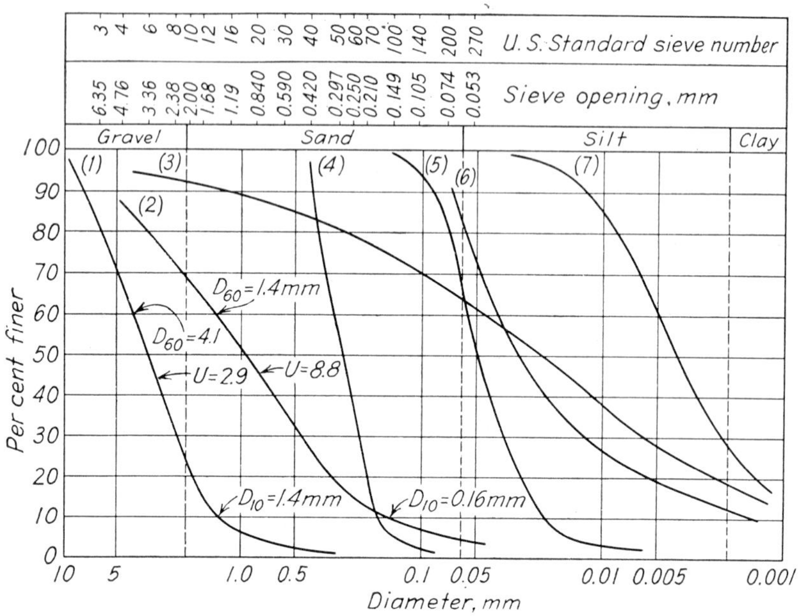

FIG. 1.1. Typical grain-size curves of natural soils. (1) Pea gravel, Castle Rock, Colo. (2) River gravel, Denver, Colo. (3) Glacial till, Peoria, Ill. (4) Sand, Grenada, Miss. (5) Glacial rock flour, Winchester, Mass. (6) Clayey silt, Smead, Mont. (7) Silty clay, Marathon, Canada.

that of the electrical forces acting at the surface of the particle. A material in which the influence of the surface charges is predominant is said to be in the colloidal state. The colloidal particles of soil usually consist of one of a very few kinds of minerals that exist only in the colloidal state. Hence, they are called the *soil colloids* or the *clay minerals.*

The three most important groups of clay minerals are *montmorillonite, illite,* and *kaolinite.* They are all crystalline hydrous aluminosilicates. The results of studies using the electron microscope and X-ray diffraction techniques show that the clay minerals have a lattice structure in which the atoms are arranged in several sheets similar to the pages of a book. The arrangement and the chemical composition of these sheets determine the type of clay mineral. All three types have a flake-shaped structure due to the sheetlike arrangement of the crystal lattice. Depending upon the arrangement of the atoms in the crystal, the individual flakes or particles may separate in the presence of water or may even expand and allow the water molecules to arrange themselves between the sheets of the crystal lattice.

All clay minerals possess the ability to adsorb positive ions from an aqueous solution. When a salt such as NaCl is dissolved in water, it is dissociated into positive ions (cations) of Na^+ and negative ions (anions) of Cl^-. The unbalanced charges that exist on the surface of the clay mineral are mostly negative and thus attract the Na^+ cations. When the cation is attached to the clay mineral, it is said to be *adsorbed*. However, this type of adsorption is not permanent because if a portion of the clay mineral containing adsorbed Na^+ ions is placed in a solution of KCl (potassium chloride) most of the Na^+ ions are replaced by K^+ ions. This process of replacement by the excess ion is called *base exchange*.

Table 1.6. Base-exchange Capacity of Clay Minerals

Mineral	Number of Positive Charges Adsorbed per 100 g
Montmorillonite	$360\text{–}600 \times 10^{20}$
Illite	120–240
Kaolinite	20– 90

The base-exchange capacity of the different types of clay minerals may be measured by washing a sample of each with a solution of a salt not commonly found in nature, such as ammonium chloride (NH_4Cl), and determining the amount of adsorbed NH_4^+ by measuring the difference between the original and final concentration of the washing solution. It is convenient to express the base-exchange capacity in terms of the total number of positive charges adsorbed per 100 g of clay mineral, since this factor is independent of the size and weight of each adsorbed ion or the number of charges associated with it. Table 1.6 shows the approximate range of values obtained for the three principal types of clay minerals. The ability to adsorb water on the surface of the clay mineral also varies in the same order.

Table 1.6 indicates that the montmorillonite minerals are about ten times as active as the kaolinite minerals. Hence, a much smaller amount of montmorillonite than of kaolinite would be required to impart typical colloidal properties to a mixed-grained soil. Few engineering laboratories are equipped to determine the nature of the clay mineral present in a given soil, but many of the index properties of soils such as the consistency, cohesion, and soil-moisture relationships are directly dependent upon the clay-mineral composition. The influence of the mineralogical composition is, therefore, reflected in the index properties.

1.6. Weight-volume Relationships of Soil Aggregate

Definitions. The looseness or denseness of a soil sample may be determined quantitatively in the laboratory. The terms porosity, void ratio, and relative density are commonly used to define the density of the sample. Figure 1.2 is a diagram of a soil sample in a

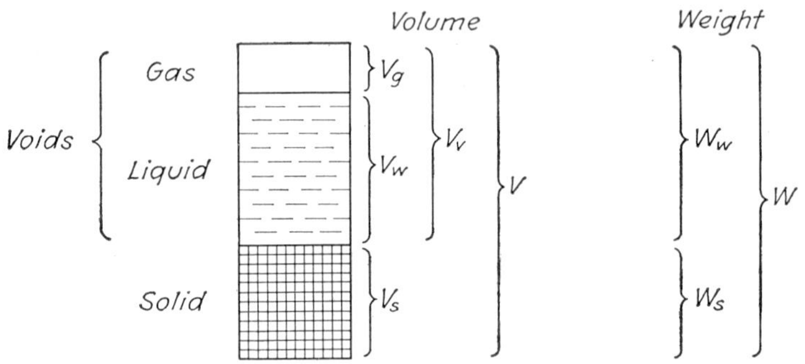

FIG. 1.2. Diagram of a sample of soil illustrating meaning of the symbols used in weight-volume relationships.

sealed container as it would look if the solid, liquid and gaseous phases could be segregated. The volume of solids is designated by the symbol V_s, the volume of the water by V_w and the volume of the gas by V_g. Since the relationship between V_g and V_w usually changes with groundwater conditions as well as under imposed loads, it is convenient to designate all the volume not occupied by solid material as void space, V_v. If the total volume of the sample is designated as V, then the *porosity* is defined by the equation

$$\text{Porosity}, n = V_v/V \tag{1.1}$$

Frequently this value is expressed as a percentage, whereupon the term *percentage of voids* is used. As a soil is compressed, the value of both the numerator and the denominator of the preceding equation changes. It is very convenient in many of the calculations necessary in settlement computations to refer the void space to an unchanging denominator. For this reason the quantity known as *void ratio* is commonly used. It is defined by

$$\text{Void ratio, } e = V_v/V_s \qquad 1.2$$

One of the most important index properties of fine-grained soils is the *natural water content, w*. It is defined by

$$\text{Water content, } w(\%) = 100W_w/W_s \qquad 1.3$$

In the above equation W_w equals the weight of water and W_s equals the weight of oven-dry solid matter. It should be noted that the weight of water is referred to the unchanging quantity W_s rather than to the total weight of the sample. When a mixture of soil and water is slowly dried, the sample continues to lose moisture up to rather high temperatures until the minerals that make up the soil break down and lose their water of constitution. For this reason an arbitrary oven temperature of 105° to 110° C has been chosen for the water-content determination.

Many soils below the water table and some fine-grained soils above it are in a saturated condition. However, the voids of most soils above the water table are filled partly with water and partly with air. Even some submerged soils have a significant air or gas content. The *degree of saturation* is defined as

$$\text{Degree of saturation, } S_r(\%) = 100V_w/V_v \qquad 1.4$$

Thus, the degree of saturation of a completely saturated soil is 100 per cent.

The weight per unit of volume or *unit weight* γ is one of the most important physical properties of a soil. It must be known, for example, before computations of earth pressure or overburden pressure can be made. By definition

$$\text{Unit weight, } \gamma = W/V \qquad 1.5$$

wherein W is the total weight of the soil including the soil moisture, and V the total volume. If the soil is saturated, its unit weight is denoted by γ_{sat}, and, if it is oven-dry, by γ_d. The latter quantity is designated as the *dry density*.

In practice, it is often inconvenient to determine the value of γ directly from measurements of the total weight and total volume. It is more commonly determined indirectly by computation based on a knowledge of the *unit weight of the solid constitutents* γ_s. This quantity is defined as

$$\text{Unit weight of solid constituents, } \gamma_s = W_s/V_s \qquad 1.6$$

It is often preferable to deal with the *specific gravity of the solid constituents* s_s, defined as

$$\text{Specific gravity of solid constituents, } s_s = \gamma_s/\gamma_w \qquad 1.7$$

where γ_w is the unit weight of water, taken as 1 g/cu cm in the metric system or 62.5 lb/cu ft in the English system. The value of γ_s or s_s may be determined by test in the laboratory, but it can usually be estimated with sufficient accuracy. For routine computations, the value of s_s for sands may be taken as 2.65. Tests on a large number of clay soils have indicated that the value of s_s usually falls in the range from 2.5 to 2.9 with an average value of about 2.7.

Table 1.7. Specific Gravity of Most Important Soil Constituents

Gypsum	2.32	Dolomite	2.87
Montmorillonite*	2.4	Aragonite	2.94
Orthoclase	2.56	Biotite	3.0–3.1
Kaolinite	2.6	Augite	3.2–3.4
Illite*	2.6	Hornblende	3.2–3.5
Chlorite	2.6–3.0	Limonite	3.8
Quartz	2.66	Hematite, hydrous	4.3±
Talc	2.7	Magnetite	5.17
Calcite	2.72	Hematite	5.2
Muscovite	2.8–2.9		

From E. S. Larsen and H. Berman, *The Microscopic Determination of the Non-Opaque Minerals,* second edition, U. S. Department of the Interior, Bull. 848, Washington, 1934.

* Theoretical values computed on the basis of the atomic weights of the constituents of the space lattice (according to R. E. Grim).

Table 1.7 gives the specific gravity of the most important soil constituents. It may be of assistance in estimating the value of s_s for a soil of known mineral composition.

Typical values of porosity, void ratio, and unit weight of various soils are listed in Table 1.8.

Density of Soil Aggregate. The void ratio or porosity of a coarse-grained soil usually does not in itself furnish a direct indication of the behavior of the soil under load or during excavation, because at the same void ratio one soil may be in a dense state whereas another may be loose. Much more significant is the *relative density,* which is defined by the equation

$$\text{Relative density, } D_r = \frac{e_{\max} - e}{e_{\max} - e_{\min}} \qquad 1.8$$

in which $e_{\max}$ is the void ratio of the soil in its loosest stable state, e is the actual void ratio and $e_{\min}$ is the void ratio in the densest possible state. Thus, $D_r = 1.0$ for a very dense soil and 0 for a very loose soil.

In practice, the relative density is usually determined indirectly by penetration or load tests (Arts. 3.4 and 3.5), because direct measurement of the void ratio of a soil in the field is not convenient. However, if e is known, the values of e_{max} and e_{min} can be determined in

Table 1.8. Porosity, Void Ratio, and Unit Weight of Typical Soils in Natural State

Description	Porosity, n	Void Ratio, e	Water Content, w*	Unit Weight g/cu cm γ_d	γ_{sat}	lb/cu ft γ_d	γ_{sat}
1. Uniform sand, loose	0.46	0.85	32	1.43	1.89	90	118
2. Uniform sand, dense	0.34	0.51	19	1.75	2.09	109	130
3. Mixed-grained sand, loose	0.40	0.67	25	1.59	1.99	99	124
4. Mixed-grained sand, dense	0.30	0.43	16	1.86	2.16	116	135
5. Glacial till, very mixed-grained	0.20	0.25	9	2.12	2.32	132	145
6. Soft glacial clay	0.55	1.2	45	...	1.77	...	110
7. Stiff glacial clay	0.37	0.6	22	...	2.07	...	129
8. Soft slightly organic clay	0.66	1.9	70	...	1.58	...	98
9. Soft very organic clay	0.75	3.0	110	...	1.43	...	89
10. Soft bentonite	0.84	5.2	194	...	1.27	..	80

*w = water content when saturated, in per cent of dry weight
γ_d = unit weight in dry state
γ_{sat} = unit weight in saturated state

the laboratory. The loosest state for a granular material can usually be created by allowing the dust-dry material to fall into a container from a funnel held in such a way that the free fall is about ½ in. If the material is silty, a looser state can sometimes be established by permitting the soil to settle through water. The densest state can usually be achieved by packing the soil into a container by means of a combination of static pressure and vibration. Several procedures should be used in order that the most effective one will be discovered.

For a soil containing appreciable amounts of silt or clay, the relative density loses its significance, because the values of e_{max} and e_{min}

have no definite meaning. Yet, many construction operations deal with such materials. Moreover, the beneficial effects of compacting such soils have been demonstrated by long experience. The need for a method of defining the degree of compaction led in 1934 to the development by R. R. Proctor of the *compaction test*. The apparatus is illustrated in Fig. 1.3. It consists of a metal cylinder having a volume of $\frac{1}{30}$ cu ft and an internal diameter of 4.0 in., together with a metal tamper having a weight of 5.5 lb and a circular face of 2-in.

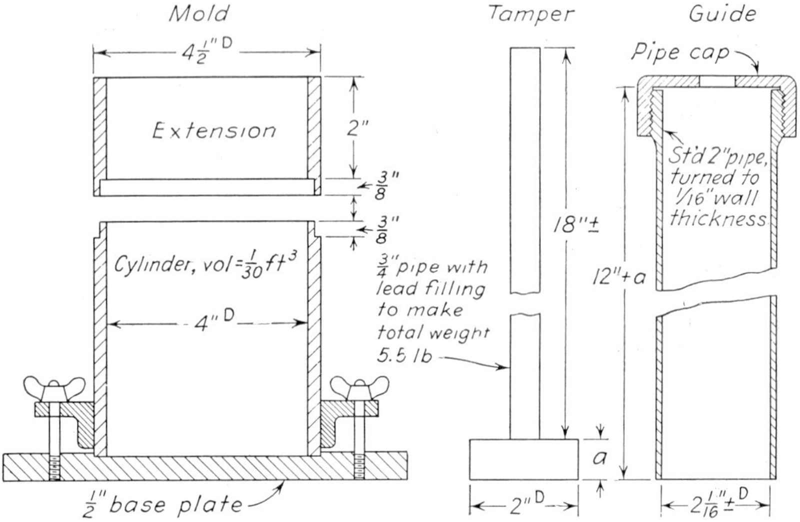

FIG. 1.3. Apparatus used in the standard Proctor compaction test.

diameter. The soil is placed in the cylinder in three approximately equal layers. Each layer is compacted by 25 blows of the tamper falling freely through a distance of 12 in. After compaction, the soil is struck off level with the top of the cylinder, and the weight of the soil in the container is determined. A sample is then extracted from the middle of the compacted cylinder of soil for a moisture-content determination.

From the weight and volume of the soil in the container, the unit weight γ of the soil is computed. The measure of compaction, however, is the *dry density* γ_d, the weight per cubic foot of the solid soil constituents in the container. The values of γ_d are determined for a series of samples of the soil, each of which has a different initial water content. Ordinarily the first determinations are made with the soil in a fairly dry state, and in successive determinations water is added until the weight of moist soil that can be packed into the mold has reached a maximum and starts to decrease. The results of the test are represented by a moisture-density curve, Fig. 1.4, in which γ_d for each determination is plotted against the corresponding value of w.

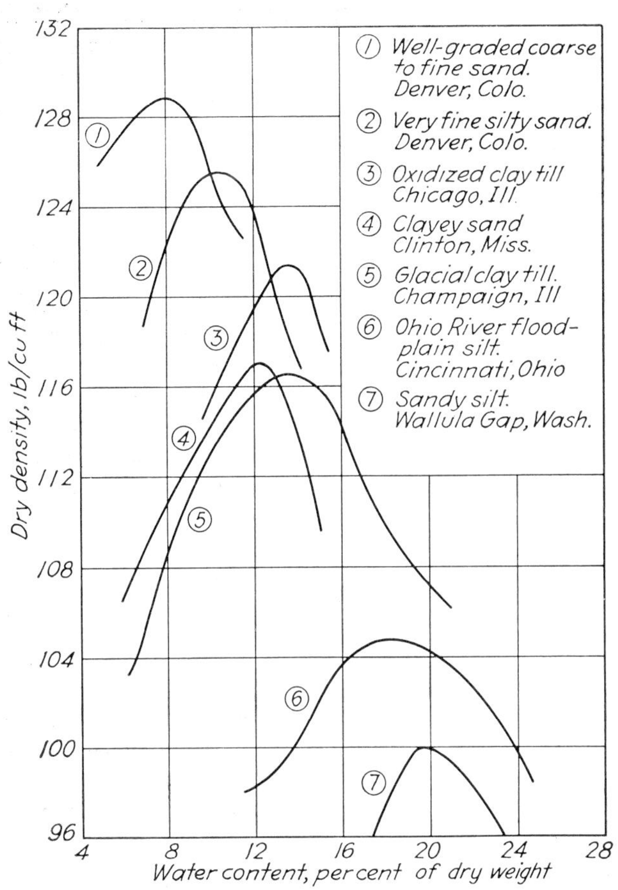

FIG. 1.4. Moisture-density relationships for various types of soils as determined by standard Proctor compaction tests.

The ordinate of the peak of the curve is the *maximum dry density* or *100 per cent compaction*, and the abscissa is the *optimum moisture content*.

The degree of compaction of the soil in the field is determined by measuring the weight of a sample of the soil, the volume occupied by the sample, and the water content. From these data the dry density in the field can be computed. The ratio between this value and that of γ_d at 100 per cent compaction, multiplied by 100, is defined as the *per cent compaction.* Specifications for placing fill in the field usually require the contractor to obtain from 90 to 100 per cent compaction.

The details of the compaction test were selected so that the value of dry density at 100 per cent compaction in the test was comparable with the highest dry density that could be achieved economically by means of the rolling equipment available in 1934. Since that time, more efficient methods of compaction have been devised, and the test may be regarded merely as an arbitrary laboratory procedure. To approximate present field conditions more closely, various changes have been suggested in the details of the compaction test. For example, many engineering organizations now use the Modified AASHO compaction test. This test is identical with the standard Proctor test, except that the sample is compacted in 5 layers, the hammer weighs 10 lb, and the height of fall is 18 in.

It is emphasized that, unlike the definition of relative density, the definition of the per cent compaction is arbitrary in that it depends on the details of the test procedure. Moreover, there is no definite relation between the per cent compaction and the strength of the compacted soil either at or subsequent to the time of compaction. Compaction tests are of value primarily in the control of placing fill so as to assure a uniform degree of density.

Relationships among Soil Aggregate Properties. The various aggregate properties discussed in this article are interrelated, and it is possible to derive algebraic expressions for computing each of the quantities in terms of one or more of the others. However, most of these expressions are not easily committed to memory. Hence, it is preferable and often more expeditious to carry out any necessary computations with the aid of the diagram, Fig. 1.2, and the defining equations for the various quantities. The simplicity of this procedure will become apparent upon study of the illustrative problems that follow.

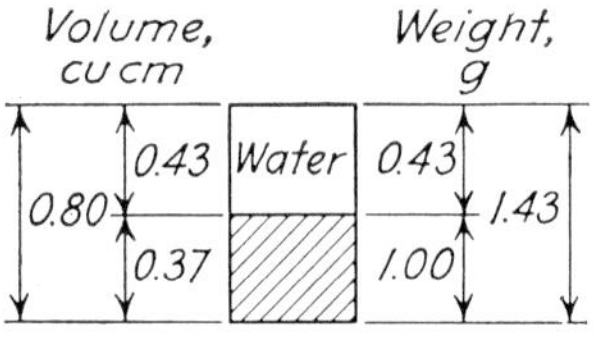

FIG. P1.1.

ILLUSTRATIVE PROBLEMS

1. A sample of soft saturated clay has a natural water content of 43 per cent. The specific gravity of the solid matter is 2.70. What are the void ratio, the porosity, and the saturated unit weight of the soil?

Solution. Inasmuch as the soil is completely saturated, the sketch Fig. P1.1 represents the weight–volume relationships. Since the water content is known, the quantities on the right-hand side of the diagram may be set down immediately, on the assumption that $W_s = 1.0\ g$.

The volume of solids is then computed and recorded on the left side. It is equal to $1.00/2.70 = 0.37$ cu cm. Since water has a specific gravity of unity, the volume of water may be recorded as 0.43 cu cm. Hence, the total volume may be taken as $0.43 + 0.37 = 0.80$ cu cm. Then, by definition,

$$e = V_v/V_s = 0.43/0.37 = 1.16$$

$$n = V_v/V = 0.43/0.80 = 0.54$$

$$\gamma_{\text{sat}} = W/V = 1.43/0.80 = 1.79 \text{ g/cu cm}$$

or

$$\gamma_{\text{sat}} = 1.79 \times 62.5 = 112 \text{ lb/cu ft}$$

2. A sample of sand above the water table was found to have a natural moisture content of 15 per cent and a unit weight of 120 lb/cu ft. Laboratory tests on a dried sample indicated values of $e_{\min} = 0.50$ and $e_{\max} = 0.85$ for the densest and loosest states, respectively. Compute the degree of saturation and the relative density. Assume $s_s = 2.65$.

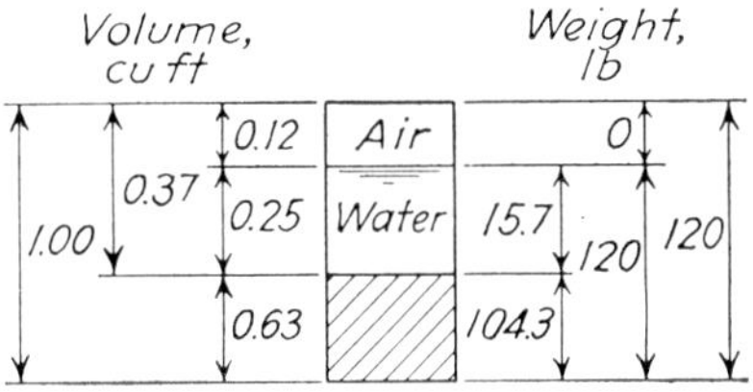

FIG. P1.2.

Solution. In this problem the total volume may conveniently be taken as 1 cu ft, whence the total weight is 120 lb. This is also the weight of soil plus water. These values may be written in the sketch, Fig. P1.2.

Since the water content is 15 per cent, we may write, by definition,

$$w = W_w/W_s = 0.15$$

or

$$W_w = 0.15W_s$$

Moreover, since $W_w + W_s = 120$ lb, we may write

$$0.15W_s + W_s = 120$$

whence

$$W_s = 120/1.15 = 104.3 \text{ lb}$$

and

$$W_w = 120 - 104.3 = 15.7 \text{ lb}$$

These values may now be inscribed on the right side of the diagram.

The volume of solids is

$$\frac{104.3}{2.65 \times 62.5} = 0.63 \text{ cu ft}$$

and the volume of water is

$$15.7/62.5 = 0.25 \text{ cu ft}$$

By difference, the volume of air is 1.0 — (0.63 + 0.25) = 0.12 ft. The volume of voids is, therefore, 0.37 cu ft.

By definition, the degree of saturation is

$$S_r = 100 \frac{V_w}{V_v} = 100 \times \frac{0.25}{0.37} = 68 \text{ per cent}$$

By definition, the void ratio is

$$e = V_v/V_s = 0.37/0.63 = 0.59$$

and the relative density is

$$D_r = \frac{e_{\max} - e}{e_{\max} - e_{\min}} = \frac{0.85 - 0.59}{0.85 - 0.50} = \frac{0.26}{0.35} = 0.74$$

3. A sample of soil compacted according to the standard Proctor test has a unit weight of 131.1 lb/cu ft at 100 per cent compaction and at the optimum moisture content of 14.0 per cent. What is the dry density? If the voids became filled with water, what would be the saturated unit weight? Assume that the solid matter has a specific gravity of 2.67.

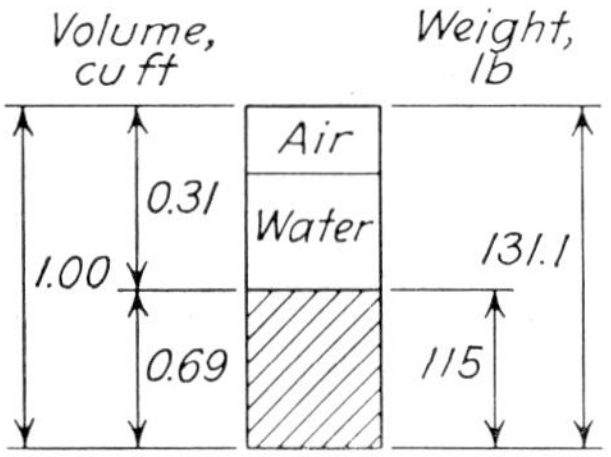

FIG. P1.3.

Solution. The conditions may be represented by the sketch, Fig. P1.3, drawn on the assumption that the total volume is 1 cu ft.

Since the water content is 14.0 per cent, we may write, by definition,

$$W_w = 0.140 W_s$$

and, since $W_w + W_s = 131.1$ lb

$$0.140 W_s + W_s = 131.1$$

$$W_s = 131.1/1.140 = 115.0 \text{ lb}$$

Then, by definition

$$\gamma_d = W_s/V = 115/1 = 115 \text{ lb/cu ft}$$

The volume of the solids is

$$\frac{115}{2.67 \times 62.5} = 0.69 \text{ cu ft}$$

The volume of voids is then 1.00 — 0.69 = 0.31 cu ft. If the voids were filled with water, the weight of the water would be

$$0.31 \times 62.5 = 19.4 \text{ lb}$$

and the saturated unit weight would be

$$\gamma_{sat} = \frac{115 + 19.4}{1} = 134.4 \text{ lb/cu ft}$$

1.7. Structure and Consistency of Soil Aggregate

Primary and Secondary Structure. The *primary structure* of a soil refers to the arrangement of the grains. This arrangement is usually developed during the process of sedimentation or rock weathering. In addition, various discontinuities may arise subsequent to the deposition or formation of the soil. These constitute the *secondary structure* of the deposit. They correspond to such phenomena as the development of systems of joints in sedimentary rocks.

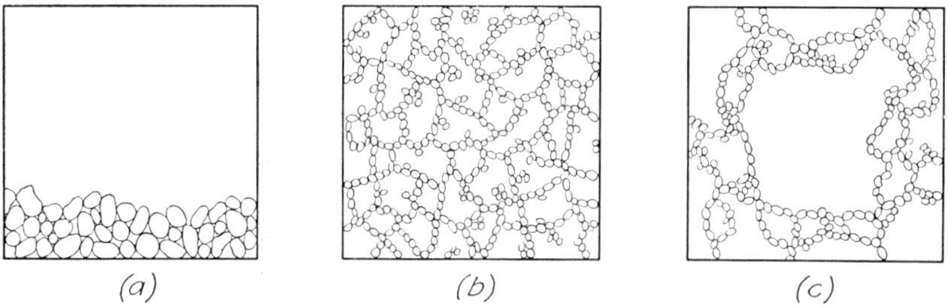

FIG. 1.5. Diagram illustrating (*a*) a dense single-grained structure, (*b*) a honeycombed structure and (*c*) a flocculent structure.

The primary structure of a soil may be described as *single-grained, honeycombed,* or *flocculent.* In a single-grained structure, Fig. 1.5*a*, each grain touches several of its neighbors in such a way that the aggregate is stable even if there are no forces of adhesion at the points of contact between the grains. The arrangement may be dense or loose, and the properties of the aggregate are greatly influenced by the denseness or looseness. A honeycombed structure consists of grains that touch each other at only a few points, Fig. 1.5*b*. The configuration would be unstable if there were no adhesion at the points of contact. In the flocculent structure, Fig. 1.5*c*, groups of grains joined into honeycombed floccules are themselves arranged in a honeycombed structure. The flocculent structure cannot be maintained without the existence of strong forces of cohesion between the particles, and it can be broken readily into a honeycombed structure. Although it is probably rare in nature, no other structure seems conceivable for stiff clays that contain as much as 90 per cent of water by volume; such clays exist, for example, beneath Mexico City.

The principal types of secondary structure are *cracks, joints, slickensides,* and *concretions.* Cracks and joints are commonly formed as a result of desiccation sometime after the deposition of the material. Slickensides are polished surfaces encountered in stiff clays that have experienced differential movement or expansion. Concretions are accumulations of carbonates or iron compounds. All these features disrupt the continuity of the soil mass and may impart to it properties significantly different from those of intact samples taken from the deposit.

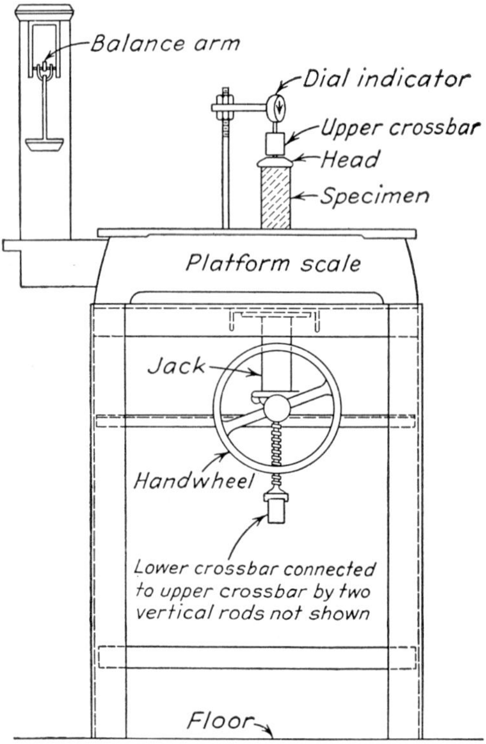

FIG. 1.6. One type of apparatus commonly used to determine the unconfined compressive strength of samples of soil.

Consistency and Sensitivity of Clays. Undoubtedly the most important index property of fine-grained soils in the natural state is the *consistency.* The consistency of natural cohesive soil deposits is expressed qualitatively by such terms as *soft, medium, stiff,* and *hard.* The exact meaning of these terms, however, varies widely in different parts of the country, depending upon whether the local soils are generally hard or generally soft. Rather than rely on such vague terms, the engineer should develop his ability to estimate the compressive strength of the soil.

Quantitatively, the consistency of an undisturbed soil may be expressed by the *unconfined compressive strength,* q_u. In the unconfined compressive strength test a prism or cylinder of soil having a height of 1½ to 2 times the average diameter is loaded to failure, in simple compression, quickly enough that the water content of the soil does not change. The equipment illustrated in Fig. 1.6 is commonly used for the test. Simultaneous observations are made of the load and of the vertical shortening. The failure load or, if the sample does not fail outright, the load required to produce 20 per cent strain is expressed as the load per unit of cross-sectional area, usually in tons per square foot. Table 1.9 indicates the relationship

Table 1.9. Qualitative and Quantitative Expressions for Consistency of Clays

Consistency	Field Identification	Unconfined Compressive Strength q_u (tons/sq ft)
Very soft	Easily penetrated several inches by fist	Less than 0.25
Soft	Easily penetrated several inches by thumb	0.25–0.5
Medium	Can be penetrated several inches by thumb with moderate effort	0.5 –1.0
Stiff	Readily indented by thumb but penetrated only with great effort	1.0 –2.0
Very stiff	Readily indented by thumbnail	2.0 –4.0
Hard	Indented with difficulty by thumbnail	Over 4.0

between the qualitative terms describing consistency and the quantitative values of the unconfined compressive strength. If equipment for making compression tests is not available, a rough estimate can be based on the simple field tests suggested in the table.

When a sample of undisturbed clay is kneaded in the hands at unaltered water content, it usually becomes softer. This characteristic may be investigated by remolding without loss of water a sample of soil previously tested in unconfined compression, packing the remolded material into a cylindrical mold, extruding the remolded sample, and testing it in unconfined compression. The ratio

$$S_t = \frac{\text{unconfined compressive strength, undisturbed}}{\text{unconfined compressive strength, remolded}} \qquad 1.9$$

is known as the *sensitivity* of the clay.

The sensitivity of many clays, including most that have a glacial origin, is not greater than 2 to 4, but some clays are known with sensitivities of 8 or more. Natural soils having values of sensitivity greater than 4 are known as *sensitive* clays, and, if the sensitivity exceeds 8, as *extrasensitive* clays. These should be treated with care during construction operations because disturbance tends to transform them, at least temporarily, into viscous liquids. Some clays with such secondary structural characteristics as cracks, joints, or slickensides may have sensitivities less than 1.

If a sample of remolded clay with a sensitivity greater than 1 is allowed to stand without further disturbance and without change in water content, it may regain at least part of its original strength and stiffness. This increase in strength is due to the gradual reorientation of the adsorbed molecules of water, and is known as *thixotropy.*

Atterberg Limits. If the water content of a thick suspension of clay is gradually reduced, the clay–water mixture passes from a liquid state through a plastic state and finally into a solid state.

It has been found that the water contents corresponding to the transitions from one state to another are usually different for clays having different physical properties in the remolded state, and are approximately equal for clays having similar physical properties. Therefore, the limiting water contents may serve as index properties useful in the classification of clays.

The significance of the limiting water contents for each physical state was first suggested by A. Atterberg in 1911. Hence, these limits are commonly known as the *Atterberg limits,* and the tests required to determine them are the *Atterberg-limit tests.* Actually, as the soil-water mixture passes from one state to another, there is no abrupt change in the physical properties. The limit tests, therefore, are arbitrary tests that have been adopted to define the limiting values.

Above the *liquid limit,* L_w, the soil–water system is a suspension. Below the liquid limit and above the *plastic limit,* P_w, the soil–water system is in a plastic state. In this state the soil may be deformed or remolded without the formation of cracks and without change in volume. The range of water content over which the soil–water system acts as a plastic material is frequently referred to as the *plastic range,* and the numerical difference between the liquid limit and the plastic limit is called the *plasticity index,* I_w:

$$\text{Plasticity index, } I_w = L_w - P_w \qquad 1.10$$

Somewhat below the plastic limit the soil–water system reaches the *shrinkage limit.* Further reduction of the water content by drying is not accompanied by decrease in volume; instead, air enters the voids of the system and the material becomes unsaturated.

The liquid limit and the plasticity index together constitute a measure of the plasticity of a soil. Soils possessing large values of L_w and I_w are said to be highly plastic or *fat*. Those with low values are described as slightly plastic or *lean*. The interpretation of liquid and plastic limit tests is greatly facilitated by the use of the *plasticity chart* developed by A. Casagrande. In this chart, Fig. 1.7*a*, the ordinates represent values of the plasticity index, and the abscissas values of the liquid limit. The chart is divided into six regions by the inclined line A having the equation $I_w = 0.73(L_w - 20)$, and the two vertical lines $L_w = 30$ and $L_w = 50$. All soils represented by points above line A are inorganic clays; the plasticity ranges from low ($L_w < 30$) to high ($L_w > 50$) with increasing values of the liquid limit. Soils represented by points below line A may be inorganic silts, organic silts, or organic clays. If they are inorganic, they are said to be of low, medium, or high compressibility, depending on whether the liquid limit is below 30, between 30 and 50, or above 50. If they are organic silts, they are represented by points in the region corresponding to a liquid limit between 30 and 50, and, if organic clays, to a liquid limit greater than 50.

The distinction between organic and inorganic soils can be made by performing two liquid-limit tests on the same material, one starting with moist or air-dried soil, and the other with oven-dried soil. Oven-drying produces irreversible changes in organic constituents that significantly lower the liquid limit. If the liquid limit of the oven-dried sample is less than about 0.7 times that for the air-dried sample, the soil may be classed as organic.

It has been found that points representing the values of L_w and I_w for soils having a common geological origin in a restricted locality usually define a straight line more or less parallel to the A line. The position of the line is characteristic for the given soils. The results of tests on a number of varieties of soils are shown in Fig. 1.7*b*. The same figure also indicates the influence of oven-drying organic soils.

The moisture contents of many natural clay deposits lie within the plastic range. Those deposits having water contents close to the liquid limit are usually much softer than those with moisture contents close to the plastic limit. One of the most important index properties of natural clay deposits is, therefore, the *liquidity index*, defined by the equation:

$$\text{Liquidity index, } I_l = \frac{w - P_w}{L_w - P_w} = \frac{w - P_w}{I_w} \qquad 1.11$$

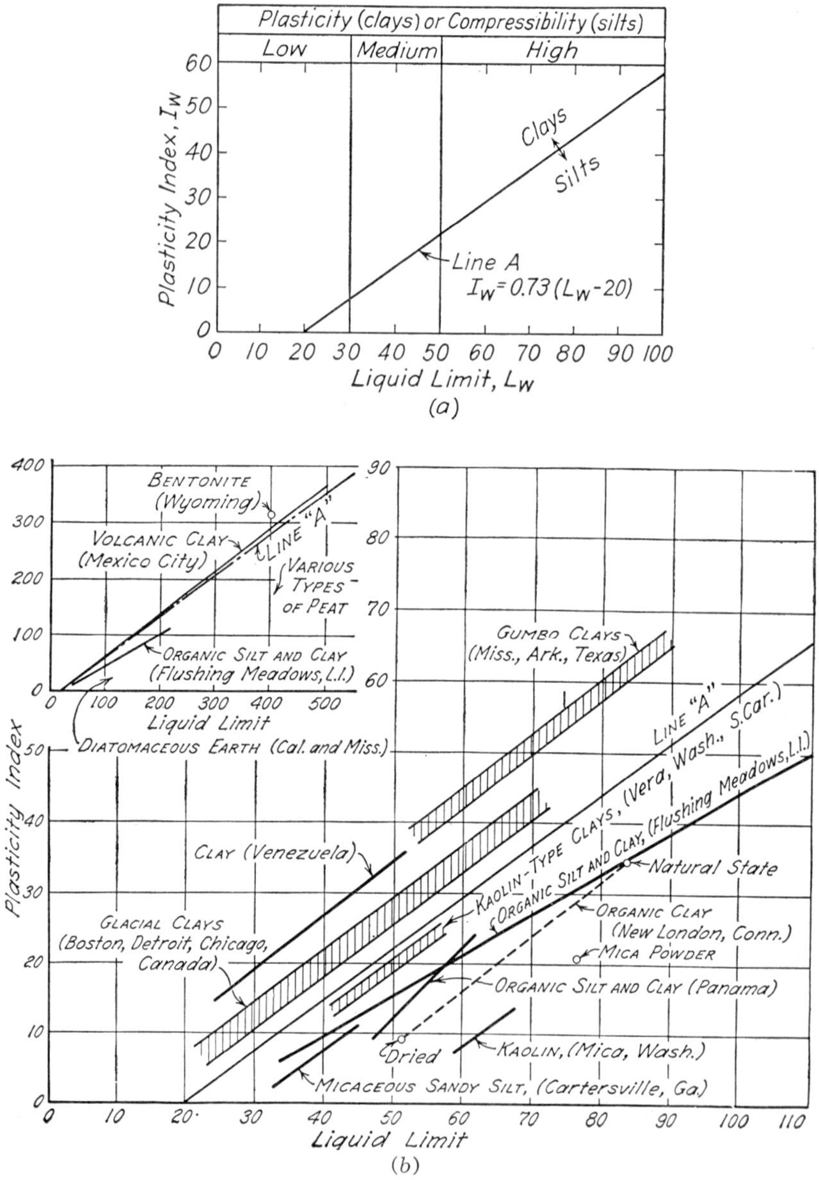

FIG. 1.7. (*a*) Plasticity chart; (*b*) relation between liquid limit and plasticity index for typical soils (after A. Casagrande).

It may be seen that I_l is negative for soils having water contents less than the plastic limit. As the water content increases from the plastic limit to the liquid limit, the value of I_l increases from 0 to 1.0. If the water content is greater than the liquid limit, the liquidity index is greater than 1.0. The consistency of a clay, at least in the remolded state, may be estimated when the natural water content and limit values are known. The relationships are illustrated in Fig. 1.8.

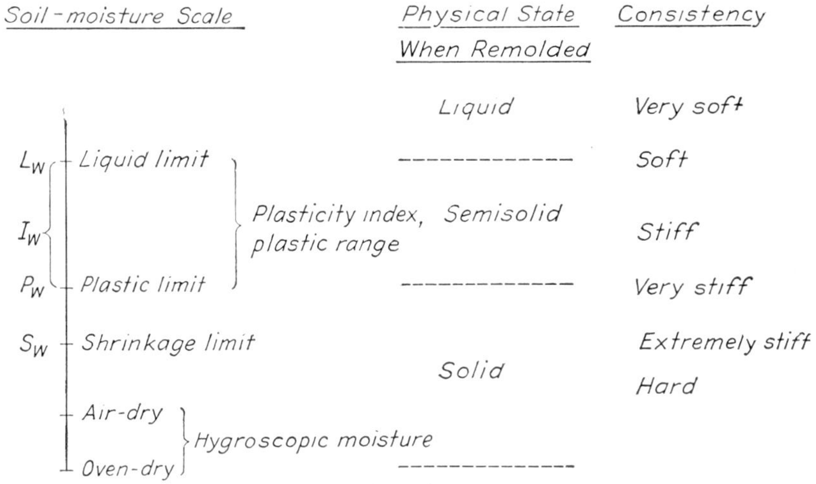

FIG. 1.8. Diagram of the soil-moisture scale showing Atterberg limits, corresponding physical state of remolded soil, and approximate consistency of the materials.

None of the Atterberg-limit tests is difficult to perform, although a certain amount of experience is required to develop the technique necessary to obtain reproducible results. The liquid-limit test is commonly made by means of the mechanical apparatus designed by A. Casagrande, Fig. 1.9. A mixture of soil and water is placed in the cup, and a groove 2 mm wide and 8 mm high is made in the center of the soil pat. The operator then turns the crank which lifts the cup to a height of 1 cm above the base and allows it to fall freely from this position. The soil is at the liquid limit if 25 blows are required to cause the lower edges of the groove to come into contact with each other for a length of about ½ in. The water content at this number of blows is the liquid limit.

The plastic-limit test is performed by rolling a sample of plastic soil into a thread with a diameter of ⅛ in. If the soil does not crumble, the thread is picked up, remolded, and rolled out again.

This process is repeated until the thread just begins to crumble when it reaches the diameter of 1/8 in. The water content at which crumbling takes place is defined as the plastic limit.

The shrinkage limit of a soil is determined by preparing a sample of known volume at a moisture content above the liquid limit and by drying the sample in an oven. The weight and volume of the oven-dry sample are measured. From these data and the initial water

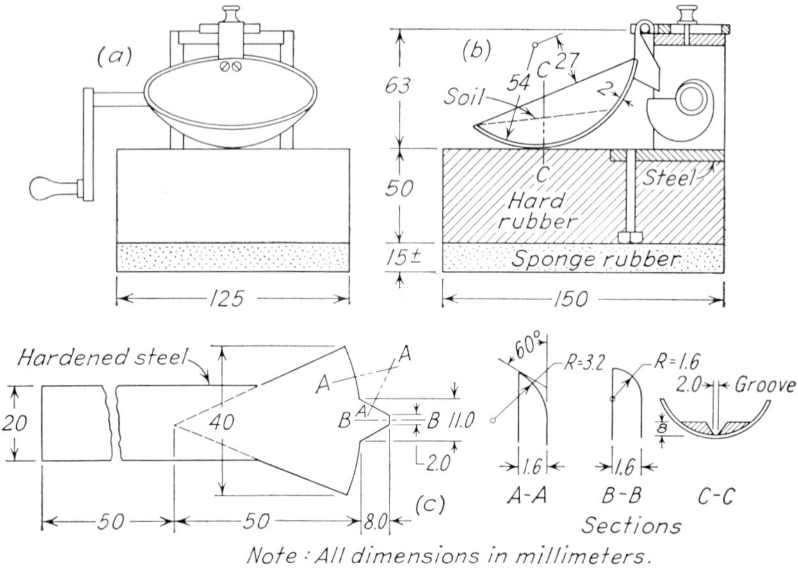

FIG. 1.9. Mechanical liquid-limit device and grooving tool.

content a computation is made of the water content at which the dried sample would be just saturated. This water content is considered to be the shrinkage limit.

1.8 Soil-classification Systems

Introduction. Because the soil deposits of the world are infinitely varied, it has not been found possible to create a universal system of soil classification for dividing soils into various groups and subgroups on the basis of their important index properties. However, useful systems based on one or two index properties have been devised. Some of these systems are in such common use by workers in various fields involving soils that the engineer must have at least a general knowledge of them. At the same time it is essential to keep in mind that no system can adequately describe any soil for all engineering purposes. Indeed, many systems ignore the properties that are the most important from the standpoint of the foundation engineer.

Textural Systems. Since the grain size is probably the most obvious characteristic of a soil, it is natural that the earliest classification systems should have been based on texture alone. Indeed, many such systems have been suggested. Several of the more common are shown in Fig. 1.10. The MIT system is preferred for engineering purposes because of its simplicity and the repetition of the numerical values chosen as the boundaries of the various ranges of size.

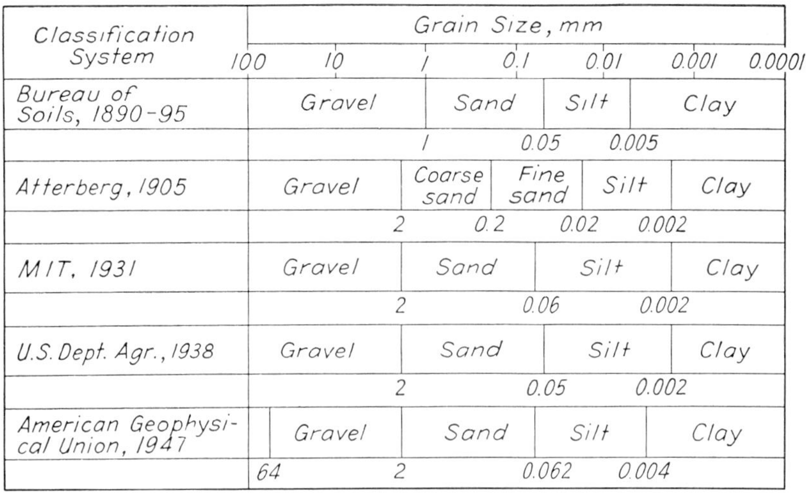

FIG. 1.10. Comparison of several common textural classification systems.

In the use of the MIT system, the grain-size distribution of the soil is first determined. Then the weight of the soil grains within each size group is computed. Finally, the percentage of the soil represented by each size group is obtained. Thus, a mixed-grained soil might be described as "3 per cent gravel, 46 per cent sand, 17 per cent silt and 34 per cent clay, according to the MIT classification."

The textural method of classification was further developed by the U. S. Bureau of Soils and Chemistry by assigning to the soil a single name to denote the texture rather than specifying the percentage of gravel, sand, silt, and clay. A triangular classification chart, Fig. 1.11, was devised by the Bureau to facilitate use of the method. After the percentages of sand-, silt-, and clay-size materials in a given sample of soil have been determined, the point on the triangular chart representing this mechanical composition is located, and the soil is given the name representative of the area in which it is located on the chart. Although the triangular chart does not reveal any properties of the soil other than grain-size distribution, it is still widely used in various modified forms by workers in the fields of agriculture and highway engineering. One major disadvantage is that the textural name as derived from the chart does not always correctly express the physical characteristics of the soil.

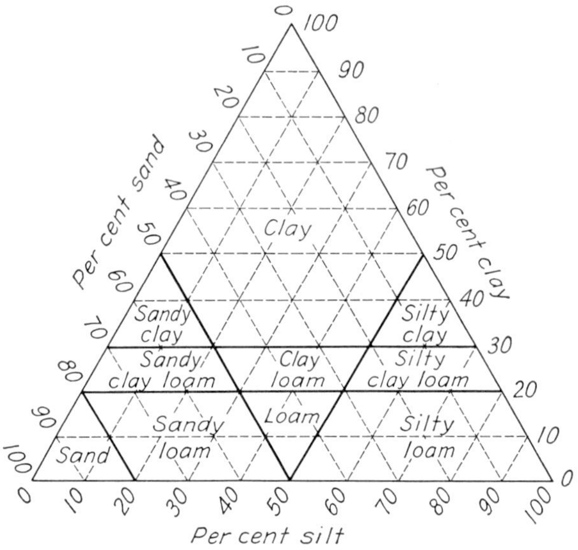

FIG. 1.11. Triangular classification chart developed by the U. S. Bureau of Soils and Chemistry.

For example, since some clay-size particles are much less active than others, a soil described as clay on the basis of this system may have physical properties more typical of silt.

Public Roads System. A soil-classification system widely used by highway engineers was originally proposed by the U. S. Bureau of Public Roads about 1928. All soils were divided into eight groups designated by the symbols *A*-1 through *A*-8. It was believed that the soil best suited for the subgrade of a highway was well-graded material composed largely of sand and gravel, but containing a small amount of excellent clay binder. A material of this type was given the designation *A*-1, and all other soils were grouped roughly in decreasing order of stability. The Public Roads system has undergone many revisions since it was originated. In the beginning, neither the percentages of the various size fractions nor the plasticity characteristics of the clay fraction were definitely specified.

In 1945 an extensive revision of the Public Roads system was made by a committee of highway engineers for the Highway Research Board. The characteristics of the various groups are defined in Table 1.10. In this modified Public Roads system the inorganic soils are classified in 7 groups corresponding to A-1 through A-7. These in turn are divided into a total of 12 subgroups. In addition, it is recommended that any soil containing appreciable fine-grained material be more completely identified by the determination of the *group index.*

The group index may be determined from the equation

$$\text{Group index} = 0.2a + 0.005ac + 0.01bd \qquad 1.12$$

in which

a = that portion of percentage passing no. 200 sieve greater than 35 and not exceeding 75, expressed as a positive whole number (0 to 40)
b = that portion of percentage passing no. 200 sieve greater than 15 per cent and not exceeding 55 per cent, expressed as a positive whole number (0 to 40)
c = that portion of the liquid limit greater than 40 and not exceeding 60, expressed as a positive whole number (0 to 20)
d = that portion of the plasticity index greater than 10 and not exceeding 30, expressed as a positive whole number (0 to 20)

If the pertinent index value for a soil falls below the minimum limit associated with a, b, c, or d, the value of the corresponding term is zero, and the term drops out of the group index equation. The group index is appended to the major group classification. For example a clay soil having a group index of 18 might be classified as A-7-6 (18).

Although the Highway Research Board revision has been widely accepted by highway engineers, some are still using the revision of February 1942 as a basis for classification. This and earlier versions of the Public Roads system included the group A-8, which is composed primarily of peat and muck soils. These soils contain an excess quantity of organic matter and possess very high water contents. They exhibit little stability under any conditions and are rarely suitable as foundation materials.

Airfield Classification System. One of the most complete soil-classification systems in use today was developed by Arthur Casagrande

Table 1.10. Public Roads Classification
1945 Highway Research Board Modification

General Classification	Granular Materials (35 per cent or Less Passing No. 200)							Silt-Clay Materials (More than 35 per cent Passing No. 200)			
Group Classification	*A*-1		*A*-3	*A*-2				*A*-4	*A*-5	*A*-6	*A*-7
	A-1-*a*	*A*-1-*b*		*A*-2-4	*A*-2-5	*A*-2-6	*A*-2-7				*A*-7-5, *A*-7-6
Sieve analysis per cent passing: No. 10 No. 40 No. 200	 50 max 30 max 15 max	 50 max 25 max	 51 min 10 max	 35 max	 35 max	 35 max	 35 max	 36 min	 36 min	 36 min	 36 min
Characteristics of fraction passing No. 40: Liquid limit Plasticity index	 6 max		 N.P.	 40 max 10 max	 41 min 10 max	 40 max 11 min	 41 min 11 min	 40 max 10 max	 41 min 10 max	 40 max 11 min	 41 min 11 min
Group index	0		0	0		4 max		8 max	12 max	16 max	20 max
Usual types of significant constituent materials	Stone fragments — gravel and sand		Fine sand	Silty or clayey gravel and sand				Silty soils		Clayey soils	
General rating as subgrade	Excellent to good					Fair to poor					

Classification procedure: With required test data in mind, proceed from left to right in chart; correct group will be found by process of elimination. The first group from the left consistent with the test data is the correct classification. The *A*-7 group is subdivided into *A*-7-5 or *A*-7-6 depending on the plastic limit. For $P_w < 30$, the classification is *A*-7-6; for $P_w \geqq 30$, *A*-7-5.

N.P. denotes nonplastic.

for the U. S. Engineer Department. This system, which has been designated as the Airfield Classification (AC) system, was designed primarily for use in the construction of military airfields. Although it has more groupings than any other classification system in general use, the system is simple because the groupings are arranged in a logical and orderly manner.

According to the AC system, the coarse-grained soils are divided into

1. Gravel and gravelly soils, symbol *G*.
2. Sands and sandy soils, symbol *S*.

The gravels and sands are each subdivided into four groups:

(*a*) Well-graded, fairly clean material, symbol *W*.
(*b*) Well-graded material with excellent clay binder, symbol *C*.
(*c*) Poorly graded, fairly clean material, symbol *P*.
(*d*) Coarse materials containing fines, not included in preceding groups, symbol *F*.

Fine-grained soils are divided into three groups:

1. Inorganic silty and very fine sandy soils, symbol *M*.
2. Inorganic clays, symbol *C*.
3. Organic silts and clays, symbol *O*.

Each of these three groups of fine-grained soils is subdivided according to its liquid limit into

(*a*) Fine-grained soils having liquid limits less than 50, that is, of low to medium compressibility, symbol *L*.
(b) Fine-grained soils having liquid limits greater than 50, that is, of high compressibility, symbol *H*.

Highly organic soils, usually fibrous, such as peat and swamp soils of very high compressibility, are not subdivided and are placed in one group, symbol *Pt*. The following summary describes the general characteristics of the soils in the various groups.

A. Coarse-grained soils: Less than 50 per cent passing no. 140 U.S. Standard sieve (0.1 mm).

GW-SW groups: Well-graded gravelly and sandy soils having few or no fines. Where subject to frost action they should contain less than 3 per cent finer than 0.02 mm.

GC-SC groups: Well-graded coarse-grained soils with excellent clay binder or other natural cementing agent. Negligible swelling and shrinkage. May be subject to frost action. Stable wet or dry.

GP-SP groups: Poorly graded gravels or sand, with little or no fines. Uniform or gap gradation. Uniform gravels or sands.

GF-SF groups: Any coarse-grained material with detrimental fines. Silty gravels or sands, poorly graded sand-clay mixtures.

B. Fine-grained soils: More than 50 per cent passing no. 140 U.S. Standard sieve (0.1 mm).

1. Fine-grained soils of low to medium compressibility; L_w less than 50 per cent.

   ML group: Inorganic silts and very fine sands, rock flour, silty or clayey fine sands with slight plasticity.

   CL group: Inorganic clays of low to medium plasticity, sandy clays, silty clays, lean clays.

   OL group: Organic silts of low compressibility.

2. Fine-grained soils of high compressibility; L_w greater than 50 per cent.

   MH group: Micaceous or diatomaceous fine sandy and silty soils, elastic silts.

   CH group: Inorganic clays of high plasticity, fat clays.

   OH group: Organic clays of medium to high compressibility.

C. Fibrous organic soils of very high compressibility

Pt group: Peat and other highly organic swamp soils.

The fine-grained soils are differentiated on the basis of a modified plasticity chart, Fig. 1.12.

Shortcomings of Engineering Classifications. The various textural systems, the Public Roads system, and the Airfield Classification system have in common the fact that they are based on the properties of either the grains themselves or of remolded material; they do not take into consideration the properties of the intact material as found in nature. It is primarily the properties of the intact material that determine the behavior of the soil during and after construction. Hence, the systems of classification can serve only as the starting points for adequate description of soils in the conditions under which they are encountered in the field. Nevertheless, even with these limitations, much information concerning the general characteristics of a soil can be inferred as a consequence of its proper classification according to one of the systems described under the preceding subheadings.

Still further useful information can be obtained from sources outside the field of civil engineering, particularly geology and pedology. The foundation engineer should possess a knowledge of at least the descriptive terminology of these two sciences.

Geological Terminology. A knowledge of the geological origin of a soil deposit often furnishes insight into the physical characteristics of the deposit. Therefore, whatever geological data are available

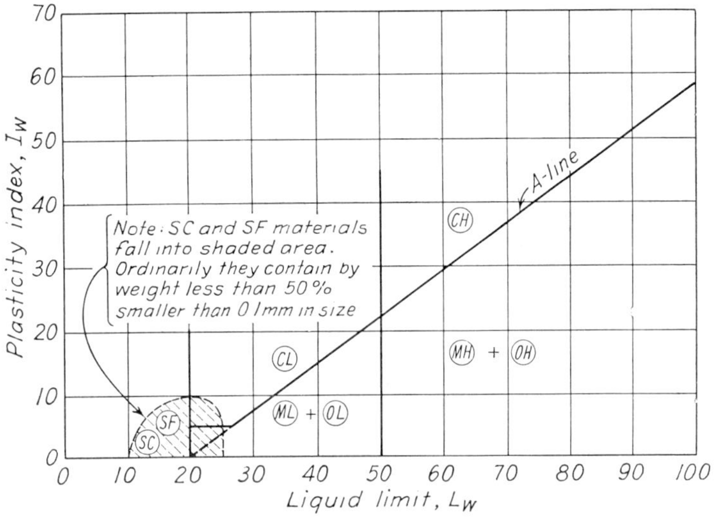

FIG. 1.12. Plasticity chart used in Airfield Classification system (after A. Casagrande).

should be consulted. In this connection it is helpful to be familiar with the definitions in the following list:

Residual soils have been developed from the parent rock over which they now lie. Because erosion is an active process, and because the climate is not generally humid, deep deposits of residual soil are seldom encountered within continental United States. As a matter of fact, residual soils are not common in the northern half of the United States or in Canada, because the continental glaciers removed whatever products of weathering had formed on the bedrock.

Transported soils no longer cover the rock material from which they were derived. They may be further classified according to the mode of transportation as follows: *alluvial* soils, which have been transported by running water; *lacustrine* soils, which have been deposited in quiet lakes; *marine* soils, which have been deposited in sea water; *aeolian* soils, which have been transported by wind; *colluvial* soils, which have been deposited primarily through the action of gravitational force, as in landslides; *glacial* soils, which have been deposited by the glaciers that once covered a large part of the northern hemisphere.

In addition, the foundation engineer should be familiar with the specific types of soil in the following list, because the soils are commonly encountered and the terminology is used by engineers and geologists alike.

Glacial till is composed of material deposited by a glacier without subsequent transportation by water. Since large masses of moving ice are able to carry materials of all sizes with equal ease, typical till is a heterogeneous mixture of soil and rock fragments varying in size from that of boulders to that of clay.

Glacial outwash is the material deposited by streams that drained the front of a melting glacier. It is usually composed of gravel, sand, and silt.

Loess is an aeolian soil consisting predominantly of particles having a diameter of about 0.06 mm. The particles are usually somewhat cemented by calcium carbonate or clay. Typical loess is usually very porous and is further characterized by its ability to stand on almost vertical slopes.

Modified loess is formed from loess by processes of immersion, saturation, erosion, or decomposition. All of these reduce the porosity and may alter the mineral composition.

Varved clay is a particular type of lacustrine deposit in which strata of clay alternate with strata of somewhat coarser material.

Lake marl or *boglime* is a white fine-grained powdery calcareous deposit precipitated by plant action. It is usually found in glaciated areas and is closely associated with peat deposits.

Marl is a calcareous silt or clay of marine origin.

Caliche is a soil in which the grains have been cemented together by the accumulation of carbonates. It is usually formed under semi-arid conditions at a depth of several feet below the surface of the ground.

Adobe is a light-colored windblown clay that has been deposited in shallow lakes.

Tuff is a fine-grained slightly cemented material composed of very small mineral fragments of volcanic origin. It may be deposited by wind or by water.

Bentonite is a natural soil material derived from the chemical alteration of volcanic ash. It is composed almost exclusively of very active colloidal particles that swell markedly in the presence of water. It exhibits a high degree of thixotropy.

Gumbo is a rather loosely used term that usually refers to very sticky, highly plastic, dark colored clays.

Hardpan is an engineering term used to describe any stratum of hard and cohesive soil that is particularly difficult to excavate or to penetrate by means of the tools commonly used in making soil borings.

Pedological System. Pedology is that branch of the science of geology which deals with the outer 4 or 5 ft of the earth's crust. Within this zone the rock or disintegrated rock material has become altered by weathering to an extent and in a manner that depends on the climate, the presence of organisms, the topography, the character of the parent material, and the length of time the processes of weathering have been active. In the terminology of the pedologist, this altered zone is called *soil.*

Wherever mature soils have developed, a succession of several distinct layers is found within a few feet of the surface. These layers are called *soil horizons* and the complete system of horizons is referred to as the *soil profile.* The uppermost horizon has usually lost by leaching a significant amount of fine-grained mineral matter although it may contain a rather large quantity of finely divided organic matter derived from decaying vegetation. It is often called the *zone of depletion.* The second horizon is the *zone of accumulation* for the fine-grained mineral matter, and in a normal profile it is more plastic than the horizons either above or below. Together these two horizons constitute the *solum,* the zone in which the soil-forming processes are the most active. Below the solum is a third horizon composed of relatively unaltered parent material. The parent material may be partially decomposed and disintegrated bedrock, or it may be soil material or rock material that has been transported and deposited at a new location. Obviously, where fresh materials are still being deposited, as on the flood plains of rivers, there is no opportunity for a typical soil profile to form.

The pedological classification of a soil is determined principally on the basis of the geology of the parent material and certain inherent characteristics of the profile. The latter include the number, color, texture, structure, thickness, chemical composition, and relative arrangement of the horizons.

All soils that have similar profiles except for the texture of the surface horizon are grouped together as a *soil series* and given a series name. These names are usually taken from the locality where the soil was first mapped. Ordinarily the series name is further modified by terms indicating the texture of the surface soil as determined by a triangular classification chart similar to Fig. 1.11. The engineering characteristics of a soil profile are primarily determined by the texture, plasticity, thickness, and drainage characteristics of the horizons. Correlations that have been made indicate that several different pedological series may have similar engineering properties.

Pedology has been utilized most extensively by agronomists, but it also has particular application to the work of the highway or airport engineer. Moreover, since the pedological description of a soil includes a description of the parent material, the foundation engineer can obtain from such a description useful general information about the nature of the soils in a locality with which he is not familiar.

Most of the pedological information is contained in county soil survey reports published by the U.S. Department of Agriculture since the early part of this century. Only those reports published during the last 20 years, however, can be depended upon for information of high accuracy. The recent reports include a description of the soil profile and the geology of each soil series in the county, as well as a map showing the boundaries of each soil series.

A relatively new development in the field of pedology is the use of aerial photographs in the preparation of soil maps. The elements revealed on the photographs often permit more accurate delineation of the soil boundaries than can be achieved by field surveys alone. Moreover, under given climatic conditions, the character of the parent material is revealed by such elements of the photographic pattern as *land form, soil color, erosion, surface drainage, vegetative cover, slope, land use,* and *microrelief.* The pattern also depends on the texture and slope of the soil profile and on the groundwater conditions. Hence, each pattern must be correlated with the corresponding soil profile as determined by ground surveys, but after this has been done the photographs permit rapid studies of very large areas.

In a similar manner, the engineering properties of the surface soils may also be correlated with the air-photo pattern. Thus, the engineer who is able to interpret photographic patterns has a valuable tool for locating the boundaries between soils of different types, for predicting the engineering characteristics of the soils, and for planning an intelligent program of soil exploration to check his predictions.

Illustrative Problems

1. A sample of inorganic soil has the following grain-size characteristics:

Size (mm)	Per Cent Passing
1.0	96
0.074 (No. 200)	71
0.05	67
0.005	31
0.002	19

The liquid limit is 53 per cent, and the plasticity index 22 per cent.

Classify this soil according to the following systems: Original U.S. Bureau of Soils; 1945 Public Roads; Airfield Classification.

Solution.

Bureau of Soils:

Clay size 31 per cent, silt size 67 — 31 = 36 per cent

Classification (Fig. 1.11): Clay

Public Roads:

Computation of group index:

$$a = 71 - 35 = 36$$

$$b = 40$$

$$c = 53 - 40 = 13$$

$$d = 22 - 10 = 12$$

$$\begin{aligned}\text{Index} &= 0.2 \times 36 + 0.005 \times 36 \times 13 + 0.01 \times 40 \times 12 \\ &= 14.3\text{, say, } 14\end{aligned}$$

On basis of limit values and Table 1.10, soil is either *A*-7-5 or *A*-7-6.

Since $P_w = L_w - I_w = 53 - 22 = 31$, which is greater than 30, classification is *A*-7-5 (14)

Airfield Classification:

On the basis of Fig. 1.12, classification is *MH*.

2. The soil represented by curve 6, Fig. 1.1, was found to be non-plastic, so that the plastic limit test could not be performed. The liquid limit was 49 per cent. Classify the soil according to the three systems mentioned in Prob. 1. Also state the percentages of sand, silt, and clay according to the MIT Classification.

Solution.

Bureau of Soils:

Clay size 19 per cent, silt size 73 — 19 = 54 per cent

Classification (Fig. 1.11): Silty loam

Public Roads:

Computation of group index (94 per cent passing no. 200)

$$a = 40$$

$$b = 40$$

$$c = 49 - 40 = 9$$

$$d = 0$$

$$\text{Index} = 0.2 \times 40 + 0.005 \times 40 \times 9 = 9.8, \text{ say, } 10$$

On the basis of Table 1.10, soil is classified as *A*-5 (10)

Airfield Classification:

On the basis of Fig. 1.12, classification is *ML*.

MIT Classification:

Soil is 13 per cent clay size, 84 — 13 = 71 per cent silt size, and 16 per cent sand size.

1.9. Summary of Procedures for Classification and Description of Foundation Materials

The first step in the adequate description of a natural rock formation or soil deposit is the proper identification, classification, and description of the rocks or soils comprising the formation or deposit. The information needed to accomplish this has been given in the preceding sections of this chapter and is summarized in the following paragraphs. More detailed data concerning the physical properties of the materials are given in Chapter 2. On any project, however, the properties of the formation or deposit itself must ultimately be evaluated. These properties may be quite different from those that might be inferred from a study of the individual specimens or samples obtained for purposes of classification and description. Hence, Chapter 4 is devoted to the characteristics of natural deposits and formations.

If the foundation material consists of rock, adequate classification and description consist of obtaining the following information for representative samples: kind of rock; principal minerals; texture; color; structural details such as stratification and schistosity; evidences of disintegration or weathering; and the results of any physical tests pertinent to the specific engineering problem, such as compression tests (see Art. 2.13).

If the foundation material consists of soil, the first requirement is to ascertain whether the soil is primarily coarse-grained, primarily fine-grained, or mixed-grained. If it is the last, the specimen should be divided into two parts, usually by means of a no. 140 U.S. Standard sieve (0.10 mm), and each part classified and described by the procedures appropriate for coarse-grained and fine-grained soils, respectively.

Second, the specimen is assigned a tentative classification in one of the major categories (gravel, sand, silt, clay, organic silt, organic clay, or peat) or, preferably, as one of the specific types listed in the geological classification, Art. 1.8. The pertinent index properties for this type of soil are then determined.

To facilitate the latter step, Table 1.11 has been prepared. It consists of a list of the major soil types and, for each type, a check list of the appropriate index properties.

After the index properties have been determined, the tentative classification of the soil should be verified by comparing the properties with those characteristic of soils in the assumed category. If they are in agreement, the tentative classification may be considered correct; if they disagree, the index properties that have been determined will serve to indicate the proper category. If the soil must be assigned to a group according to one of the standard systems, the classification can be made most reliably at this stage.

With experience, most of the foregoing steps can be eliminated, and many of the index properties can be estimated; indeed, every engineer should develop the habit of expressing his descriptions of soils in terms of estimated numerical values of the index properties rather than in qualitative terms. However, even the experienced technician may need to follow the detailed procedure for soils difficult to classify, and the beginner will find most of the steps essential.

The complete description of the soil includes the results of all the appropriate examinations and tests listed in Table 1.11, together with the results of all additional physical tests or field tests carried out during the soil investigation. These data should ultimately be supplemented by whatever is learned concerning the characteristics of the deposit which the samples represent, including a summary of the geological features (see Chapter 4).

Table 1.11. Minimum Requirements for Adequate Soil Description

			Sand and Gravel	Hardpan[14]	Glacial Till	Tuff, Fine-grained	Inorganic Silt	Loess[15]	Modified Loess	Adobe	Clay	Gumbo	Lake Marl	Marl	Organic Silt	Organic Clay	Peat
General Information from Field Examination		Color	●	●	●	●	●	●	●	●	●	●	●	●	●	●	●
		Odor[2]													●	●	●
		Texture and structure[3]		●		●	●	●	●	●	●	●	●	●	●	●	●
		Dilatancy[4]					●	●	●	●			●	●	●		
		Grain properties[5]	●	●	●												●
		Plasticity				●	●	●	●	●	●	●	●	●	●	●	
		Dry strength[6]				●	●	●	●	●	●	●	●	●	●	●	
Results of Classification Tests	Intact Samples[1]	Natural water content, w		●	●	●	●	●	●	●	●	●	●	●	●	●	●
		Natural void ratio,[7] e	●		●			●	●	●							
		Unconfined compressive strength, q_u		●	●	●	●	●	●	●	●	●	●	●	●	●	
		Sensitivity,[8] S_t				●	●		●	●	●	●	●	●	●	●	
	Representative Samples	Unit weight of solid constituents, γ_s		●	●	●	●	●	●	●	●	●	●	●	●	●	
		Maximum void ratio,[9] e_{max}	●		●			●	●								
		Minimum void ratio[9], e_{min}	●		●			●	●								
		Liquid limit, L_w				●	●	●	●	●	●	●	●	●	●	●	
		Plastic limit,[10] P_w				●	●	●	●	●	●	●	●	●	●	●	
		Shrinkage limit, S_w				●			●	●	●	●	●	●	●	●	
		Mechanical analysis[11]	●		●	●	●	●	●	●		●	●	●	●		
		Carbonate content[12]			●		●	●	●	●	●	●	●	●	●	●	
		Organic matter content[13]										●	●	●	●	●	

Footnotes to this table are on page 49.

[1] If no undisturbed or tube samples were obtained, use the spoon samples.

[2] If the odor is faint, heat the sample slightly. This intensifies the odor.

[3] Describe appearance of fresh fracture of intact sample (granular, dull, smooth, glossy). Then rub small quantity of soil between the fingers, and describe sensation (floury, smooth, gritty, sharp). If large specimens break up readily into smaller fragments, describe appearance of walls of cracks (dull, slickensided) and average spacing of cracks.

[4] Perform shaking test. Describe results (conspicuous, weak, none), depending on intensity of phenomena observed.

[5] Describe shape (angular, subangular, subrounded, rounded, well rounded) and mineralogical characteristics of macroscopic soil particles only. Mineralogical characteristics include types of rocks and minerals represented among the grains so far as they can be discerned by inspection under the hand lens. Describe rock fragments (fresh, slightly weathered, or thoroughly decomposed; hard or friable). If a sand contains mica flakes, indicate mica content (slightly, moderately, or very micaceous). In connection with peat, the term grain properties refers to the type and state of preservation of the predominant visible remnants of plants such as fibers, twigs, or leaves.

[6] Crush dry fragment between fingers, and indicate hardness (very low, low, medium, high, very high).

[7] If no undisturbed samples have been obtained, substitute results of standard penetration test or equivalent.

[8] Applies only to clay and fine silt at a water content above the plastic limit.

[9] Prepare sample as described in Art. 1.6.

[10] In addition to numerical value of P_w, state whether threads were tough, firm, medium, or weak.

[11] Present results either in form of semilogarithmic graph, or else by numerical values of D_{10} and $U = D_{60}/D_{10}$ accompanied by adjectives indicating the type of grain-size grading.

[12] Calcium carbonate content can be detected by moistening the dry material with dilute HCl. Describe result of test (strong, weak, or no effervescence).

[13] To determine presence of organic matter, determine L_w first in fresh state and then after drying in oven at 105° C. Describe results of test (highly or slightly organic).

[14] Add to data on texture a description of general appearance, structure, and degree of cohesiveness of chunks in fresh state and after soaking in water.

[15] Add to data on texture a description of the macroscopic features of the loess, such as diameter and spacing of root holes.

Wolmar Fellenius (1876–   )

Chairman and active leader of the Swedish Geotechnical Commission, appointed in 1913 to study the causes of catastrophic landslides on the Swedish State Railways. The report of this Commission, published in 1924, is justly regarded as a milestone in the application of soil mechanics to practical problems. The investigators were among the first to perform laboratory and field tests for determining the shearing strength of soils, and to recognize the essential differences in the properties of undisturbed and remolded clays.

PLATE 2.

CHAPTER 2

Physical Properties of Soil and Rock

2.1. Importance of Hydraulic and Mechanical Properties

In the preceding chapter only those properties necessary for the identification and classification of soils and rocks have been discussed. However, the foundation engineer must also have a quantitative knowledge of the physical properties of the materials with which he deals. If the construction of a foundation requires that the water table be lowered, the engineer must be informed concerning the hydraulic properties and the drainage characteristics of the subsurface materials. Since all earth materials are compressible to some degree, the engineer must have a basis for estimating the amount of compression that will take place under a given load. This requires consideration of the stress-deformation characteristics of the material. Finally, to avoid the complete failure of his foundations, the engineer must have information in regard to the ultimate strength of the supporting material. These basic properties of soil and rock will be considered in this chapter.

2.2. Permeability of Soil

Definitions and Darcy's Law. A material is said to be permeable if it contains continuous voids. Every soil and every rock satisfy this condition. Nevertheless, there are large differences in the degree of permeability of the various earth materials. The quantity of flow through a dense rock may be so small as to pass unnoticed because evaporation prevents the accumulation of water on the exposed face; yet the flow of water through such a material may produce seepage pressures between the mineral grains as large as those exerted in more permeable materials under the same conditions of hydraulic head.

In order to understand the forces that govern the flow of water through earth materials, it is necessary to know the meaning of certain terms used in the field of hydraulics. In Fig. 2.1 points a and b represent the extremities of the path along which part of the water flows through a sample of earth material. At each extremity a standpipe known as a *piezometric tube* has been installed to permit observation of the level to which the water rises at these points. The water level in the tube at b is known as the *piezometric level* at b and the vertical distance from this level to point b is the *piezometric head* at b. The vertical distance between a and b represents the *position head*, ΔH. If the piezometric level at a is equal to the piezometric level at b, the system is in a state of rest, regardless of the magnitude of ΔH, and there is no flow from a to b. Flow occurs only if there

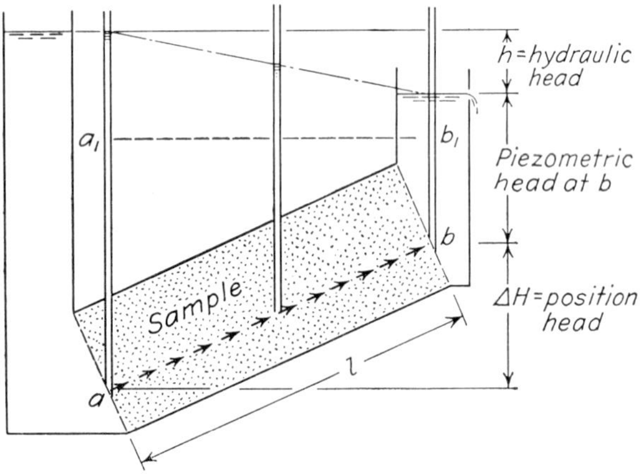

FIG. 2.1. Diagram illustrating the meaning of hydraulic head, piezometric head, and position head.

is a difference in the piezometric levels at a and b. This difference is known as the *hydraulic head* h, or as the *difference in piezometric level* between a and b. In the figure the two points a_1 and b_1 are at the same elevation. Under the conditions represented, the pressure at a_1 must exceed that at b_1 by a quantity equal to the unit weight of water times the difference in piezometric level. This quantity, $\gamma_w h$, is referred to as the *excess hydrostatic pressure* and is designated by the symbol u. This pressure causes the water to move through the soil from a to b. The ratio

$$i_p = \gamma_w(h/l) = u/l \qquad 2.1$$

is known as the *pressure gradient* from a to b, and the ratio

$$i = i_p/\gamma_w = h/l \qquad 2.2$$

is defined as the *hydraulic gradient* between a and b. It is a pure number.

The flow of water through permeable substances is governed by an empirical relationship

$$v = ki \tag{2.3}$$

first stated by H. Darcy in 1856. In this expression, v is the *discharge velocity,* defined as the quantity of water that percolates in a unit time across a unit area of a section oriented at right angles to the direction of flow; i is the hydraulic gradient; and k is a coefficient known as the *coefficient of permeability.* The value of k, which has the units of a velocity, depends primarily on the characteristics of the permeable substance, but it is also a function of the unit weight and the viscosity of the fluid. Since water is the only fluid with which the foundation engineer is concerned, the influence of differences in the unit weight of the fluid is negligible. Moreover, the changes in viscosity within the ordinary range of temperature of groundwater are relatively unimportant and can usually be ignored in the solution of practical problems. Hence, it is customary and justifiable for civil engineers to regard the coefficient of permeability as a property of the soil or rock.

In a general way, the coefficient of permeability increases with increasing size of voids, which in turn increases with increasing grain size. However, the shape of the void spaces also has a marked influence on the permeability. As a consequence, no simple relationships have been found between permeability and grain size except for fairly coarse soils with rounded grains. For example, in his studies of the permeability of loose filter sands, Allen Hazen found that

$$k = CD_{10}^2 \tag{2.4}$$

where C is equal approximately to 100/cm sec and D_{10} is expressed in centimeters.

Similarly, it has been found that the coefficient of permeability of coarse-grained soils varies approximately as the square of the void ratio. No such simple relation has been established for soils containing flake-shaped particles.

Permeability Tests. The lack of simple and general relationships between the coefficient of permeability and the results of classification tests often leads to the necessity for performing permeability tests, either in the field or in the laboratory. Permeability tests on soil samples are usually made with a falling-head permeameter or a constant-head permeameter. The constant-head permeameter, Fig. 2.2*a*, gives reliable results for highly permeable materials such as clean sands and gravels. The value of k is computed by means of the equation

$$k = QL/hAt \tag{2.5}$$

in which Q is the quantity of discharge, L the length of the sample in the direction of flow, A the cross-sectional area of the sample, h the hydraulic head, and t the time. All these quantities are measured in the test.

The falling-head permeameter, Fig. 2.2c, is more suitable for tests on materials of low permeability because the dimensions of the ap-

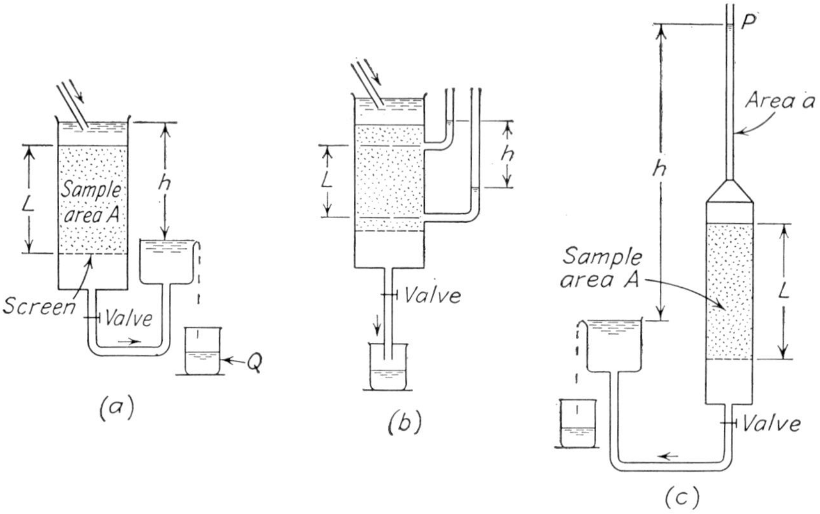

FIG. 2.2. (a and b) Constant-head permeameters; (c) falling-head permeameter.

paratus can be adjusted so that the measurements of head and time may be carried out with high accuracy over a wide range in the values of the coefficient of permeability. The value of k may be calculated from the quantities measured during the test by means of the equation

$$k = 2.3 \frac{aL}{At} \log_{10} \frac{h_0}{h_1} \tag{2.6}$$

In this equation, a is the cross-sectional area of the standpipe, L the length of the sample, A the cross-sectional area of the sample, t the time, and h_0 and h_1 are the original and final hydraulic heads, respectively.

The results of permeability tests on cohesionless materials are often misleading because it is impracticable to obtain representative samples of permeable material and to place them in the permeameter without disturbance. Not only is it difficult to obtain undisturbed samples, but since most granular deposits are quite erratic, it is difficult even to obtain representative ones.

Permeability tests are also subject to various types of experimental errors. One of the most important of these arises from the formation of a filter skin of fine material on the surface of the sample. Segregation, for example, of the silty particles in sand may be caused by careless placement of the sample in the permeameter or by erosion of the fine material while the test is being conducted. The segregated layer of fine material greatly reduces the measured permeability. Figure 2.2*b* shows a type of constant-head apparatus that may be used to eliminate the effect of the surface skin. In this apparatus the loss in head is measured through a distance in the interior of the sample, and the drop in head across the filter skin has no effect on the results.

In conducting permeability tests the technician must be careful to saturate the sample completely and to make sure that no air bubbles will be released by the water during the test. The air bubbles block the voids and thus lower the permeability. At best, permeability tests on laboratory specimens can serve only as a basis for a rough estimate of the permeability of an actual soil deposit. For this reason an estimate of the permeability of a coarse-grained deposit based on grain size may often be as useful as the results of laboratory tests.

The permeability of clay samples can best be determined indirectly from the data obtained by performing consolidation tests. The basis for the computations will be discussed more fully in Art. 2.8.

Permeability of Stratified Deposits. Many deposits consist of layers or lenses of materials that differ in grain size and permeability. The average coefficient of permeability of such deposits differs significantly in the horizontal and vertical directions. In the horizontal direction the average permeability may be almost as great as the permeability of the most permeable layer or lens, whereas in the vertical direction the average may be almost as small as the permeability of the least permeable layer or lens. The ratio of the average coefficients of permeability in the horizontal and vertical directions ranges for most natural deposits between 2 or 3 and several hundred.

Field Pumping Tests. The most reliable information concerning the permeability of a deposit of coarse-grained material below the water table can usually be obtained by conducting pumping tests in the field. Although such tests have their most extensive application in connection with dam foundations, they may also prove advisable on large bridge or building foundation jobs where the water table must be lowered. A test is made by drilling one well from which water is pumped and several others in which the position of the water table is observed. The observation wells are drilled at various distances from the pumping well along two straight lines, one oriented approximately in the direction of the groundwater flow and the other at right angles to it. Once a steady state of flow has been established and the water levels in all the wells are nearly stationary, the coefficient of permeability may be calculated from the equation

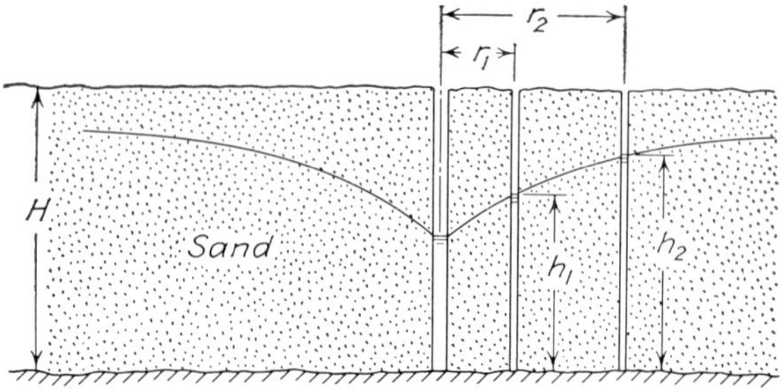

FIG. 2.3. Illustration of flow of water toward a well during a pumping test.

$$k = \frac{2.3q}{2\pi H(h_2 - h_1)} \log_{10} \frac{r_2}{r_1} \qquad 2.7$$

in which q equals the rate of flow across the boundary of any cylindrical section having a radius r and, therefore, the quantity of water pumped from the well per unit of time. The quantities in the equation are defined in Fig. 2.3.

Approximate values of the coefficient of permeability for various types of soils and the recommended method of determining these values are assembled in Table 2.1. As the table indicates, it is difficult to obtain reliable values for the coefficient of permeability of soils that contain appreciable quantities of silt or very fine sand. There is no indirect method of computing the permeability of such materials, and laboratory tests are likely to be extremely unreliable unless made by experienced technicians.

ILLUSTRATIVE PROBLEMS

1. A loose uniform sand with rounded grains has an effective grain size D_{10} equal to 0.3 mm. Estimate the coefficient of permeability.

Solution. The estimate can be made by means of eq. 2.4. Thus,

$$k = CD_{10}^2 = 100 \times (0.03)^2 = 9 \times 10^{-2} \text{ cm/sec}$$

2. In a constant-head permeability test a sample 8 cm long was tested in a permeameter with an inside diameter of 5 cm. A state of steady flow was established under a head of 50 cm, and the discharge was collected for 30 sec. It amounted to 120 cu cm. Compute the value of k.

Table 2.1. Coefficient of Permeability of Various Soils

k(cm/sec)	Drainage	Soil Type	Determination of k			
10^2	Good	Clean gravels	Pumping tests. Reliable if properly conducted.	Constant head permeameter-reliable.		
10^1	Good	Clean gravels				
1.0	Good	Clean sands				Computation from grain size
10^{-1}	Good	Clean sand and gravel mixtures			Falling-head permeameter: Reliable	
10^{-2}	Good					
10^{-3}	Good					
10^{-4}	Good					
10^{-5}	Poor	Very fine sands			Falling-head permeameter: Unreliable	
10^{-6}	Poor	Organic and inorganic silts, mixtures of sand silt and clay, glacial till, stratified clay deposits.				
10^{-7}	Practically impervious	Impervious soils, e.g., homogeneous clays below zone of weathering.			Falling-head permeameter: Fairly reliable	Computation from consolidation data (Reliable)
10^{-8}						
10^{-9}						

After Casagrande and Fadum

Solution: According to eq. 2.5,

$$k = \frac{QL}{hAt} = \frac{120 \times 8}{50 \times \pi \times \frac{5^2}{4} \times 30} = 3.3 \times 10^{-2} \text{ cm/sec}$$

3. A falling-head permeability test was performed in a permeameter with an inside diameter of 5 cm. The inside diameter of the standpipe was 2 mm. The sample had a length of 8 cm. During a period of 6 min the head on the sample decreased from 100 to 50 cm. Compute the value of k.

Solution. According to eq. 2.6,

$$k = 2.3 \frac{aL}{At} \log_{10} \frac{h_0}{h_1} = 2.3 \frac{0.2^2 \times 8}{5^2 \times 6 \times 60} \log \frac{100}{50}$$

$$= 8.2 \times 10^{-5} \log 2.0 = 2.5 \times 10^{-5} \text{ cm/sec}$$

2.3. Permeability of Rock

The permeability of sound rock is of little more than academic interest to the engineer engaged in the design and construction of building or bridge foundations, because almost every deposit of rock contains extremely permeable channels such as joints, bedding planes, or solution cavities. Moreover, it is probable that most reported values of the coefficient of permeability are too low because the formation of a relatively impervious skin on the surface of a specimen of rock is almost unavoidable when the specimen is being ground to a shape suitable for testing. Nevertheless, a study of the values in Table 2.2 indicates that the permeability of rocks is comparable

Table 2.2. Permeability of Sound Rocks

cm/sec

Type of Rock	Range in Values of k
Granite	7.1×10^{-11} to 2.5×10^{-10}
Slate	9.7×10^{-11}
Marble	8×10^{-10} to 2.5×10^{-8}
Limestone	8.4×10^{-10} to 1.3×10^{-6}
Sandstone	1.95×10^{-7} to 3.7×10^{-6}

After Kessler and Sligh.

with that of fine-grained soils, and that even sound granite cannot be regarded as impervious.

2.4. Intergranular and Porewater Pressures

Definitions. The stresses that act within a saturated mass of soil or rock may be divided into two kinds; those that are transmitted directly from grain to grain of the solid constituents, and those that act within the fluid that fills the voids. The former are called *intergranular pressures* or *effective stresses,* and the latter *porewater pressures* or *neutral stresses.* This is an important distinction in foundation engineering, because only the intergranular pressures can induce changes in volume of a soil mass. Likewise, only intergranular pressures can produce frictional resistance in soils or rocks.

The distinction between intergranular and porewater pressures can be illustrated by means of Fig. 2.4, which shows a container partly filled with granular material and completely filled with water. To the bottom of the container is attached a flexible tube connected to a reservoir. In Fig. 2.4*a* the water level in the reservoir is at the same elevation as that in the container, so that no flow takes place. On plane *ab* at depth $(H_1 + z)$ below the top of the container, the vertical pressure p is

$$p = H_1\gamma_w + z\gamma_{\text{sat}} \tag{2.8}$$

where γ_w is the unit weight of water, and γ_{sat} the unit weight of the saturated soil. Since p depends on the weight of all the overlying soil and water, it is referred to as the *total pressure*. It consists of the intergranular pressure $\bar{p}$ and the porewater pressure u_w. Thus

$$p = \bar{p} + u_w \tag{2.9}$$

The water above *ab* extends in continuous voids to the height z and in a continuous mass above the soil for a distance H_1. Hence, according to the laws of hydraulics, the porewater pressure u_w at *ab* must be

$$u_w = (H_1 + z)\gamma_w \tag{2.10}$$

The intergranular pressure $\bar{p}$ is then

$$\bar{p} = p - u_w = H_1\gamma_w + z\gamma_{\text{sat}} - (H_1 + z)\gamma_w$$

or

$$\bar{p} = z(\gamma_{\text{sat}} - \gamma_w) \tag{2.11}$$

The quantity $\gamma_{\text{sat}} - \gamma_w$ is known as the *submerged unit weight*, γ'. Therefore, for the conditions illustrated in Fig. 2.4*a*,

$$\bar{p} = z\gamma' \tag{2.12}$$

Thus, the intergranular pressure is independent of the depth H_1 of water above the submerged soil or rock.

Under the conditions of Fig. 2.4*a* there is no flow of water through the voids. In the pressure diagram to the right of the figure the change in width of the unshaded area represents the distribution of the porewater pressure with depth, and the change in width of the shaded area represents the distribution of the intergranular pressure. If the level of the water in the reservoir is different from that in the container then flow occurs and the conditions are similar to those illustrated in Fig. 2.4*b* or 2.4*c*. Thus, eq. 2.10 is no longer valid, and eq. 2.11 and 2.12 are no longer applicable.

Seepage Pressures and Critical Hydraulic Gradient. If the free water surface is maintained at the top of the container and the reservoir is lowered, a state of steady flow is soon established in accordance with the conditions illustrated in Fig. 2.4*b*. Since the reservoir and the flexible tube represent a standpipe freely communicating with the bottom of the container, the porewater pressure at the bottom of the container may be expressed as

$$u_w = (H_1 + H - h)\gamma_w \tag{2.13}$$

This represents a decrease of $h\gamma_w$ in the porewater pressure as compared to the condition shown in Fig. 2.4*a*. It may be noted that this decrease cannot be attributed to the velocity of the flowing water, because the velocity head $v^2/2g$ is negligible for velocities comparable to the highest ones encountered in natural soil deposits. Hence, the total pressure on the bottom of the container is determined solely by the weight of the soil and water above it. As a result the intergranular pressure must be increased by $h\gamma_w$ over that indicated in Fig. 2.4*a*. At plane *ab* the proportionate increase in the intergranular pressure is therefore equal to $\frac{z}{H}h\gamma_w$. This increase in intergranular pressure due to the flow of water through the voids is known as *seepage pressure*. It is the result of the frictional drag of the flowing water on the soil grains.

It may be noted that the loss in head between the top of the granular material and the depth z is hz/H. The corresponding hydraulic gradient i is then h/H. Thus, the seepage pressure may be expressed as $iz\gamma_w$, and the intergranular pressure at depth z is

$$\bar{p} = z\gamma' + iz\gamma_w \tag{2.14}$$

If the reservoir is raised above the container, so that water flows from the reservoir upward through the granular material, Fig. 2.4*c*, the porewater pressure at the bottom of the container is increased by the amount $h\gamma_w$. Thus the intergranular pressure on plane *ab* is reduced to

$$\bar{p} = z\gamma' - iz\gamma_w \tag{2.15}$$

By increasing the upward hydraulic gradient the seepage pressure can be increased to $z\gamma'$, whereupon the intergranular pressure $\bar{p}$ becomes zero. This occurs when

$$z\gamma' - i_c z\gamma_w = 0 \tag{2.16}$$

and

$$i_c = \gamma'/\gamma_w \tag{2.17}$$

The hydraulic gradient at which the intergranular pressure becomes zero is known as the *critical hyraulic gradient* i_c. Under these conditions a cohesionless soil cannot support any weight on its surface.

Moreover, as the hydraulic gradient approaches the critical value, the soil becomes much looser and the coefficient of permeability k increases. Hence, if an excavation is made into a cohesionless soil

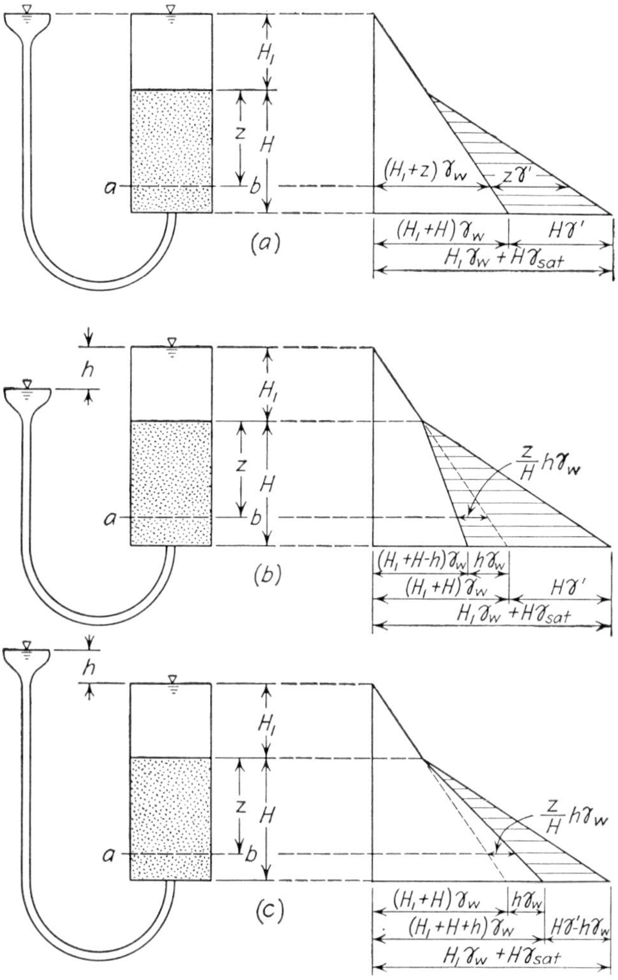

FIG. 2.4. Diagram illustrating meaning of intergranular and porewater pressure.

below groundwater level to such a depth that the intergranular pressure is reduced to zero, a visible agitation of the grains of soil is observed. This phenomenon is known as *boiling* or the *quick condition*. Most quicksand deposits are the result of these particular hydraulic conditions.

ILLUSTRATIVE PROBLEMS

1. Compute the vertical intergranular pressure at a depth of 30 ft in the deposit represented by Fig. P2.1.

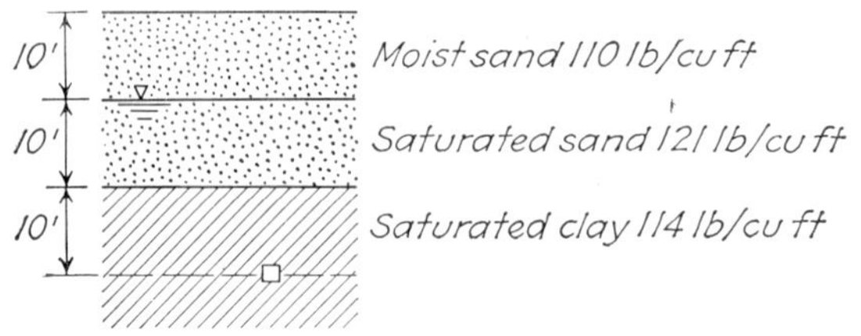

FIG. P2.1.

Solution. On the basis of submerged weights, eq. 2.12, we may write

$$
\begin{aligned}
10 \times 110 &= 1100 \text{ lb/sq ft} \\
10 \times (121 - 62.5) &= 585 \\
10 \times (114 - 62.5) &= \underline{515} \\
\bar{p} &= 2200 \text{ lb/sq ft}
\end{aligned}
$$

On the basis of eq. 2.9, we may write

$$
\begin{aligned}
p &= 10 \times 110 + 10 \times 121 + 10 \times 114 = 3450 \text{ lb/sq ft} \\
u_w &= 20 \times 62.5 \qquad\qquad\qquad\qquad\quad = \underline{1250} \\
\bar{p} &= \qquad\qquad\qquad\qquad\qquad\qquad\quad\; 2200 \text{ lb/sq ft}
\end{aligned}
$$

2. In the deposit shown in Fig. P2.2 the groundwater level was originally at the ground surface. The water table was lowered by

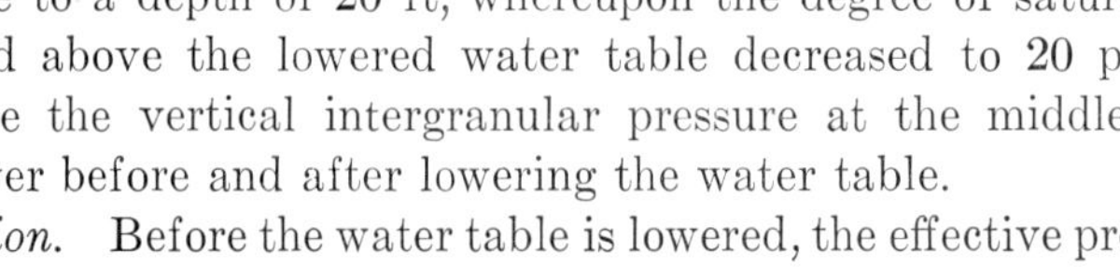

FIG. P2.2.

drainage to a depth of 20 ft, whereupon the degree of saturation of the sand above the lowered water table decreased to 20 per cent. Compute the vertical intergranular pressure at the middle of the clay layer before and after lowering the water table.

Solution. Before the water table is lowered, the effective pressure is

$$
\begin{aligned}
50 \times (135 - 62.5) &= 3625 \text{ lb/sq ft} \\
12.5 \times (120 - 62.5) &= \underline{719} \\
\bar{p} &= 4344 \text{ lb/sq ft}
\end{aligned}
$$

After the water table is lowered, the moist weight of the sand in the upper 20 ft becomes

$$116 + 0.2(135 - 116) = 116 + 3.8 = 119.8 \text{ lb/cu ft}$$

Therefore, the effective pressure at 20 ft becomes

$$\begin{aligned} 20 \times 119.8 \quad\quad &= 2396 \\ 30 \times (135 - 62.5) &= 2175 \\ 12.5 \times (120 - 62.5) &= \underline{719} \\ \bar{p} &= 5290 \text{ lb/sq ft} \end{aligned}$$

Note that lowering the water table *increases* the intergranular pressure in the layer of clay.

2.5. Soil Moisture, Drainage, and Frost Action

Capillary Phenomena. The level assumed by the water in an observation well extending into a deposit of soil is known as the *groundwater level, free water surface,* or *water table.* This definition is valid irrespective of the coefficient of permeability of the deposit, although the determination of the groundwater level in very fine-grained soils by means of observation wells may be very inaccurate unless specialized techniques are used.

Below the water table, most soils are completely or almost completely saturated. Above the water table, the degree of saturation depends on the climatic conditions, the grain size of the soil, and the distance to the water table. Coarse-grained soils are only partially saturated even at elevations close to the water level, whereas fine-grained soils may be saturated for a considerable distance above it. In the latter instance, the free water surface may be defined as the level at which the porewater pressure is equal to atmospheric pressure; that is, $u_w = 0$.

If gravity were the only force acting upon the water in the soil, the soil above groundwater level would always be completely dry except at such times as recent precipitation was flowing through it. Water, however, exhibits the force of surface tension due to the molecular attraction between its molecules at an air-water interface. The surface tension combined with the attraction between water and most solid substances, as demonstrated by the ability of water to wet these substances, constitutes a force that is opposed to gravity and tends to draw or retain moisture above the water table. This force is known as *capillarity.*

Capillarity may be demonstrated by immersing the lower end of a glass tube of small diameter into a vessel containing water. The water will rise in the tube to a height governed primarily by the diameter of the tube and the cleanliness of its inner surface. This height is known as the *height of capillary rise.* It is given in centimeters by the expression

$$\text{Height of capillary rise, } h_c = \frac{2T_s}{r\gamma_w} \cos \alpha \qquad 2.18$$

in which T_s is the surface tension of the water in grams per centimeter, r the radius of the tube in centimeters, and α the contact angle between the surface of the water and the wall of the tube. Above the free water surface the pressure in the water is negative with respect to atmospheric pressure. Thus, at elevation z above the water table,

$$u_w = -z\gamma_w \qquad 2.19$$

The conditions for capillary rise in a mass of soil are not strictly analogous to those in a glass tube because the voids in the soil have variable diameters. Nevertheless, the average diameter of the voids is related to the average grain size and thus to the height of capillary rise. Hence, as a rough approximation for the maximum height h_c (cm) to which water can be lifted by capillarity in a given soil, we may write

$$h_c = C/eD_{10} \qquad 2.20$$

in which e is the void ratio, D_{10} is Hazen's effective size in centimeters, and C is an empirical constant that depends on the shape of the grains and the surface impurities. In general, C has a value between 0.1 and 0.5 sq cm.

The height of capillary rise is greatest for very fine-grained soil materials, but the rate of rise in such materials is slow because of their low permeability. Thus, the capillary rise that occurs in a given length of time, such as 24 hours, is a maximum for materials of intermediate grain size such as silts and very fine sands. Figure 2.5 shows the relationship between the grain size of a uniform quartz powder and the height of capillary rise for a 24-hour period.

The water that rises by capillarity above the free water surface attains the maximum height h_c only in the smaller voids. A few large voids may effectively stop capillary rise in certain parts of the soil mass. Hence the height h_{cc}, Fig. 2.6, to which the soil is completely saturated by capillary rise, is likely to be considerably smaller than h_c. Since the threads of water are continuous, however, the stress in the water is everywhere governed by eq. 2.19.

Some of the water that percolates into the ground from the surface does not reach the free water table but is held by surface tension either in the smaller voids or as annular rings surrounding the points of contact between soil grains. Additional moisture may rise above the free water table by the processes of evaporation and condensa-

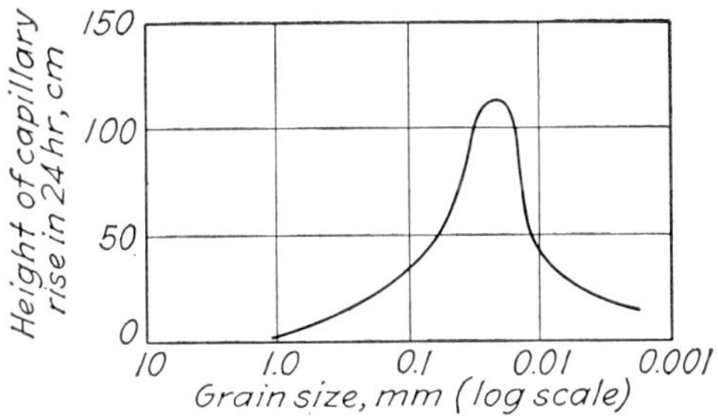

FIG. 2.5. Relation between grain size of uniform quartz powder and height of capillary rise for 24-hr period (after A. Atterberg).

tion. Soil moisture from these sources is known as *contact moisture.* The surface tension in the films of water tends to pull the soil grains together and gives rise to the *apparent cohesion* of sands and silts. The cohesion is called apparent because immersion destroys the films of water and the tendency of the grains to stick together.

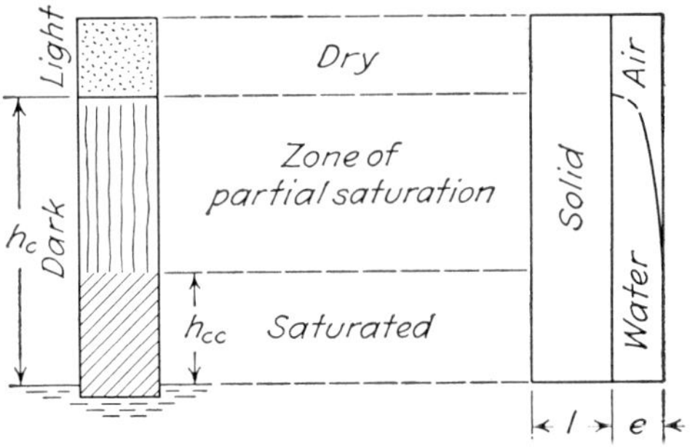

FIG. 2.6. Capillary rise of water into dry sand.

Drainage by Gravity. The difficulties of construction below water level in soil deposits may be avoided, and the stability of the materials increased, by lowering the water table. In coarse-grained materials this can usually be done by draining the water into ditches or galleries or by pumping from sumps or wells. Since gravity is the force that causes the flow of the water toward the drains, the process is known as *drainage by gravity.*

As the water flows toward the drains, it produces seepage pressures that tend to move the grains toward the outlet. If the soil is a mixed-grained material containing fine sand or silt, the finer constituents are likely to be washed out. This may lead to settlement or to the development of erosion tunnels in the drained soil and, in addition, may clog the drains. Hence, it is usually necessary to prevent the migration. This can be accomplished most effectively by covering the drained material where the water emerges with a coarser granular material, known as a *filter*, which does not impede the flow of water but which contains voids too small to be invaded by the fine particles of the material being drained.

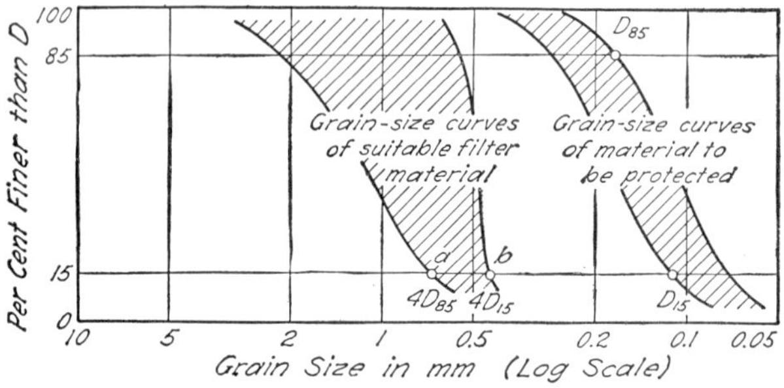

FIG. 2.7. Diagram illustrating specification for grain size suitable for filter. Right-hand shaded area encloses all grain-size curves for material to be protected; left-hand area indicates range within which curves for filter material must lie.

The grain-size requirements for satisfactory filter materials are defined by the graph shown in Fig. 2.7. Experiments have shown that the 15 per cent size D_{15} of the filter material must be at least four times as large as the 15 per cent size of the coarsest layer of soil in contact with the filter and not more than four times as large as the 85 per cent size D_{85} of the finest adjoining layer of soil. This may be expressed by the inequality:

$$\frac{D_{15}\text{ (of filter)}}{D_{85}\text{ (of protected soil)}} < 4 < \frac{D_{15}\text{ (of filter)}}{D_{15}\text{ (of protected soil)}} \qquad 2.21$$

A filter meeting these requirements will prevent the fine material in the protected soil from being washed out; however, in some cases the filter itself may be subject to the loss of fines. It is then necessary to construct a system of filters that become increasingly coarser as they approach the drains. Each layer of the system must meet the grain-size requirements for a filter with respect to the layer of material adjacent to it. Such a system is called a *graded filter*.

Not all the water can be removed from a soil by gravitational processes. The amount that remains is known as the *water-holding capacity* of the soil. Although several laboratory tests have been devised to determine the water-holding capacity, there is no close correlation between the results of the tests and the water-holding capacity in the field. Moreover, the amount of water that remains in the voids is of little practical importance compared with the rate at which the drainage can be accomplished. For clean sands and gravels only a few hours or days are required to lower the water table several feet, whereas for silty soils several weeks may be needed.

As the water flows out of a coarse-grained soil, part of the void space is filled with air. This process is known as *drainage by air invasion.* In fine-grained soils, however, the void space continually decreases, and there is practically no air invasion until the shrinkage limit is reached. This process is known as *drainage by consolidation.*

Drainage by Desiccation. It is not possible to drain fine-grained soils in a reasonable length of time by gravity alone. However, in nature, drainage may be effected by desiccation. During this process, moisture is gradually lost by evaporation. As the water evaporates, the solid soil particles are drawn into closer and closer contact on account of the surface tension of the water. At the shrinkage limit the strength of the soil is great enough to withstand the forces of surface tension without further consolidation, and, if further desiccation takes place, the process of air invasion begins. The moisture content does not decrease to zero but attains a value in equilibrium with the relative humidity of the soil atmosphere. Clay soils may retain as much as 7 per cent moisture even in the air-dry state.

The process of desiccation increases the strength of fine-grained soils because the increasing tension in the water pulls the colloidal particles into close contact. The relationship between the water content and the unconfined compressive strength of typical soils is shown in Fig. 2.8. Clay soils that have become desiccated in the field and are subsequently saturated usually retain a large portion of the strength imparted by the process of desiccation. Layers of such clay are not uncommon and may provide adequate foundations for light structures.

When an air-dried or oven-dried sample of fine-grained soil is placed in water, the soil *slakes.* The force of surface tension draws the water into the voids and compresses the air trapped in the interior. The pressure in the air may become so high that the corresponding tension in the soil skeleton exceeds the tensile strength of the soil and causes the slaking.

Frost Action. If the temperature of the surface of the soil is lowered below the freezing point, moisture migrating from the lower layers may collect and freeze beneath the upper crust of ice. If the soil profile is such that water cannot rise by capillarity from the groundwater table, consolidation takes place in the region from which the moisture migrates. If, as is more common, the soil may draw moisture from the water table, a continuous process is set up in which moisture rises from the water table toward the cold surface by capillary action, evaporation, and condensation until lenses of ice are formed in the soil near the surface of the ground. This phenomenon, known as *frost action,* is most prevalent in very fine sands and silts, because these soils are capable of transmitting the greatest amount of water in the shortest time by means of capillary action. In the northern part of the United States it is not uncommon for lenses of ice to reach an aggregate thickness of 6 to 18 in. during one winter season. Frost action is responsible for much damage to highway pavements. When the lenses melt, the surface of the ground subsides and the soil becomes a thick suspension with negligible bearing capacity. Most highway departments in the northern states require the removal of all frost-heaving materials, from the subgrade down to the average depth of frost penetration. The discarded material is replaced by frost-free sands or gravels or mixed-grained soils having less than 3 per cent of the particles finer than 0.02 mm.

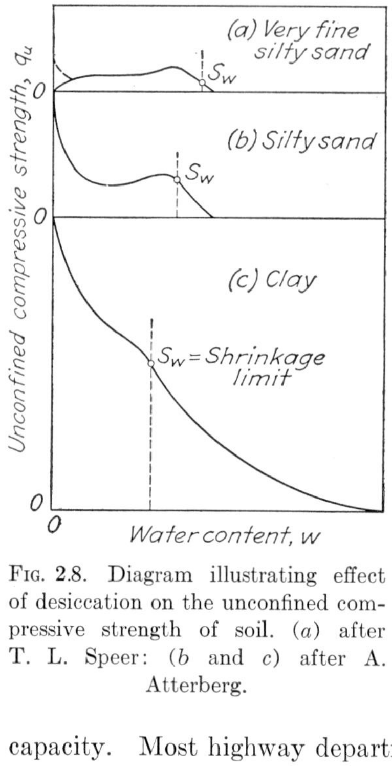

FIG. 2.8. Diagram illustrating effect of desiccation on the unconfined compressive strength of soil. (*a*) after T. L. Speer: (*b* and *c*) after A. Atterberg.

2.6. Importance of Stress–strain Characteristics of Earth Materials

The stress–strain characteristics of a soil or rock determine the settlement that a given structure will experience. In some instances, they may also serve as an indication of construction difficulties that may arise during excavation into soil masses.

The computation of the settlement of a structure due to the deformations of the subsoil is analogous to the computation of the deflection of a truss. The solution of the latter problem requires a knowledge of the stress in each member, the dimensions of the member, and the stress–strain characteristics of the material of which it is made.

One of the materials best known to the structural engineer is steel. Within the range of working stresses, the stress–strain curve for steel is practically a straight line, and within this range the steel is said to be *elastic*. Its properties in simple compression or tension are fully described by two constants

$$E\text{, the modulus of elasticity} = p/\epsilon = 30{,}000{,}000 \text{ lb/sq in.}$$

$$= 2{,}000{,}000 \text{ tons/ sq ft,}$$

and

$$\mu\text{, Poisson's ratio} = \epsilon_L/\epsilon = 0.3$$

where

p = intensity of direct stress in a given direction

ϵ = strain in direction of p

ϵ_L = strain at right angles to direction of p

A sample of an elastic material under a vertical pressure p experiences a reduction in volume equal to $p(1 - 2\mu)/E$ per unit of volume. Hence, if $\mu = 0.5$, no volume change occurs; the sample merely shortens vertically and spreads laterally. On the other hand, if $\mu = 0$, no lateral spread whatsoever accompanies the vertical shortening. For steel, $\mu = 0.3$. This indicates that lateral spreading and volume change both take place.

The stress–strain relationships for soils and rocks are more complex than those for a nearly elastic material such as steel within the range of working stresses. One of the most important differences is that the stress–strain curves are commonly not linear. Hence, the quantities E and μ have no meaning unless they are defined arbitrarily. Moreover, the values of these quantities are dependent on the lateral pressure p_c. Inasmuch as real soils are acted upon by lateral pressures due to the overburden and to other loads supported by the soil, it is often necessary to take this dependence into account.

The stress–strain characteristics of cohesive soils may be investigated by means of unconfined compression tests in which there is no lateral pressure acting on the sample, or by triaxial compression tests in which the sample is usually subjected to a constant lateral pressure while the vertical pressure is increased. Cohesionless soils cannot be tested without lateral restraint because they do not remain intact under their own weight. Therefore, triaxial tests are required for investigating the stress–strain characteristics of sand samples.

2.7. Stress–strain Characteristics of Sands

For a given lateral pressure p_c, the stress–strain characteristics of dry sand depend primarily on the relative density of the sand, and to a much smaller degree on the shape and size of the grains. Typical stress–strain curves are shown in Fig. 2.9.

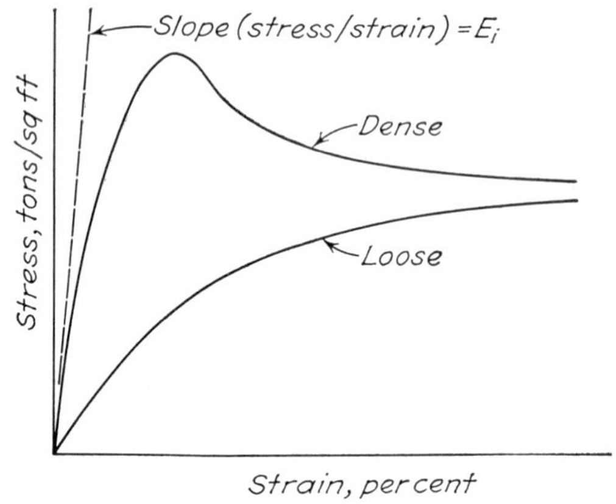

FIG. 2.9. Typical stress-strain curves for sand under constant lateral pressure.

It is apparent that neither of these curves even approximates a straight line through the origin. Hence, the term modulus of elasticity has no meaning in connection with such soils. However, the ratio of stress to strain for very low stresses may be represented in a crude way by the tangent to the stress–strain curve at the origin. The slope of this tangent (stress per unit of strain) is designated as the *initial tangent modulus*, E_i. The value of E_i is several times greater for dense than for loose sand. Therefore, the settlement of a structure exerting low pressures on a dense sand is likely to be much smaller than that of a similar structure on loose sand.

The values of E_i for both loose and dense sand increase markedly with increasing lateral pressure. This is shown by Fig. 2.10 which represents the results of triaxial compression tests on samples of a typical sand in the loose and dense states. For loose sands E_i increases directly with p_c. For dense sands, E_i increases rapidly with p_c when p_c is small. For large lateral pressures, the rate of increase drops to a value comparable with that for loose sands.

The quantity $\mu = \epsilon_L/\epsilon$ is not constant but varies during the progress of a test. The volume of a specimen of loose sand decreases in a general way during the test, whereas that of a dense sand first decreases and then increases as failure is approached.

Saturation does not appreciably alter the relationships described above, provided the water content of the sand can change freely.

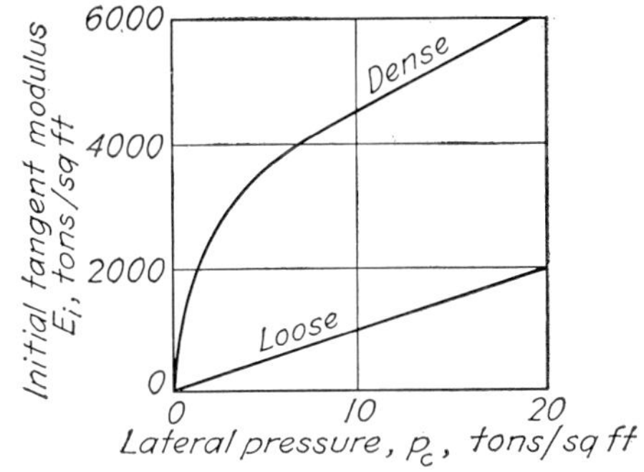

FIG. 2.10. Relation between initial tangent modulus and confining pressure for sands.

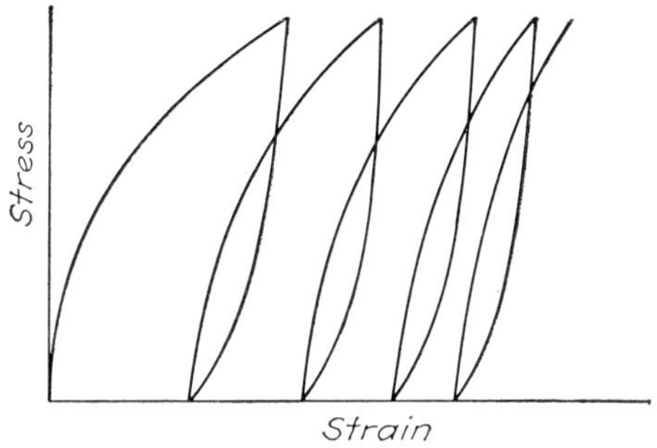

FIG. 2.11. Relation between stress and strain for sand subjected to repeatedly applied vertical load in triaxial compression test.

This is usually the case in the field because the permeability of most sands is fairly high. However, in very fine-grained or silty sands the water content may remain almost unchanged during a rapid change in stress. Under this condition, the values of E_i for loose saturated sands are commonly smaller than for dry sands of the same relative density, whereas for dense sands they are larger.

The repeated application and removal of vertical stress on a sample of laterally confined sand leads to a stress–strain diagram like that shown in Fig. 2.11. There is an increase of total deformation after each application of the load, but the magnitude of the increase becomes successively smaller.

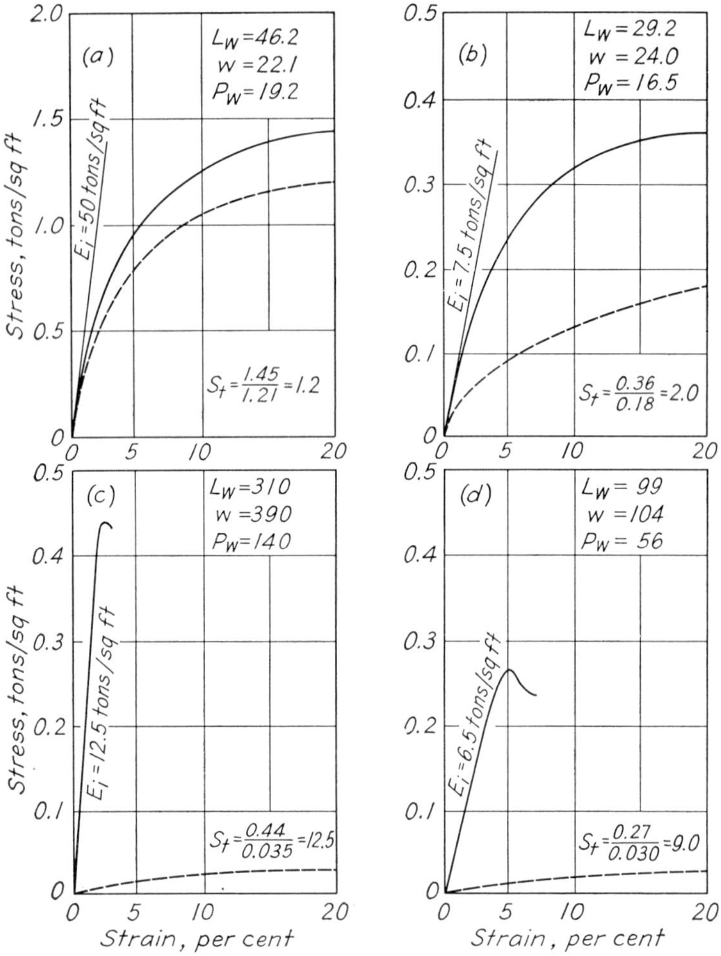

FIG. 2.12. Stress-strain curves in unconfined compression for undisturbed (full lines) and remolded (dash lines) samples of four typical clays: (*a*) stiff insensitive clay from St. Louis, Mo.; (*b*) soft insensitive glacial clay from Detroit, Mich.; (*c*) soft sensitive lacustrine volcanic clay from Mexico City; (*d*) soft sensitive organic silty clay from New Haven, Conn.

2.8. Stress–strain Characteristics of Clays

Unconfined Tests. The simplest procedure for investigating the stress–strain characteristics of a clay is the performance of an unconfined compression test (Art. 1.7). Stress–strain curves for four cohesive soils are shown in Fig. 2.12. Curves *a* and *b* are representative of common types; curves *c* and *d* represent rather unusual materials.

The dash lines in Fig. 2.12 indicate the results of tests on completely remolded samples of the materials at their original water content. The values of E_i are significantly smaller for the remolded than for the undisturbed samples. Values of the sensitivity S_t, as defined in Art. 1.7, are also indicated on the diagrams.

It may be observed that the stress–strain diagrams for the clays of low sensitivity *a* and *b* are curved throughout their entire range, whereas those for the clays of high sensitivity *c* and *d* are linear almost to the point of failure. The latter behave in their undisturbed state like brittle elastic materials and upon remolding assume the character of viscous fluids.

Although the stiffest clay represented by the diagrams in Fig. 2.12 has a value of E_i equal to 50 tons/sq ft, this value is relatively low in comparison with 2,000,000 tons/sq ft for steel and approximately 200,000 tons/sq ft for concrete.

Influence of Lateral Pressure. If a specimen of clay is subjected to an all-around pressure p_c and then to an additional vertical pressure p, the corresponding stress–strain diagram is similar to that for an unconfined specimen of the same material, except that E_i and the failure load are both larger. If the all-around pressure is at least as great as the lateral pressure that acted on the sample before it was removed from the ground, the value of E_i is given approximately by the equation

$$E_i = Cp_c \tag{2.22}$$

If the water content of the clay is permitted to change until it reaches equilibrium under the pressure p_c, and then the test is completed with no further change in water content, the value of C ranges from about 10 for clays of high plasticity to about 100 for clays of low plasticity. If the test is completed so slowly that the water content can change freely, the value will be much smaller.

Influence of Complete Confinement. In many instances the settlement of a structure is due to the presence of one or more layers of soft clay located between layers of sand or stiffer clay. The adhesion between the soft and stiff layers almost completely prevents the lateral bulging of the soft layers. In the laboratory this condition is simulated most closely by the *confined compression* or *consolidation test.*

During a consolidation test the specimen is completely confined by a metal ring, Fig. 2.13. The load is transmitted to the upper and lower faces of the specimen through two porous disks that permit water to flow into or out of the clay. The deformation is measured by means of a dial gage.

The pressure p is increased in steps. After each increment the load is allowed to remain constant until the deformation practically ceases. This usually requires several hours even for a specimen having a thickness of as little as ¾ in., because the deformation takes place only as fast as water can be squeezed out of the clay.

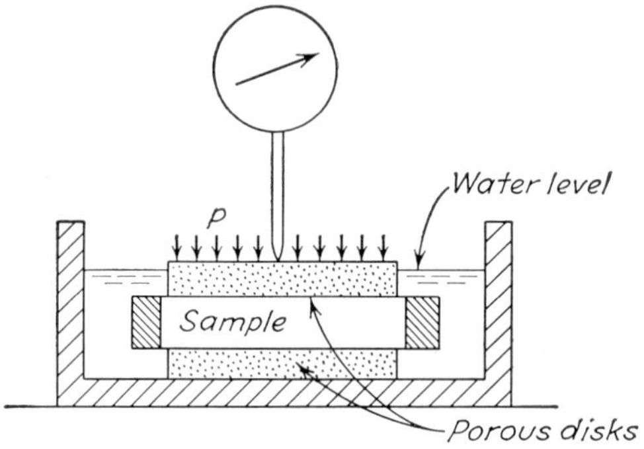

FIG. 2.13. Diagram showing apparatus for consolidation test.

When the rate of deformation under a given pressure has become very small, a new increment is applied and the procedure is repeated.

The results are plotted in the form of a curve representing the final void ratio corresponding to each increment of pressure as a function of the total pressure. It is convenient to plot the pressure to a logarithmic scale. The diagram is then known as an e-log p curve.

Two different conditions of practical importance must be recognized, namely, whether the sample of clay was taken from a normally loaded stratum or from a preloaded one. A stratum is said to be *normally loaded* if it has never been acted on by vertical pressures greater than those existing at present. On the other hand, a *preloaded* stratum was at one time in its history subjected to greater vertical pressures than are now active.

A typical e-log p curve for an undisturbed sample of a normally loaded clay of low sensitivity is shown in Fig. 2.14. It is designated by K_u. The results of a consolidation test on the same material remolded at a water content near the liquid limit are indicated by K_r.

The curve K_u characteristically consists of two branches, a fairly horizontal initial portion and a nearly straight inclined portion. The curve K_r is more or less parabolic at low pressures, and it merges into a straight line with a slope flatter than that of K_u.

The coordinates of point *a*, Fig. 2.14, represent the void ratio and effective pressure corresponding to the state of the clay in the field. When a sample is extracted by means of the best techniques, the water content of the clay does not increase significantly. Hence, the void

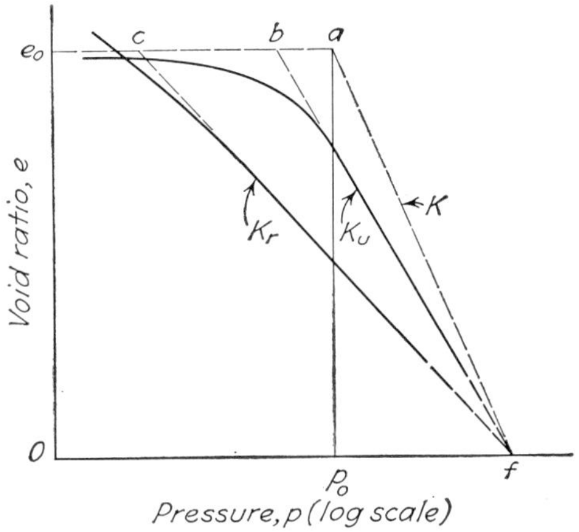

Fig. 2.14. Typical *e*-log *p* curves for undisturbed (K_u) and remolded (K_r) samples of insensitive normally loaded clay.

ratio e_0 at the start of the test is practically identical with that of the clay in the ground. When the pressure on the sample reaches p_0, the *e*-log *p* curve should pass through point *a* unless the test conditions differ in some manner from those in the field. In reality, K_u always passes below point *a*, because even the best sample is at least slightly disturbed.

The engineer is interested in the *e*-log *p* relationship for the clay in the field, not in the laboratory. Therefore, he requires some procedure for extrapolating from the results of the laboratory tests to those representative of conditions in the field. One procedure for accomplishing this end can be developed on the basis of a comparison between curves K_u and K_r. It has been observed that the extension of the straight lower branch K_r intersects the axis $e = 0$ at approximately the same point as the extension of the inclined branch of K_u. This point is denoted by *f*. If the great degree of disturbance represented by remolding a carefully taken sample does not appreciably change the location of f, it may be reasoned that the slight disturbance due to sampling should have virtually no effect. Hence, it is assumed that the field e-log p relation may be represented by a straight line K extending from a to f.

If a clay deposit is known to be normally loaded, the value of p_0 can be computed from a knowledge of the water content, the unit weight of the solid constituents, and the degree of saturation of the overlying strata, provided the position of the water table is known. The initial void ratio of the clay can also be computed readily. Hence, the coordinates of point a can be determined. The procedures for computing p_0 and e_0 have been discussed in Arts. 1.6 and 2.4.

Point f can be determined by performing a consolidation test on an undisturbed sample or, if the expense of obtaining one is too great, on a partly disturbed or remolded sample. Points a and f can then be connected by a straight line to establish the field relationship K upon which a computation of settlement can be based.

The slope of K on the semilogarithmic diagram is designated as the *compression index* C_c, defined by the equation

$$C_c = \frac{e_0 - e_1}{\log_{10} p_1 - \log_{10} p_0} = \frac{e_0 - e_1}{\log_{10} \dfrac{p_1}{p_0}} \qquad 2.23$$

The numerical value of C_c can be determined readily from the diagram. If e_1 corresponds to any arbitrary pressure p_1 and e_2 to the pressure $p_2 = 10p_1$, then $C_c = e_1 - e_2$.

It has been found that the value C_c is closely related to the liquid limit of normally loaded clays. The relationship between the two quantities is approximately

$$C_c = 0.009(L_w - 10) \qquad 2.24$$

where L_w is expressed as a percentage. Equation 2.24 is of great practical importance because it permits one to compute the approximate settlement of a structure above a deposit of fairly insensitive normally loaded clay if only the liquid limit is known, and even if no consolidation tests have been performed.

The results of a consolidation test on a carefully taken sample of normally loaded clay of high sensitivity (S_t greater than about 8) resemble curve K_u, Fig. 2.15. At a pressure near p_0 the structure of the clay appears to break down, and a slight increase of pressure leads to a large decrease in void ratio. If the statistical relationship, eq. 2.24, is used to determine the compressibility of such a material, the computed settlement is too small. This constitutes an error on the unsafe side. Hence, the curve K representing the field e-log p relationship should itself be used.

The approximate shape of the curve K for a highly sensitive clay can be obtained in the following manner. A tangent is drawn to the steep part of the curve at its point of inflection c. This intersects the line e_0 at b. The lower branch of the curve is extended to $e = 0$ to locate point f. The vertical line Af is then drawn, and curve K so located that for any value of e the distance from K to Af divided by the distance from K_u to Af is equal to aA divided by bA.

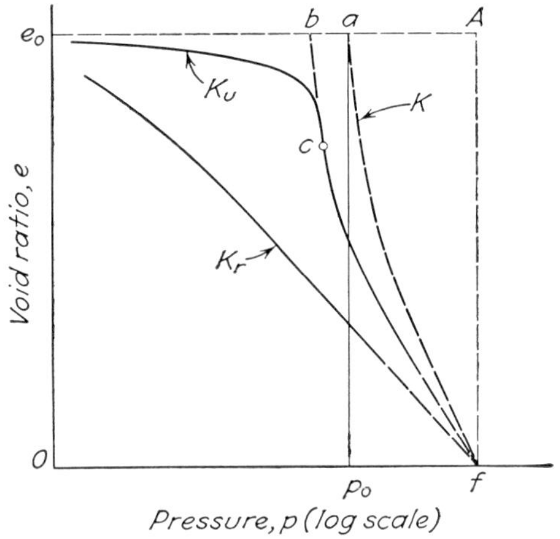

FIG. 2.15. Typical e-log p curves for undisturbed (K_u) and remolded (K_r) samples of highly sensitive normally loaded clay.

The e-log p diagram for a typical preloaded clay is indicated by curve K_u in Fig. 2.16. Point a' corresponds to the void ratio e_0 and pressure p_0' of the clay in the field at the time when the stratum was under its maximum load. By some process, such as erosion, the pressure was reduced and the void ratio increased slightly by swelling until the present state, represented by a, was reached.

The curve K_u passes below point a, whereas the field curve K should pass through it. Moreover, the field curve K should pass close to point a', because before the erosion the field relationship should have been represented by $a'f$. We cannot construct this curve because the pressure p_0' corresponding to point a' is no longer known; the present overburden pressure is only p_0. If we fail to recognize that the clay is preloaded and base a settlement computation on af, the computed settlement will be far too great. The change Δe_n in void ratio computed for an increase Δp in pressure is likely to be 4 to

10 times as great as the real change Δe_p, provided Δp is not greater than about half of $p_0' - p_0$. As Δp approaches $p_0' - p_0$, the error becomes smaller.

Occasionally, deposits of extrasensitive clay are encountered for which the e-log p curve, Fig. 2.17, is typical. Point b, representing the upward extension of K_u, lies to the right of point a corresponding to the existing overburden pressure. The quantity $p_0' - p_0$ may

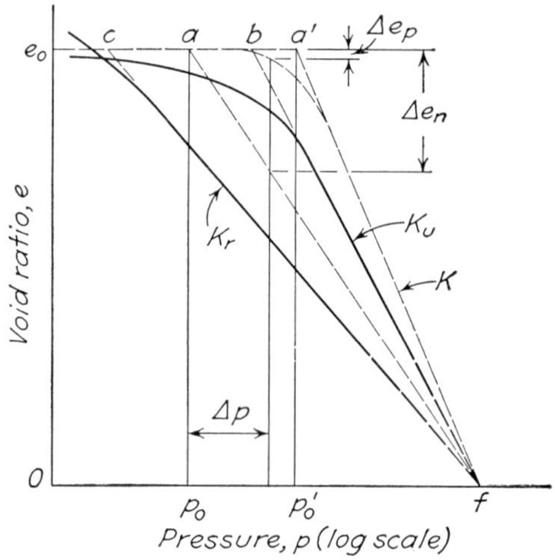

FIG. 2.16. Typical e-log p curves for undisturbed (K_u) and remolded (K_r) samples of preloaded clay of ordinary sensitivity.

represent the influence of preloading, or it may possibly represent an increase in the strength or rigidity of the clay due to such processes as thixotropy. These matters have not yet received adequate study. In any event, when such a material is encountered, it seems reasonable to assume that the pressure p_0' can be approached before the structure of the clay will break down, and that the settlements associated with values of Δp less than about half of $p_0' - p_0$ will, like those for a preloaded deposit of ordinary sensitivity, be relatively small. For such materials, estimates of settlement should be based on tests on the best undisturbed samples that can be obtained.

In connection with practical problems, the most important consideration is to be able to recognize whether or not a clay is preloaded. It almost certainly is if the water content is closer to the plastic limit than to the liquid limit. Moreover, by performing a consolidation test on a very carefully taken sample, one can often obtain the data required to make the decision. In Fig. 2.14 it may be noticed that the upward projection of the straight part of K_u intersects the line $e = e_0$ at point b which is located on the left of point a. This is always true of a normally loaded clay of ordinary sensitivity. On the other hand, in Fig. 2.16 which refers to a preloaded clay of ordinary sensitivity, b is located on the right side of a. If the sample is somewhat disturbed, however, b may be located farther to the left and may even reach c. Therefore, the influence of disturbance tends to destroy the evidence of preloading. Hence, if it is probable that

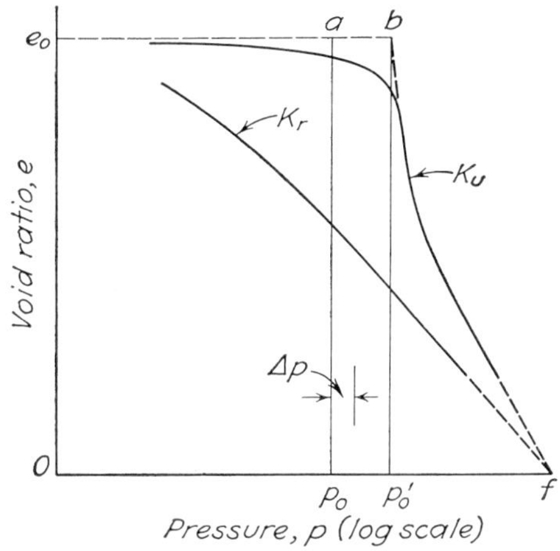

FIG. 2.17. Typical e-log p curves for undisturbed (K_u) and remolded (K_r) samples of highly sensitive slightly preloaded clay.

a clay deposit may be preloaded and samples are to be obtained for consolidation tests, the best possible techniques for sampling should be used.

If good samples are available, the pressure p_0' can be estimated by a procedure developed by A. Casagrande, Fig. 2.18. On the curve K_u, c is located at the point of maximum curvature. From c, a tangent is drawn to K_u, and a horizontal line is also constructed. The angle between these two lines is then bisected. The point of intersection of this bisector with the upward extension of a tangent to the straight part of K_u is denoted by d. The abscissa of d corresponds to the pressure p_0'.

This procedure was developed primarily on the basis of laboratory evidence and the limitations on its validity are not yet fully known. Moreover, it is seldom necessary to resort to such refinement. Usually it is sufficient to know whether a clay is strongly preloaded or not. If it is, its contribution to settlement is likely to be negligible.

The preloaded state may have been due to a former overburden of soil, to the weight of glacial ice, or to stresses induced by desiccation. Lenses or layers of preloaded clay in a normally loaded deposit are usually the result of the latter cause.

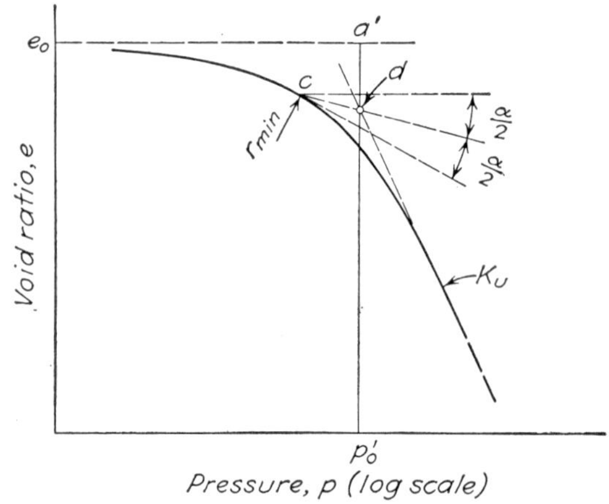

FIG. 2.18. Graphical construction for determining former maximum pressure p_0' from e-log p curve (after A. Casagrande).

Rate of Consolidation. The consolidation of clay under a load does not take place instantaneously. For each increment of load that is applied to the sample in a consolidation test, it is possible to plot a time–deformation curve such as that shown in Fig. 2.19.

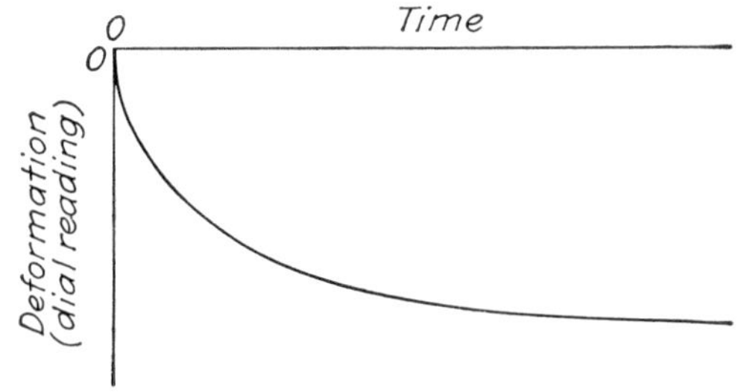

FIG. 2.19. Typical time-deformation curve for sample under constant load in consolidation test.

Clays are so impervious that the water is almost trapped in the pores. When an increment of load is applied, the porewater cannot escape immediately. Since the clay particles tend to squeeze together, pressure develops in the porewater in a manner similar to the development of pressure in the oil filling a hydraulic jack when a weight is placed on the ram. This pressure tends to make the fluid flow out. The flow is rapid at first, but, as it continues, the pressure drops and the rate of flow decreases. As the water is forced out of a clay sample, the grains of clay can move closer together. Therefore, the surface of the specimen settles. The rate of settlement is rapid at first, and then decreases to a small fairly constant value.

The progress of consolidation can be observed by measuring the decrease of porewater pressure. This decrease occurs at different rates in different parts of the sample. It occurs more rapidly at the upper and lower boundaries where drainage is facilitated than it does in the middle.

On the assumption that the laws of hydraulics govern the decrease of porewater pressure, a theory has been developed for computing the rate of settlement. This is known as the *theory of consolidation.* It leads to the following relationship between the degree of consolidation and the time

$$U\% = f(T_v) \qquad 2.25$$

where

$$T_v = \frac{c_v}{H^2} t \qquad 2.26$$

In this expression, U is the *degree of consolidation*, or the amount of settlement that has occurred at a given time compared with the total amount that will take place; T_v is a dimensionless number called the time factor; c_v is a property of the soil known as the coefficient of consolidation; H is the half-thickness of the sample; and t is the time corresponding to U. From this expression it may be observed that the time required to reach a given degree of consolidation for given drainage conditions varies as the square of H.

The theoretical relationship stated symbolically in eq. 2.25 is shown graphically in Fig. 2.20, plotted to a semilogarithmic scale. If the values of c_v and H are known for a given layer of clay, the theoretical curve can be replotted in the form of a per cent consolidation–time curve on the basis of eq. 2.26. Moreover, if the ultimate settlement due to the consolidation of the clay layer has been computed, the curve can be further modified into a settlement–time curve.

The factor H in eq. 2.26 is defined as the half-thickness of a layer drained on both sides as in a consolidation test. Thus, the time required to reach a given degree of consolidation is four times as great for a layer of given thickness drained on one side only as it is for one drained on both sides.

The coefficient of consolidation c_v can be evaluated by means of a consolidation test, by fitting the experimental deformation–time curve to the theoretical curve for eq. 2.25. The first step in the procedure is the determination of the deformations corresponding to $U = 0$ per cent and $U = 100$ per cent. This can most readily be done by procedures developed by A. Casagrande. The first step is to plot the

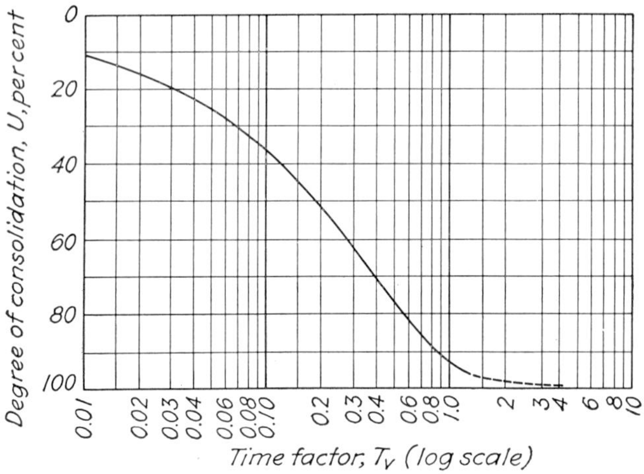

FIG. 2.20. Theoretical relationship between degree of consolidation U and time factor T_v.

deformation (dial reading) against the logarithm of the time for one of the load increments in the consolidation test. The curve has the shape shown in Fig. 2.21.

To obtain the location of the line $U = 0$ per cent, select any time t_1 at which less than half the consolidation for the entire load increment has taken place. Locate the corresponding point c on the curve. Then locate point d corresponding to a time $t_1/4$. The vertical distance between c and d is a. Lay off a horizontal line at distance a above d. The ordinate of this line is $U = 0$ per cent.

To obtain the location of the line $U = 100$ per cent, select the point of inflection e of the experimental curve, and draw a tangent downward from this point. Draw to the left a tangent to the lower, straight-line part of the experimental curve. The two tangents intersect at f, which corresponds to $U = 100$ per cent.

It may be observed that the laboratory curve approaches an inclined tangent, whereas the theoretical curve approaches a horizontal one. The difference between the two curves is called the *secondary consolidation.* The laws that govern secondary consolidation are not yet known. However, the phenomenon is of importance because it may account for a large fraction of the settlement of an actual structure. On account of secondary consolidation, the theoretical and experimental curves coincide only up to about $U = 60$ per cent. Therefore, in fitting the two curves to determine c_v, we must use values of t corresponding to less than $U = 60$ per cent.

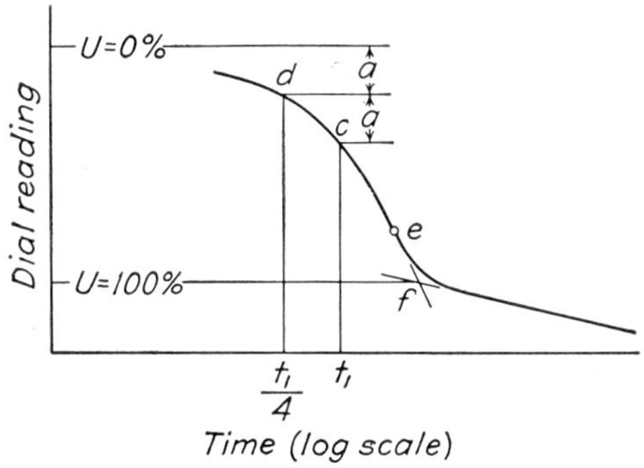

FIG. 2.21. Graphical construction for determining deformations corresponding to $U = 0$ per cent and $U = 100$ per cent in consolidation test (after A. Casagrande).

Commonly we select the value of t corresponding to $U = 50$ per cent. The corresponding value of $T_v = 0.197$. Hence, by eq. 2.26, we can evaluate c_v as

$$c_v = \frac{T_v}{t} H^2 \qquad 2.27$$

where H is the half-thickness of the test specimen.

Once c_v for the clay is known, eq. 2.26 and Fig. 2.20 can be used to plot the theoretical time-settlement relation for any structure located above a stratum of the clay.

On the basis of the consolidation test data the coefficient of permeability k of the sample during any given load increment may be computed by means of the equation

$$k = c_v \gamma_w m_v \qquad 2.28$$

In eq. 2.28 the *coefficient of volume compressibility* m_v is defined by the expression

$$m_v = \frac{a_v}{1 + e_0} \qquad 2.29$$

in which the coefficient of compressibility a_v is the ratio between the change in void ratio and the change in load for the given increment.

$$a_v = \frac{e_0 - e_1}{p_1 - p_0} \tag{2.30}$$

ILLUSTRATIVE PROBLEMS

1. A normally loaded soft clay of low sensitivity has a liquid limit of 57 per cent. Estimate the compression index.

Solution. Since the clay is soft and normally loaded, its water content is likely to be close to the liquid limit, and eq. 2.24 is applicable. Therefore,

$$C_c = 0.009(L_w - 10) = 0.009 \times (57 - 10) = 0.42$$

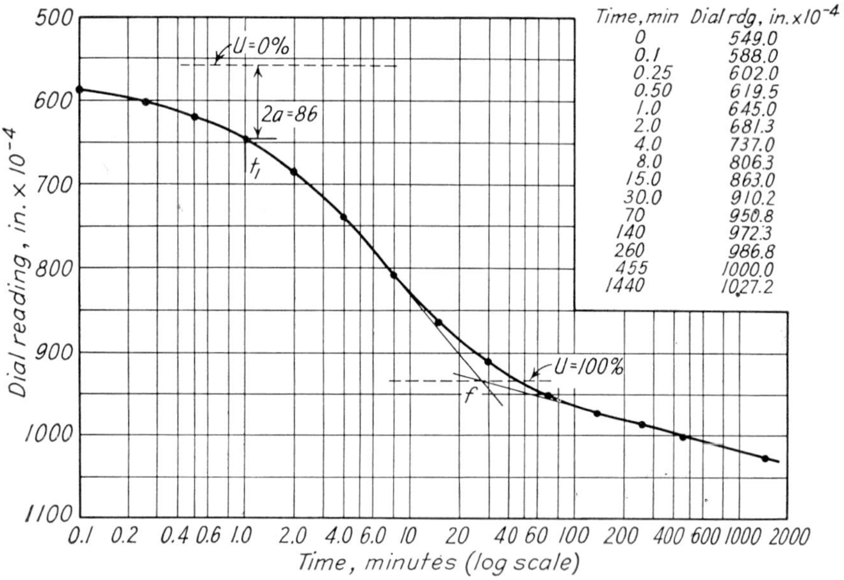

FIG. P2.3.

2. The accompanying time–dial reading curve, Fig. P2.3, was obtained during a consolidation test on a soft glacial clay ($L_w = 43\%$, $P_w = 21\%$, $w = 39\%$) when the pressure was increased from 1.66 kg/sq cm to 3.33 kg/sq cm. The void ratio after 100 per cent consolidation under 1.66 kg/sq cm was 0.945 and that under 3.33 kg/sq cm was 0.812. The dial was set at zero at the beginning of the test, and the initial height of the sample was 0.75 in. Drainage was permitted

at both faces of the sample. Compute the coefficient of permeability corresponding to the stated increment of pressure.

Solution. To determine dial reading R_d at 0 per cent consolidation, choose $t_1 = 1.0$ min

For $t_1 = 1.0$, $R_d = 645$

For $t_1/4 = 0.25$, $R_d = 602$

$a = 43$; $2a = 86$ (see Fig. 2.21)

For $U = 0\%$, $R_d = 645 - 86 = 559$

For $U = 100\%$, $R_d = 936$ (from graphical construction, Fig. 2.21)

For $U = 50\%$, $R_d = 559 + (936 - 559)/2 = 748$

For $U = 50\%$, $t = 4.5$ min $= 270$ sec (from Fig. P2.3)

For $U = 50\%$, $T_v = 0.197$ (from Fig. 2.20)

For $U = 50\%$, $2H = 0.7500 - 0.0748 = 0.6752$ in.; $H = 0.3376$ in.

Hence,

$$c_v = \frac{T_v H^2}{t} = \frac{0.197}{270} \times (0.3376 \times 2.54)^2$$

$$= 5.38 \times 10^{-4} \text{ sq cm/sec}$$

$$a_v = \frac{e_0 - e_1}{p_1 - p_0} = \frac{0.945 - 0.812}{3333 - 1666} = \frac{0.133}{1667} = 7.94 \times 10^{-5} \text{ sq cm/g}$$

$$m_v = \frac{a_v}{1 + e_0} = \frac{7.94 \times 10^{-5}}{1.945} = 4.08 \times 10^{-5} \text{ sq cm/g}$$

$$k = c_v \gamma_w m_v = 5.38 \times 10^{-4} \times 1.0 \times 4.08 \times 10^{-5}$$

$$= 2.20 \times 10^{-8} \text{ cm/sec}$$

3. In a laboratory consolidation test a sample of clay with a thickness of 1 in. reached 50 per cent consolidation in 8 minutes. The sample was drained top and bottom. The clay layer from which the sample was taken is 26 ft thick. It is covered by a layer of sand through which water can escape and is underlain by a practically impervious bed of intact shale. How long will the clay layer require to reach 50 per cent consolidation?

Solution. According to eq. 2.26

$$\frac{t_s}{t_l} = \frac{H_s^2}{H_l^2}; \qquad t_l = t_s \frac{H_l^2}{H_s^2}$$

where s refers to the sample and l to the layer. For the sample, $H_s = 0.5$ in.; for the layer, $H_l = 26$ ft. Then

$$t_l = 8 \times \frac{(26 \times 12)^2}{0.5^2} \times \frac{1}{60 \times 24 \times 365} = 5.92 \text{ years}$$

2.9. Methods of Investigating Strength of Soil and Rock

The maximum load that can be transmitted to the subsoil by a foundation depends upon the resistance of the underlying soil or rock to shearing deformations. Therefore, it is of prime importance to investigate the factors that control the shearing strength of these materials. In some instances, the shearing strength is investigated by means of a *direct-shear test* in apparatus of the type shown in Fig. 2.22. More commonly, the shearing strength is investigated by means of compression tests in which an axial load is applied to the

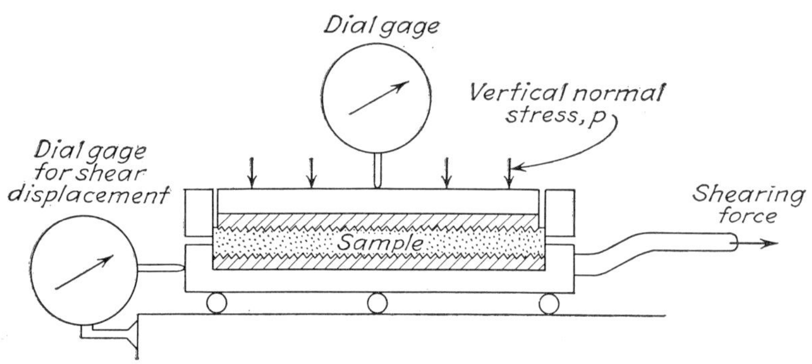

FIG. 2.22. Apparatus for direct-shear test.

specimen and increased until failure occurs. The specimen may or may not be subjected to a lateral pressure. If it is, the test is known as a *triaxial test.* Otherwise, it is known as an *unconfined compression test.*

The use of compression tests to investigate the shearing strength of materials depends upon the facts that failure in such tests takes place by shear on one or more inclined planes and that it is possible to compute the normal pressure and the shearing stress on such a plane at the instant of failure.

If we perform a series of shear tests on identical samples of soil, but subject each specimen to a different normal pressure, we can plot the results in the form of a diagram showing the normal stress p as a function of the shearing stress s at the instant of failure. The results for most soils approach a straight line as shown in Fig. 2.23. The equation of this line is

$$s = c + p \tan \phi \qquad 2.31$$

If these results have been obtained by a series of triaxial compression tests in which the vertical pressure at failure is p_1 and the

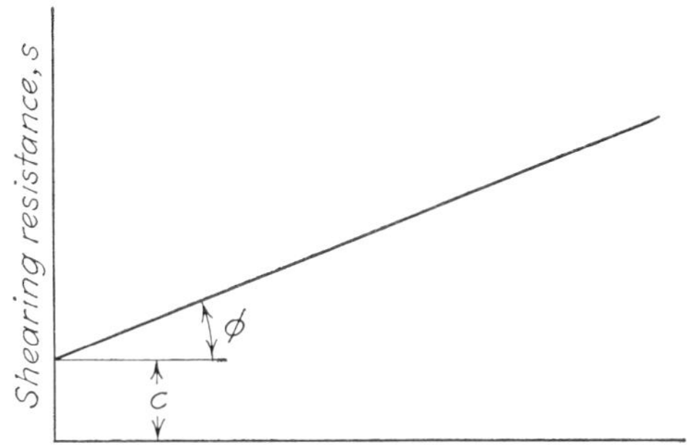

FIG. 2.23. Relation between normal pressure and shearing resistance.

lateral pressure p_3, there is a specific relationship between p_1 and p_3, namely,

$$p_1 = p_3 \tan^2 (45° + \phi/2) + 2c \tan (45° + \phi/2) \qquad 2.32$$

The quantity $\tan^2 (45° + \phi/2)$ is a constant that depends only on the value ϕ. It is designated as N_ϕ. Hence,

$$p_1 = p_3 N_\phi + 2c\sqrt{N_\phi} \qquad 2.33$$

In the preceding equations, the value c, which is equal to the shearing resistance on the failure plane if the normal pressure on that plane is zero, is called the *cohesion.* The angle ϕ is known as the *angle of internal friction.* Equation 2.33 is valid not only for the results of triaxial tests in which p_1 and p_3 are the vertical and lateral pressures, respectively, but also for any point in a material governed by the failure law eq. 2.31, if p_1 and p_3 are the major and minor principal stresses at that point.

It may also be demonstrated that the angle α between the plane of failure and the plane on which the major principal stress p_1 acts is equal to

$$\alpha = 45° + \phi/2 \qquad 2.34$$

2.10. Shearing Resistance of Sand

Dry Sand. Typical stress–strain curves showing the results of triaxial tests on dry sand are shown in Fig. 2.9. If the sand is loose, the stress increases continuously with strain until the sample reaches a state of failure. If the sand is dense, a peak shearing resistance is developed at a relatively small strain, whereupon the stress decreases to an ultimate value close to that for loose sand.

If we perform a series of triaxial tests at different lateral pressures, we obtain a relation between normal pressure and shearing stress on

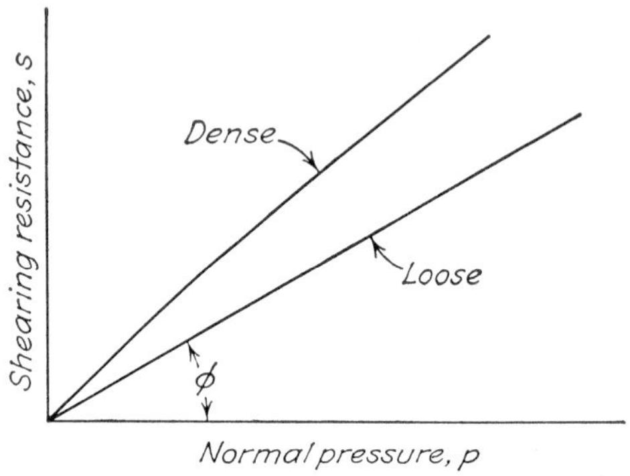

FIG. 2.24. Relation between normal pressure and shearing resistance for dry sand.

the surface of sliding as indicated in Fig. 2.24. The line representing the relationship for a dense sand is slightly curved whereas that for a loose sand is almost straight. Both relationships may be expressed with sufficient accuracy for practical purposes on the assumption that they are straight lines with the equation

$$s = p \tan \phi \qquad 2.35$$

Hence,

$$\frac{p_1}{p_3} = N_\phi \qquad 2.36$$

represents the maximum value possible for the ratio between the principal stresses. When this value is reached, the sand fails.

The value of ϕ depends upon the relative density of the sand, on the grain-size distribution, and on the shape of the grains. It may be estimated with the aid of Table 2.3.

Saturated Sand. If the sand is saturated, the values of ϕ are approximately one degree smaller than for dry sand at the same relative density. Equations 2.35 and 2.36 are still valid provided the pressures p_1 and p_3 are the actual pressures transmitted from grain to grain. In other words, the equations are valid if p represents the intergranular pressure, $\bar{p}$.

Table 2.3. Representative Values of ϕ for Dry Sand

	Degrees	
	Round Grains Uniform	Angular Grains Well Graded
Loose	28.5	34
Dense	35	46

In many practical problems the actual pressure on the plane of failure consists partly of intergranular pressure and partly of porewater pressure. With the aid of eq. 2.9, eq. 2.35 may be rewritten in the form

$$s = (p - u_w) \tan \phi \qquad 2.37$$

Whenever p and u_w can be evaluated, eq. 2.37 can be utilized for computation of the shearing resistance of the sand. Ordinarily, the total pressure p can be estimated without difficulty. If the porewater pressure is merely the consequence of submergence a distance H below the free water surface, u_w is equal to $H\gamma_w$. If in addition to the pressure due to the static head there is a seepage pressure, the value of u_w must take account of the seepage pressure. In some instances it is expedient to estimate u_w on the basis of theory; in others, observations of the porewater pressure can be made in the field.

On the other hand, in connection with one class of problems encountered under field conditions it is impracticable to compute or measure u_w at the time of failure, and eq. 2.37 cannot be used directly. This situation arises if the porewater pressure on the plane of failure changes on account of the tendency of the volume of the sand to change while shearing is in progress.

Unless the pressure $\bar{p}$ is very small, a loose sand tends to become denser when it is sheared and its pores tend to contract. If water can squeeze out of the pores readily, there is practically no change in the porewater pressure. However, if the drainage is prevented, the pressure in the porewater increases. According to eq. 2.37, if the pressure u_w increases, $p - u_w$ decreases, and s becomes smaller. On the other hand, if the sand is dense, it tends to expand during shear unless the pressure $\bar{p}$ is very large. If water can enter the pores readily, there is practically no change in the porewater pressure.

However, if the additional water cannot be supplied, the pressure in the porewater decreases, $p - u_w$ increases, and s becomes greater.

These phenomena may be investigated by laboratory tests in which no drainage, and hence no volume change, is permitted and in which the values of s and of the total pressure p at failure are determined. The initial porewater pressure is zero. Typical results of two series of such tests are shown in Fig. 2.25. For dense sands, at all values of total pressure p except the largest ones, the values of s for undrained sand are greater than the corresponding values for drained sand,

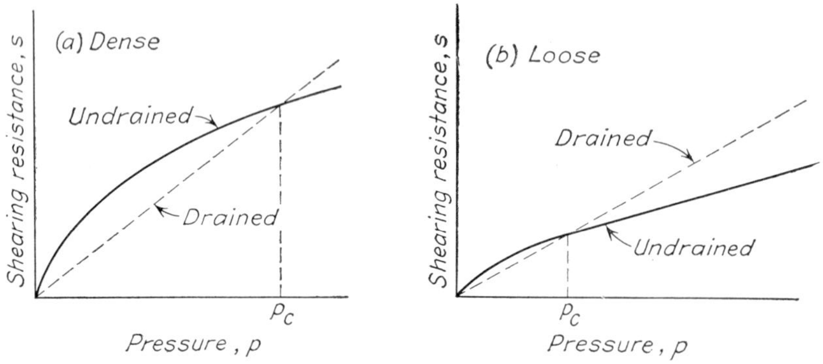

FIG. 2.25. Relation between shearing resistance and total normal pressure in drained and undrained saturated sand in (*a*) dense and (*b*) loose states (after A. Casagrande).

indicated in the figure by the dash line. For loose sands, on the other hand, the values of s for undrained sand are smaller than the corresponding ones for drained sand except at very low pressures.

If a shear failure is likely to take place in the field so rapidly that practically no flow of water can occur to or from the voids of the sand near the surface of sliding, the change in porewater pressure is likely to correspond to that in undrained tests. The shearing resistance corresponding to the total pressure p can then be determined by means of such tests. If in the field the porewater pressure is not initially zero, the value of the initial intergranular pressure $\bar{p}$ should be substituted for p before the curves representing the test results (Fig. 2.25) are utilized to evaluate s.

Coarse sands in the field are so pervious that changes in the porewater pressure on account of volume changes during shear are relatively improbable. The time required for drainage of large masses of fine and very fine sand, however, may be appreciable, and the undrained condition may be approached in the field. Under most circumstances the value of s in the field is intermediate between those corresponding to the drained and undrained conditions.

2.11. Shearing Resistance of Silt and Silty Sand

The shearing resistance of dry silt and silty sand is given by eq. 2.35 but the values of ϕ are somewhat smaller than those for sand. They range between 27° and 30° for loose silt and between 30° and 35° for dense silt.

Saturated silt has relatively low permeability. Hence, failure in the field usually occurs under undrained conditions. Moreover, the results of undrained tests on silt can be expressed with fair accuracy by the linear relation

$$s = p \tan \phi_{cq} \qquad 2.38$$

where p is the total stress on the plane of failure and ϕ_{cq} is known as the *consolidated-quick angle of shearing resistance.* Since silt usually experiences volume decrease during shear, ϕ_{cq} is usually less than ϕ. It commonly lies between 20° and 22° and is sometimes as small as 17°.

As in undrained tests of sand, the porewater pressure for undrained tests of silt is initially zero whereas in the field it may have an initial value. Therefore, if the porewater pressure in the field is not initially zero, the corresponding value of $\bar{p}$ should be substituted for p in eq. 2.38.

2.12. Shearing Resistance of Clay

Saturated Natural Clay. Experience has shown that natural masses of soft clay in the field, if loaded so rapidly that little drainage can occur, usually behave as if $\phi = 0$. On this assumption, $N_\phi = 1.0$, and eq. 2.31 reduces to

$$s = c$$

Therefore, on the basis of eq. 2.32,

$$p_1 = p_3 + 2c$$

If a compression test is made with no lateral pressure ($p_3 = 0$), the value of p_1 at failure is designated as the *unconfined compressive strength* q_u. Hence,

$$q_u = 2c \qquad 2.39$$

Since unconfined compression tests can be performed easily and cheaply, they are used extensively to evaluate the shearing strength of saturated clays.

Under some conditions, when clays have an opportunity to consolidate in the field, they behave as if they possessed frictional resistance. The conditions under which clays exhibit frictional resistance are not yet fully understood. The use of triaxial tests in an attempt to duplicate field conditions has not been found consistently reliable. At the present state of knowledge, it is recommended that the conservative value $\phi = 0$ be assumed and that the shearing strength of the clay be based on the unconfined compressive strength test.

Many stiff saturated clays contain networks of hair cracks or slickensides. The shearing strength of deposits of this kind depends upon the influence of these defects. Triaxial tests on large specimens that include a representative number of defects have been found useful in determining the shearing strength of the mass.

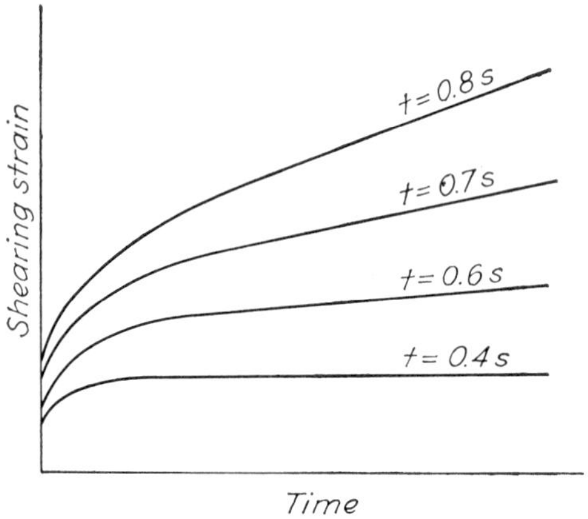

FIG. 2.26. Time-deformation curves for samples of remolded clay subject to shearing stresses t representing different fractions of the shearing resistance s.

Unsaturated Clays in the Field. Unsaturated clays and rolled clay fills usually behave as if they possessed some friction. The shearing strength may be expressed roughly by

$$s = c + p \tan \phi_a \qquad 2.40$$

in which ϕ_a depends upon the air content. It varies from about 0° for saturated clays to about 30° for almost dry clays.

Triaxial tests are commonly used to evaluate c and ϕ_a, but the limitations and reliability of the method have not been clearly established.

Creep. If the shearing stress acting on a sample of undisturbed clay is less than a value known as the *creep strength,* the clay deforms almost instantaneously upon application of the shearing stress and thereafter experiences no progressive deformation. On the other hand, if the creep strength is exceeded, the clay deforms continuously under constant shearing stress. The rate of creep increases with increasing values of the shearing stress, as shown in Fig. 2.26.

The ratio between creep strength and ultimate shearing resistance has not been extensively investigated. For some insensitive clays it appears to be as small as 0.3, whereas for brittle clays it may be on the order of 0.8. Stresses in excess of the creep strength have been an important cause of progressive lateral movement of such structures as retaining walls and abutments.

Illustrative Problems

1. A drained triaxial test is to be performed on a uniform dense sand with rounded grains. The all-around pressure p_3 is to be 2 tons/sq ft. At about what vertical pressure should the sample fail?

Solution. According to Table 2.3, the value of ϕ is likely to be about 35°. Therefore,

$$N_\phi = \tan^2 (45° + \phi/2) = \tan^2 (45° + 17.5°)$$

$$= 1.921^2 = 3.68$$

According to eq. 2.36,

$$p_1 = p_3 N_\phi = 2.0 \times 3.68 = 7.36 \text{ tons/sq ft}$$

Therefore, the sample should fail at a vertical pressure of about 7.36 — 2.00 = 5.36 tons/sq ft in excess of the all-around pressure.

2. The results of a series of triaxial tests on a soil have been expressed by means of the values $c = 1.2$ tons/sq ft and $\phi = 20°$. On the assumption that eq. 2.33 is valid, what vertical pressure would be required to produce failure of a mass of the soil if the lateral pressure were equal to 0.3 ton/sq ft?

Solution. The values of $\sqrt{N_\phi}$ and N_ϕ are

$$\sqrt{N_\phi} = \tan (45° + \phi/2) = \tan (45° + 10°) = 1.428$$

$$N_\phi = 1.428^2 = 2.039$$

$$p_1 = p_3 N_\phi + 2c\sqrt{N_\phi}$$

$$= 0.3 \times 2.039 + 2 \times 1.2 \times 1.428 = 0.61 + 3.43$$

$$= 4.04 \text{ tons/sq ft}$$

3. Compute the shearing resistance against sliding along a horizontal plane at a depth of 20 ft in the deposit of sand shown in Fig. P2.4. Assume that the sand can drain freely and that ϕ for the wet sand is 32°.

FIG. P2.4.

Solution. The total stress p at depth 20 ft is

$$7 \times 118 = 826 \text{ lb/sq ft}$$

$$13 \times 128 = \underline{1664}$$

$$p = 2490 \text{ lb/sq ft}$$

The neutral stress is

$$u_w = 13 \times 62.5 = 812.5 \text{ lb/sq ft}$$

Therefore, from eq. 2.37,

$$s = (p - u_w) \tan \phi = (2490 - 812) \tan 32° = 1050 \text{ lb/sq ft}$$

2.13. Shearing Resistance and Crushing Strength of Rock

The relationship between normal pressure and shearing resistance is quite different for different rocks but can usually be represented

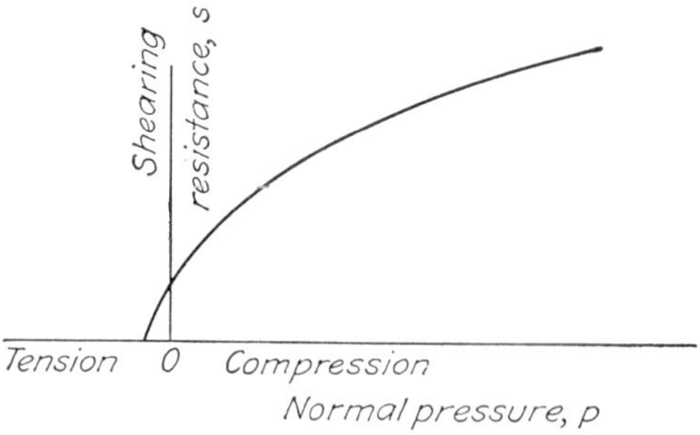

FIG. 2.27. Relation between normal pressure and shearing resistance for rock.

by a curved line as shown in Fig. 2.27. Values of shearing strength are commonly not determined because direct-shear tests are difficult to perform with accuracy. The unconfined compressive strength is more commonly employed. Table 2.4 gives values obtained from carefully conducted tests reported by R. R. Philippe and F. M. Mellinger.

Table 2.4. Unconfined Compressive Strength of Foundation Rocks

tons/sq ft

Location	Type	Strength
Mining City, Ky.	Sandstone	220
Mining City, Ky.	Sandstone	440
Sutton, W. Va.	Hard, gray sandstone	680
Berlin, Ohio	Brown, fine-grained sandstone	520
Berlin, Ohio	Gray, medium-grained, friable sandstone	410
Center Hill, Tenn.	Hard, fossiliferous limestone	600
Wolf Creek, Ky.	Hard, gray limestone	940
Falmouth, Ky.	Medium, crystalline, very fossiliferous limestone	1090
Buggs Island, Va.	Gray, granitoid gneiss	1340
Conemaugh, Pa.	Sandy, medium hard siltstone	430
Allatoona, Ga.	Dolomite	1830
Allatoona, Ga.	Quartzite	1530

M. Juul Hvorslev (1895–   )

Foremost authority on methods of boring and sounding, and of obtaining samples of soil for engineering purposes. His report on this subject, prepared after more than a decade of intensive and meticulous research, is an invaluable contribution to the literature of foundation engineering. His earlier studies of the shearing resistance of clays significantly advanced knowledge in this aspect of soil mechanics.

PLATE 3.

CHAPTER 3

Techniques of Subsurface Investigation

3.1. Methods of Exploration

Before the engineer can design a foundation in an intelligent and satisfactory manner, he must have a reasonably accurate conception of the physical properties and the arrangement of the underlying materials. The field and laboratory investigations required to obtain this essential information are called the *soil exploration* or the *exploratory program*. Because of the complexity of natural deposits, no one method of exploration is best for all situations.

The method most suitable under a wide variety of conditions consists of drilling holes into the ground and extracting samples for identification and, in some instances, for testing. Several procedures are in common use for drilling the holes. Likewise, a variety of methods is available for obtaining the samples. The choice depends upon the nature of the material and on the purpose of the exploratory program. Since the method used for drilling the holes does not necessarily dictate the method of sampling, the descriptions of procedures for drilling and for sampling will be discussed in separate articles.

After preliminary borings have disclosed the general characteristics of the underlying material, a more extensive program of boring and sampling may be indicated. Or it may prove more expedient to investigate the consistency or the relative density of the softer parts of the deposit by means of penetration tests or other direct methods that do not require sampling. The specific procedure must be chosen in consideration of the character of the soil deposit and the type of information needed for design or construction.

Other procedures less common than those mentioned above may be used under suitable conditions. For example, the character of the subsoil is occasionally investigated by direct inspection of the materials disclosed on the walls of test pits, shafts, or tunnels. In some instances, load tests are performed at the bottom of the test pits. When general information is needed concerning the location of boundaries between hard material such as bedrock and overlying softer deposits, geophysical methods may sometimes be used to advantage.

In the following articles, the various procedures for soil exploration will be discussed in greater detail.

3.2. Exploratory Borings

Auger Borings. The most primitive method for making a hole in the ground is the use of an auger. Two varieties of hand-operated augers are illustrated in Fig. 3.1. Although hand-drilled auger holes can be made to depths of more than 100 ft, they are most commonly made in connection with soil investigations for railroad, highway, or airport construction where it is usually unnecessary to drill deeper than about 12 ft. However, portable power-driven helical augers are available in diameters ranging from 3 to 12 in. These are in common use for making deeper holes in soils having sufficient cohesion to prevent caving as the material is removed.

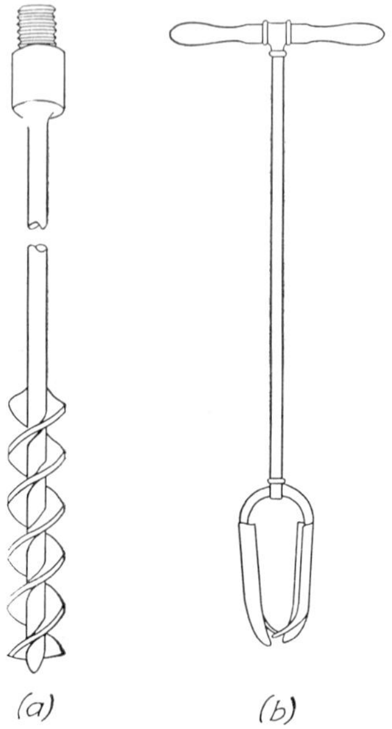

FIG. 3.1. Hand-operated augers. (*a*) Helical auger; (*b*) post-hole auger.

If the walls of the hole will not stand unsupported, the soil may be prevented from caving by means of a pipe known as *casing*. The casing is driven for a short distance into the ground and cleaned out by means of the auger. Additional lengths of pipe are added to the casing as the auger advances. Casing is rarely used with hand-operated augers, and its use is inconvenient in connection with power-operated augers because the auger must usually be removed while the casing is being driven. Therefore, auger borings are not commonly made in material that requires lateral support.

Wash Borings. Probably the most common procedure for making relatively deep holes in soil deposits is wash boring. A simple form of the necessary equipment is shown in Fig. 3.2.

The hole is started by driving a piece of casing to a depth of 5 or 10 ft. The casing is then cleaned out by means of a chopping bit fastened to the lower end of a wash pipe that is inserted inside the casing. Water is forced down through the wash pipe and emerges at high velocity through small openings in the bit. It then rises, carrying fragments of soil, through the annular space between the casing and the wash pipe. It overflows at the top of the casing through a T connection into a tub from which it is again pumped through a hose into the wash pipe. The connection between the water hose and the wash pipe is provided with a swivel joint so that the wash pipe and

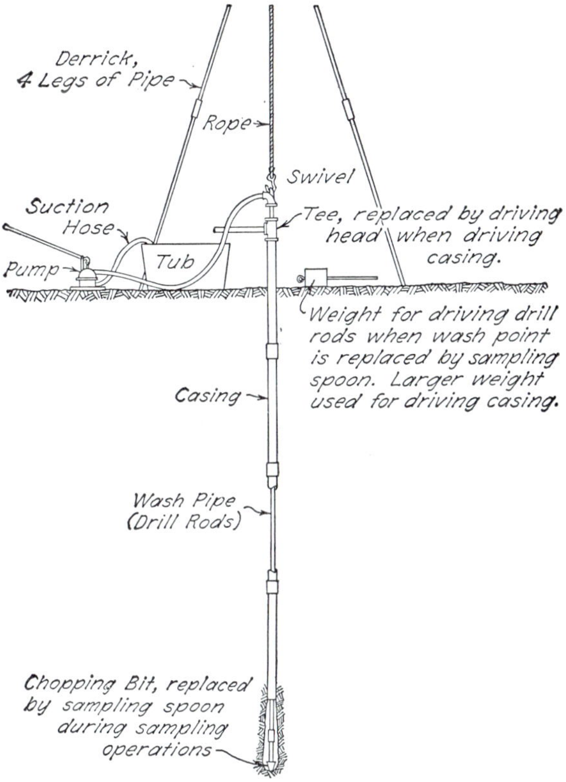

FIG. 3.2. Apparatus for making wash boring (after H. A. Mohr).

the chopping bit can be twisted as they are raised and dropped on the soil at the bottom of the hole. This facilitates the cutting. The wash pipe is extended, and additional casing is driven as the hole progresses. If the material will stand without caving, however, the casing need not extend more than 10 or 15 ft below the surface of the ground.

By this procedure, holes can usually be advanced rapidly through all but the most resistant soil strata. For ordinary exploratory borings the size of the casing is usually 2 or 2½ in. However, holes of larger diameter can be made if desired.

Percussion Drilling. If the drill hole must be advanced through exceptionally hard strata of soil or through rock, auger borings or wash borings cannot be used. One method for drilling holes through such deposits is known as *percussion* or *cable-tool drilling.* In this method a heavy drilling bit is alternately raised and dropped in such a manner that it grinds the underlying material to the consistency of a sand or silt. If possible, the bore hole is kept dry except for a

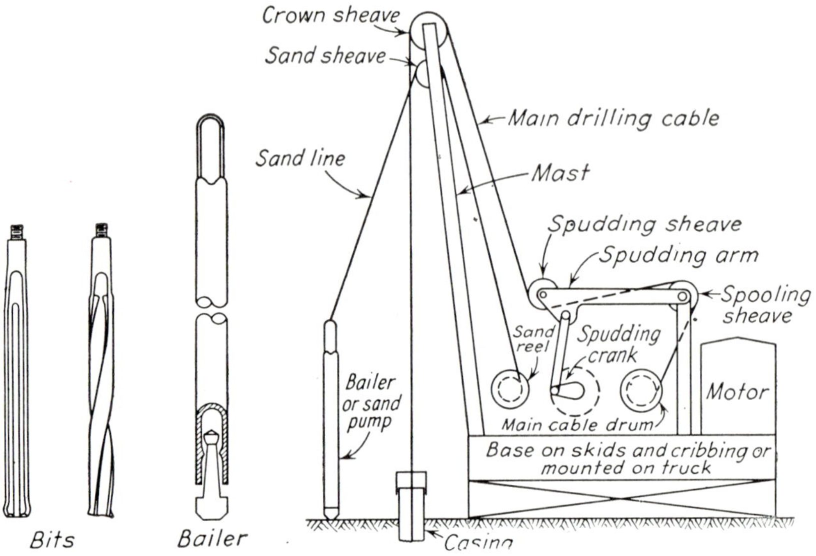

FIG. 3.3. Percussion drilling rig with typical tools (after M. J. Hvorslev).

small amount of water that forms a slurry with the material ground up by the bit. When the accumulation of slurry interferes with drilling, the drilling tools are removed from the hole and the slurry cleaned out with a bailer. The hole may be cased if the formation will not stand without collapse.

A drilling rig with typical drilling tools is shown in Fig. 3.3.

Rotary Drilling. Rotary drilling is the most rapid method for penetrating highly resistant materials unless the deposit is very loose or badly fissured. It may be used in rock, in clay, or even in sand. In this method, a rapidly rotating drilling bit cuts or grinds the material at the bottom of the hole into small particles. The particles are removed by circulating water or drilling fluid in a manner similar to that in wash borings. A sketch of a rotary drilling rig is shown in Fig. 3.4.

Casing is usually unnecessary in rotary drilling. If the hole tends to collapse, a drilling fluid is used consisting of a slurry of thixotropic volcanic clay in water. This slurry, known as *drilling mud*, coats and supports the sides of the drill hole, and seals off permeable strata.

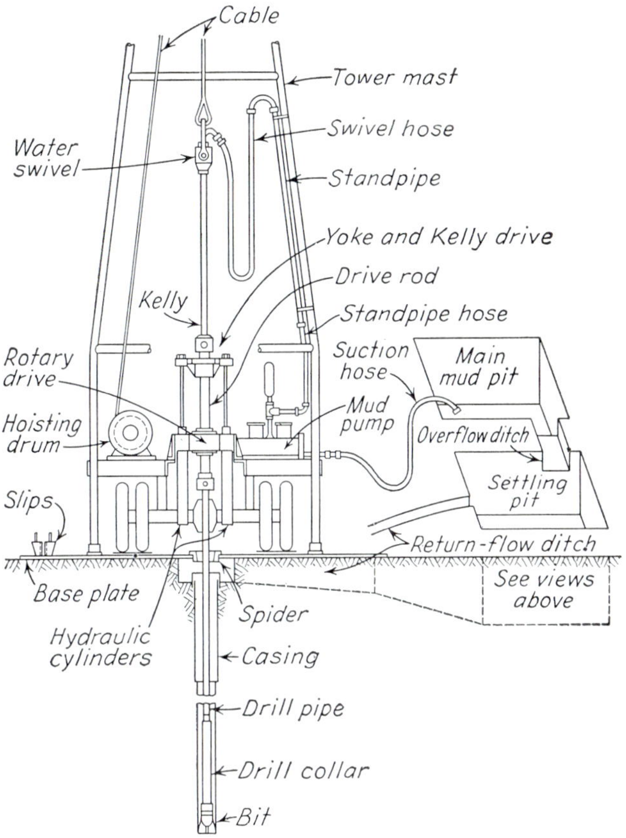

Fig. 3.4. Rotary drilling rig (after M. J. Hvorslev).

If drilling stops, the drilling mud becomes a gel and holds the rock cuttings in suspension so that they do not accumulate and jam the bit in the bottom of the hole.

Rotary borings in rock vary in diameter from about 1 in. to more than 36 in. depending upon the purpose for which the hole is made.

3.3. Sampling

Types of Samples and Sampling Spoons. The kind of samples that should be obtained from an exploratory drill hole depends upon the purpose for which the exploration is made. If it is necessary to obtain only a qualitative conception of the constituents of a deposit, sufficient information may sometimes be obtained from the cuttings of auger borings or the fragments of material washed up with the water in the wash boring or rotary drilling processes. However, shavings or cuttings of this type may not contain all the soil or rock constituents in their proper proportions. Obviously the physical state of the material is completely altered by the treatment to which it has been subjected. Hence, such fragments are rarely satisfactory for the purposes of the foundation engineer, and conclusions based upon them may be very misleading.

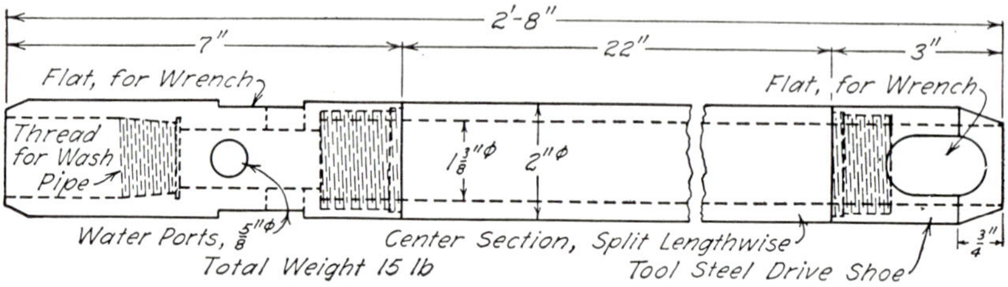

FIG. 3.5. Split-spoon sampler used in standard penetration test (courtesy Raymond Concrete Pile Co.).

For proper identification and classification of a soil or rock, *representative* samples are required. They should contain all the constituents in their proper proportions. Such samples are adequate for visual classification, for the performance of mechanical analyses, and for determining the Atterberg limits, the unit weight of the solid constituents, the carbonate content, and the quantity of organic matter. The mechanical properties of the soil, however, may be appreciably altered by the sampling process. If so, such samples are not suitable for the determination of the stress–strain characteristics or relative density of the materials.

If information of the latter type is required, the samples must be taken in such a manner that they experience negligible deformation during the process of sampling. Such samples are called *undisturbed* although a certain amount of disturbance must be regarded as inevitable.

Most samples are obtained by driving or pushing into the ground an open-ended cylindrical tube known as a *sampling spoon.* Spoons having inside diameters from 1⅜ to 2½ in. usually consist of four parts, a cutting shoe at the bottom, a barrel consisting of a length of pipe split longitudinally into two halves, and a coupling at the top for connection to the drill rods. Such a device is known as a *split spoon,* Fig. 3.5. After a sample has been taken, the cutting shoe and the coupling are unscrewed and the two halves of the barrel separated to expose the material.

If spoon samples are to be transported to the laboratory without examination in the field, the barrel is often cored out to hold a cylindrical thin-walled tube known as a *liner.* After a sample has been

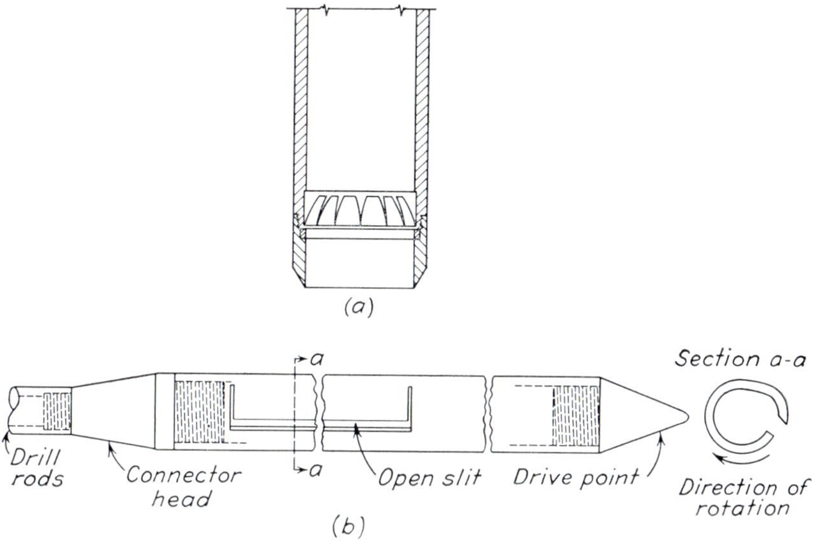

FIG. 3.6. (*a*) Spring core catcher; (*b*) scraper bucket.

obtained, the liner and the sample it contains are removed from the spoon, and the ends are sealed with caps or with metal disks and wax. Liner spoons are available for samples up to about 8 inches in diameter, but the largest samples commonly taken in this manner have a diameter of 5 in.

The degree of disturbance of spoon samples depends upon the way in which the force is applied to the spoon, whether by pushing or driving, on the rate of penetration, and on the dimensions of the sampler. If other conditions are equal, the degree of disturbance is roughly proportional to the area ratio,

$$A_r\ (\%) = 100\,\frac{D_e^2 - D_i^2}{D_i^2} \tag{3.1}$$

where D_e is the external diameter and D_i the internal diameter of the cutting shoe through which the sample must pass. If the area ratio is not greater than about 10 per cent, the distortion of the sample is small in almost any type of soil. The degree of disturbance also is less if the sample is advanced with a rapid steady motion rather than by intermittent pushing or driving.

Samples of cohesionless materials, such as sand below the water table, cannot be retained in sampling spoons without the addition of a spring core catcher, Fig. 3.6*a*. In deposits of very fine sand, or in sand containing small pebbles that may prevent the springs from closing, no recovery may be possible. Representative samples, but in a completely altered state, can be obtained by means of a *scraper bucket,* Fig. 3.6*b*. The bucket is driven below the casing and rotated. As it turns, scrapings from the sides of the hole fall into the lower compartment.

The cost of undisturbed samples increases rapidly with increasing diameter. It has been found by experience that most soft or moderately stiff cohesive soils can be sampled without excessive disturbance in thin-walled seamless steel tubes having a diameter of not less than 2 in. and an area ratio of about 10 per cent. Tubing with a diameter of 2 in. is commonly used, in lengths varying from 2 to 3 ft. The lower end of the tube is sharpened to form the cutting edge, and the upper end is machined for attachment to the drill rods. The entire tube is pushed into the ground at the bottom of the hole and is removed with the sample inside. The two ends of the tube are sealed, and the sample is shipped to the laboratory. In the laboratory the tube is usually cut into lengths of about 6 in. so that the sample can be extruded without appreciable disturbance.

It is often possible to improve the quality of samples and to increase the recovery of soft or slightly cohesive soils by the use of a *piston sampler*. This type of sampler, Fig. 3.7*a*, consists of a thin-walled tube fitted with a piston attached to a rod that extends to the ground surface inside the drill rods. While the tube is being lowered to the bottom of the drill hole, the piston rod is clamped to the drill rods, and the piston is held at the bottom of the sampler to prevent the entrance of undesirable material. Before the start of the sampling stroke the piston rod is released and clamped to the casing. The sampling tube is then pushed past the piston which remains on the surface of the soil at the bottom of the hole as shown in Fig. 3.7*b*. The presence of the piston prevents soft soils from squeezing rapidly into the tube and thus eliminates most of the distortion of the sample. The piston also helps to increase the length of sample that can be recovered, because, if the top of the column of soil begins to separate from the piston, a slight vacuum is created which tends to pull the sample into the tube. At the end of the sampling stroke the piston is locked in position at the top of the sample. During the withdrawal of the device from the drill hole, the piston prevents water

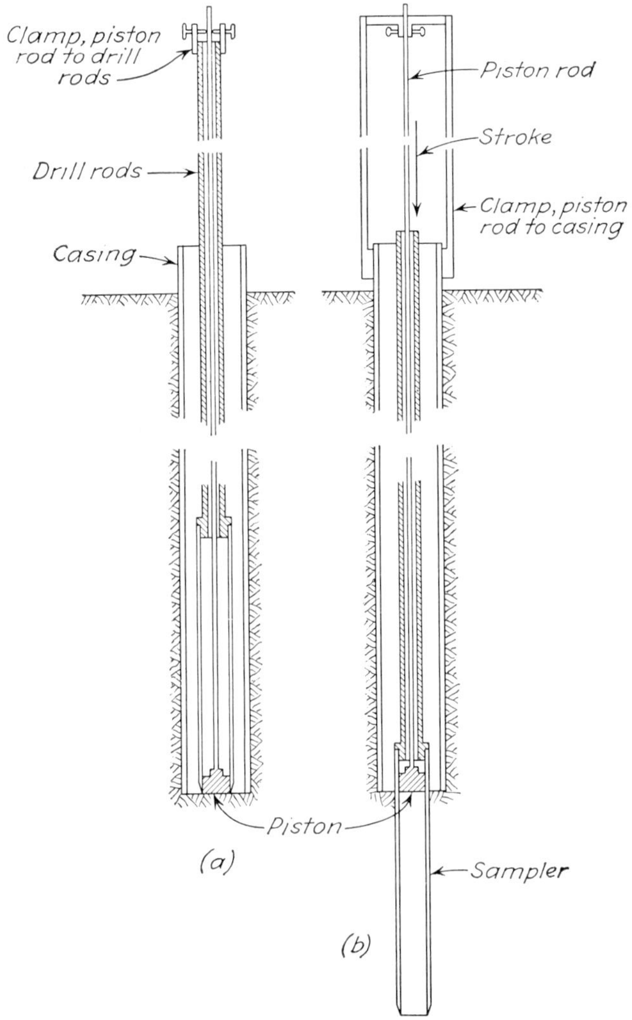

FIG. 3.7. Schematic diagram of piston sampler. (*a*) Arrangement just prior to sampling stroke; (*b*) arrangement during sampling stroke.

pressure from acting upon the top of the sample and thus increases the chances for recovery. The design of piston samplers has been refined to the extent that it is possible to take undisturbed samples of sand from below the water table. The technique, however, requires that a heavy slurry of bentonitic drilling mud be used in the drill hole.

Cores. Short intact samples of rock can be obtained during percussion drilling by means of a special core barrel. However, the necessity for taking a core greatly slows drilling progress. Therefore, percussion drilling is rarely used in connection with sampling operations for exploratory purposes.

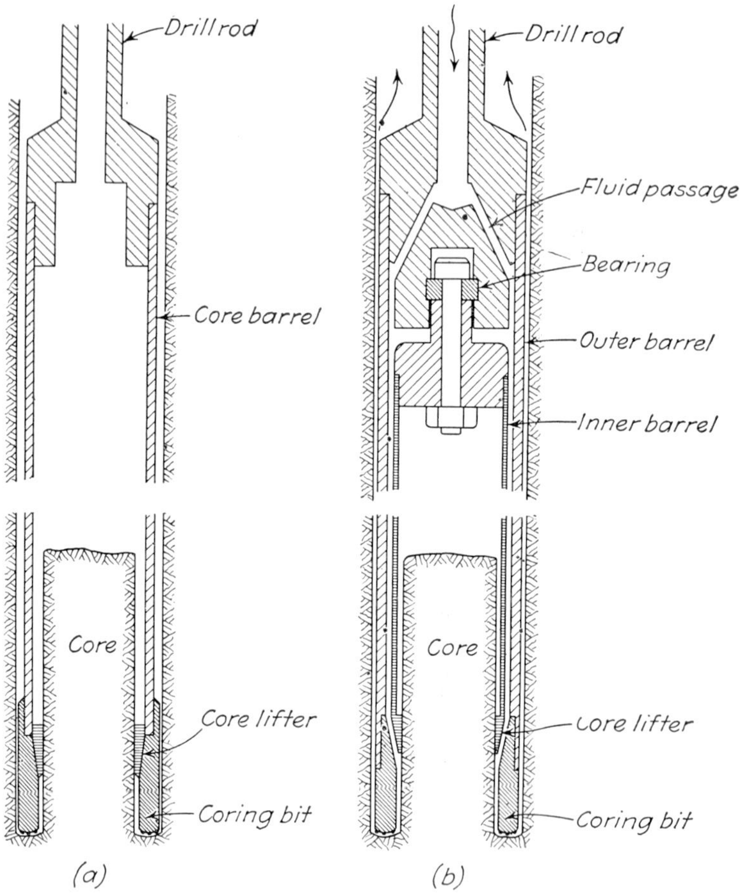

FIG. 3.8. Schematic diagram of core barrels. (*a*) Single tube; (*b*) double tube.

In rotary drilling, the drilling bit may be replaced by a core drill attached to the lower end of a core barrel as shown in Fig. 3.8. The drill cuts an annular ring in the rock and leaves a central core that enters and is retained in the core barrel. The cutting element may consist of diamonds, chilled shot, tungsten carbide inserts, or steel cutters. Core drilling does not greatly increase the time for advancing the hole and is widely used for sampling resistant soils and rocks.

The cuttings from the annular ring are commonly washed away with the drilling fluid, but in some materials the action of the fluid alters the character of the material. Under these circumstances, it may be possible to remove the cuttings with an air jet or a ring-shaped auger located directly above the cutting tool.

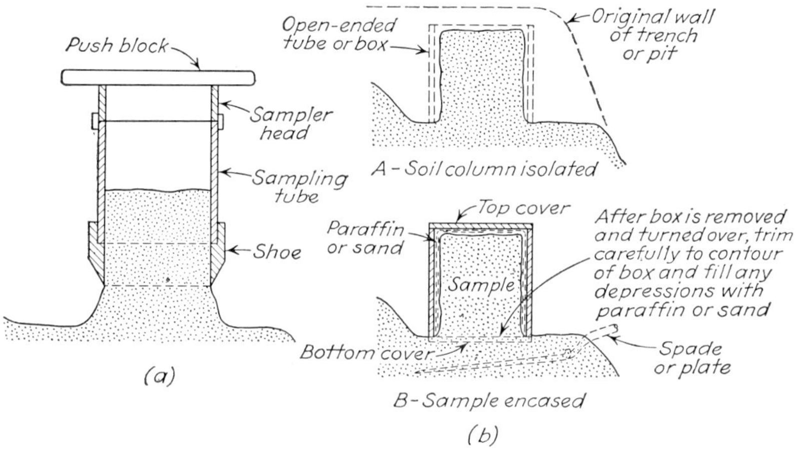

FIG. 3.9. Surface sampling (*a*) by advance trimming and (*b*) by block sampling.

The core barrel may consist of a single tube, Fig. 3.8*a*, or a double tube, Fig. 3.8*b*. Samples taken in a single-tube barrel are likely to experience considerable disturbance due to torsion, to swelling, and to contamination with the drilling fluid. The double-tube barrel is designed to protect the core against the action of the circulating fluid. By the addition of a liner, it may be adapted to sampling soils or nonuniform, fissured, friable or soft rocks.

Hand-carved Samples. Practically undisturbed samples may be carved from soils possessing at least a trace of cohesion, provided the material is exposed in a test pit, shaft, or tunnel. Two methods of obtaining such samples are illustrated in Fig. 3.9.

ILLUSTRATIVE PROBLEMS

1. Compute the area ratio for the split spoon used in the standard penetration test.

Solution. According to Fig. 3.5, the standard spoon has an inside diameter of 1⅜ in. and an outside diameter of 2 in. Therefore, from eq. 3.1,

$$A_r = 100 \times \frac{2.0^2 - 1.375^2}{1.375^2} = 112 \text{ per cent}$$

2. Compute the area ratio for a thin-walled tube sampler having an external diameter of 2 in. and a wall thickness of 1/16 in.

Solution.

$$A_r = 100 \times \frac{2.0^2 - 1.875^2}{1.875^2} = 13.8 \text{ per cent}$$

3.4. Direct Measurements of Consistency and Relative Density

Penetrometers. Several methods have been developed for determining the consistency of cohesive deposits or the relative density of cohesionless deposits without the necessity for making a drill hole and extracting samples. Most of these procedures are based on measuring the resistance offered by the soil to the advancement into the ground of a device known as a *penetrometer.* If the penetrometer is pushed steadily into the soil, the procedure is known as a *static penetration test.* If driven into the soil, it is known as a *dynamic penetration test.* As a rule, static tests are preferable in connection with soft cohesive deposits and dynamic tests with cohesionless or very hard deposits.

Many varieties of penetrometers have been designed, each especially adapted to certain kinds of material. Two of these are shown in Fig. 3.10. However, a crude but useful penetration test can be made by driving an ordinary sampling spoon into the ground and counting the number of blows of a drop hammer required to produce a given penetration. This procedure may be combined cheaply with boring and sampling operations to provide additional information concerning the characteristics of the subsoil.

Standard Penetration Test and N Values. In the United States, the most widespread test of this kind is known as the *standard penetration test.* It is made by dropping a hammer weighing 140 lb onto the drill rods from a height of 30 in. The number of blows N necessary to produce a penetration of 1 ft is regarded as the penetration resistance. The sampler has the dimensions shown in Fig. 3.5.

The results of the standard penetration test can usually be correlated in a general way with the pertinent physical properties of the soil. Table 3.1 shows such a correlation. However, the scattering of individual results from the conservative values given in the table may be quite large, and it is preferable to make direct comparisons with the results of other appropriate tests in any given locality. The correlation for clays can be regarded as no more than a crude approximation, but that for sands is quite reliable. However, if the sand is very fine or contains a large amount of silt, and in addition is located beneath the water table, Table 3.1 may indicate a relative density

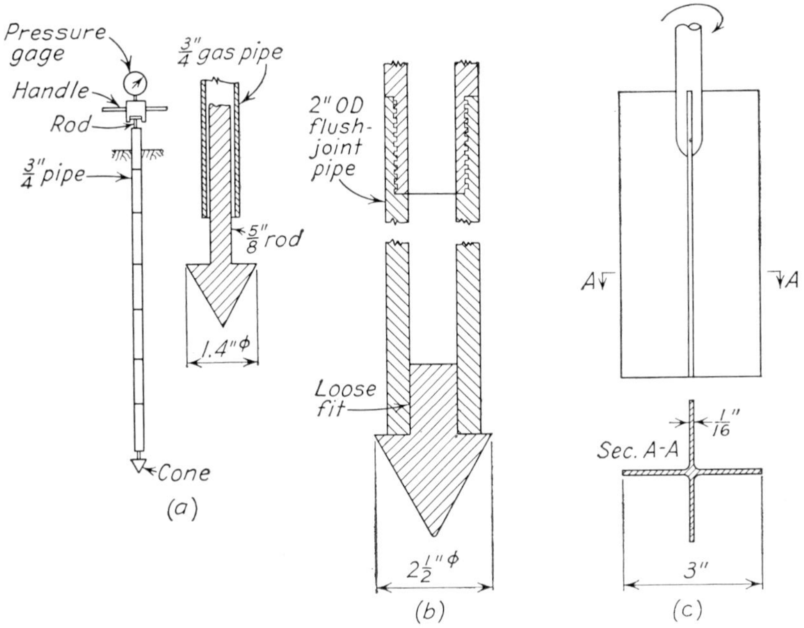

FIG. 3.10. Devices for direct measurement of consistency of soil under field conditions. (*a*) Dutch penetrometer; (*b*) conical drive point for deposits of sand and gravel; (*c*) vane auger.

Table 3.1. Penetration Resistance and Soil Properties on Basis of the Standard Penetration Test

Sands (Fairly Reliable)		Clays (Rather Unreliable)	
No. of Blows per ft, N	Relative Density	No. of Blows per ft, N	Consistency
		Below 2	Very soft
0–4	Very loose	2–4	Soft
4–10	Loose	4–8	Medium
10–30	Medium	8–15	Stiff
30–50	Dense	15–30	Very stiff
Over 50	Very dense	Over 30	Hard

considerably greater than the actual relative density of the deposit. Under these conditions the N value should be corrected as indicated in Art. 14.2 before Table 3.1 is used. Moreover, in deposits containing many boulders the results may be unreliable because of the small size of the sampling spoon compared to that of the boulders.

Although the standard penetration test cannot be regarded as a highly refined and completely reliable method of investigation, the N values give a useful preliminary indication of the consistency or relative density of most soil deposits. The information is in many instances sufficient for final design. In any event, it provides data for more intelligent planning of whatever additional exploration may be desirable at the site.

Cone Penetration Tests. Of the various penetrometers designed for use in exceptionally soft soils, the one illustrated in Fig. 3.10*a* is most widely used. It was developed by the Soil Mechanics Laboratory at Delft, the Netherlands. The penetrometer is a cone that is pushed into the soil by hand or by mechanical means. The pressure required to advance the cone at a slow constant rate is observed and is known as the penetration resistance. The shaft supporting the cone is surrounded by a casing that is advanced alternately with the point. Additional information concerning the soil properties is obtained by measuring the force necessary to advance the casing. The results have been correlated with the shearing strength of Dutch soils on the basis of numerous soil tests, and empirical rules have been devised for predicting the settlement of embankments and other structures in Holland on the basis of the tests.

A simple penetrometer that can be used for investigating the relative density of sand and gravel deposits without the use of casing is shown in Fig. 3.10*b*. It consists of a cone that fits loosely into the bottom of a string of pipes driven into the ground by means of a drop hammer with constant height of fall. A continuous record is made of the number of blows required to advance the point through each foot of depth. When the point has reached its final elevation, the pipe is withdrawn and the cone left at the bottom of the hole. Friction on the pipe is minimized by making the diameter of the cone somewhat greater than the outside diameter of the pipe. Tests of this type can be made more rapidly and cheaply than standard penetration tests. The results can be correlated with the N values, Table 3.1, by performing several series of standard penetration tests in borings adjacent to the locations of the cone penetration tests.

Direct Measurement of Shearing Strength. Various devices have been developed for direct measurement of the shearing resistance of cohesive soils. One of the more recent and most promising developments is known as the *vane auger,* Fig. 3.10*c*. In its simplest form it consists of a four-bladed vane fastened to the bottom of a vertical rod. The vane and the rod can be pushed into the soil without appreciably disturbing the material. The assembly is then rotated and the torque required to turn the vane is measured. Since the soil fails along a cylindrical surface passing through the outer edges of the vane, the shearing resistance can be computed if the dimensions of the vane and the torque are known. If the vane is rotated rapidly through several revolutions, the soil becomes remolded and the shearing strength can again be determined. Hence, it is possible to measure not only the shearing resistance but also the sensitivity of the clay. This device is best adapted to clays having an unconfined compressive strength of less than 1.0 ton/sq ft. The vertical rod is usually protected by a casing to eliminate the skin friction. The vane auger is not suitable for use in soil deposits that contain sand layers or many large pebbles.

3.5. Miscellaneous Methods of Soil Exploration

Inspection Shafts. In connection with large projects, it is sometimes economical to inspect the subsurface formations in their natural state. This may be done by excavating shafts or driving tunnels through the materials. A visual examination should be made not only by engineers but also by geologists. Ordinarily, undisturbed samples may readily be obtained from such excavations.

Relatively inexpensive inspection holes 3 to 4 ft in diameter can be drilled through rock by means of shot core barrels. Holes of this type have been used chiefly for exploration of dam foundations, but under some circumstances they may be advisable in connection with other types of structures.

Load Tests. The stress–strain characteristics of soils and soft rocks are sometimes investigated by means of field load tests. A square or circular bearing plate is established at a suitable elevation on the bottom of a test pit or shaft. Vertical load is applied to the plate in increments, and the settlement is observed after each application of load. The results are plotted in the form of a load-settlement diagram, Fig. 3.11*a*.

The results of load tests require careful interpretation and may in many instances be misleading. Unless the subsoil is uniform for a considerable depth below the base of a proposed foundation, the results of load tests are likely to give a false conception of the supporting capacity of the underlying material. The limitations of such tests will be discussed in connection with the various types of foundations, Part C.

However, load tests represent a convenient method for investigating the relative density of sand deposits, especially for the calibration of the results of penetration tests. Load tests made for this purpose and carried out in the following manner are known as *standard load tests*. The test pit should be at least 5 ft square. The loading plate should be 1 ft square, and no surcharge should be placed on

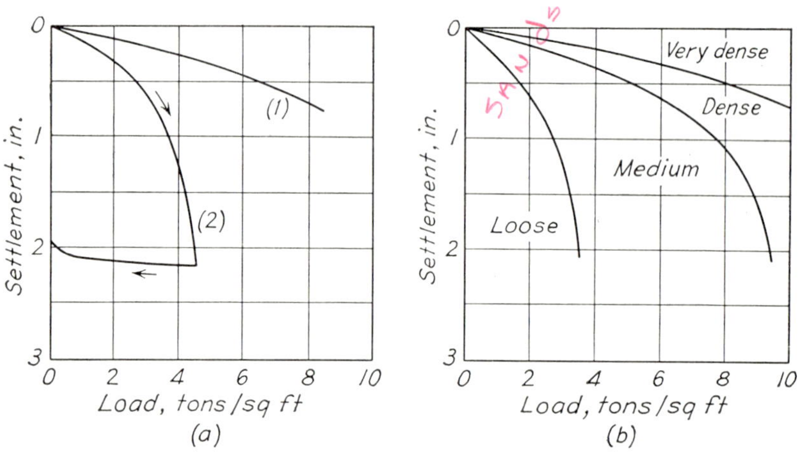

FIG. 3.11. (*a*) Load-settlement diagram for (1) a dense clean sand in a caisson 26 ft below the bottom of a river and (2) a sand of medium density at the bottom of a 30-ft shaft; (*b*) chart for estimating relative density on basis of results of standard load test on bearing plate 1 ft square.

the ground within a distance of 2 ft from the plate. Inasmuch as the capillary moisture found within 3 or 4 ft of the water table in most sand deposits is responsible for an apparent cohesion that has an influence on the test results, standard load tests should not be performed within the capillary zone. They should be carried out at elevations not less than about 4 ft above the water table or else directly at the water level.

The results of a standard load test may be interpreted by means of Fig. 3.11*b*, which shows the boundaries of the areas occupied by load-settlement curves for sands of different relative densities. If the test has been performed at the water table, the values of settlement should be reduced 50 per cent before the load-settlement curve is plotted for comparison with those in Fig. 3.11*b*.

In some instances, it is convenient to drive piles as a means for determining the resistance of the soil. The piles may then be loaded to determine their supporting capacity. The appropriate use of such tests is described in Part C.

Geophysical Methods. The boundaries between different elements of the subsoil may sometimes be located by geophysical methods. These procedures are based on the fact that the gravitational, magnetic, electrical, radioactive, or elastic properties of the different elements of the subsoil may be different. Depending on the property that is used as a basis for differentiating between the elements, the geophysical methods may be classified as gravitational, magnetic, electrical, radioactive, or seismic. Geophysical methods were developed primarily for the mining industry, and most of the procedures are not well suited for the purposes of civil engineering. Differences in unit weight, magnetic characteristics, and radioactivity of deposits near the surface of the earth are rarely great enough to permit use of the corresponding geophysical methods of exploration.

Differences in the elastic properties are the basis for seismic methods in which an explosive charge is fired at a given location and measurements made to determine the time for the resulting elastic waves to reach various recording stations. The time depends on the distances through which the waves must travel and upon the velocity of the wave through the media. In the zone of weathering at the top of the soil horizon, the velocity is on the order of 100 ft/sec. In clay it varies from 1300 to 2200 ft/sec and in sound rock from 13,000 to 25,000 ft/sec. Because of these differences, the seismic method is of some practical value in locating the surface of the bedrock beneath soft deposits. If the surface is overlain by boulders, the seismic method may be more reliable than boring.

Methods have also been developed for inducing continuous vibrations in a mass of soil and measuring the effects of these vibrations at various points. In some instances the results can be correlated with the bearing capacity of the subsoil.

Possibly the geophysical method most adaptable to the purposes of civil engineers at the present time is the electrical resistivity method. Two electrodes are driven into the ground at a considerable distance from each other, and an electrical potential is established between them. The potential at intermediate points is measured. The spacing between electrodes is then changed, and the observations are repeated. After a relationship between potential drop and electrode spacing has been established, the results can be interpreted to indicate the boundaries of bodies of earth materials with different characteristics. For example, it is usually possible to determine the elevation of bedrock, the boundaries of gravel deposits in masses of clay, and the location of the water table. In conjunction with occasional borings, the method has been found of practical value for general reconnaissance of subsurface conditions over large areas. The detailed conditions at a given site must be explored by more direct methods.

3.6. Record of Field Exploration

The records of subsurface explorations and sampling operations should be clear and accurate. Field notes should contain the date the work was done, the location with respect to a permanent system of coordinates, the elevation of the ground surface with respect to a permanent bench mark, the elevation of the water table, the elevation of the upper boundary of each successive stratum of soil or rock, a field classification of the strata encountered, and whatever values of penetration resistance or other measures of consistency were obtained. The records of special investigations during which large-diameter undisturbed samples or hand-carved samples were obtained should be as complete as possible.

The engineer or boring foreman in charge of any exploration should always keep in mind that seemingly insignificant details of the procedure required to advance the hole and keep it open may yield information as valuable as that obtained from the samples. For this reason the records should include the types of tools used in the investigation, together with any changes made. The depth at which a change was made and the reason for the change should also be noted. Furthermore, the methods used to stabilize the walls of the bore hole or to sheet the test pit should be recorded. All this information should be condensed and presented for every exploratory hole whether successfully completed or not.

After samples have been examined and tested in the laboratory, the materials should be classified according to the most appropriate method, and the field notes together with the laboratory results should be assembled in such a form that the boundaries between different materials are plotted at their correct elevations on a suitable vertical scale.

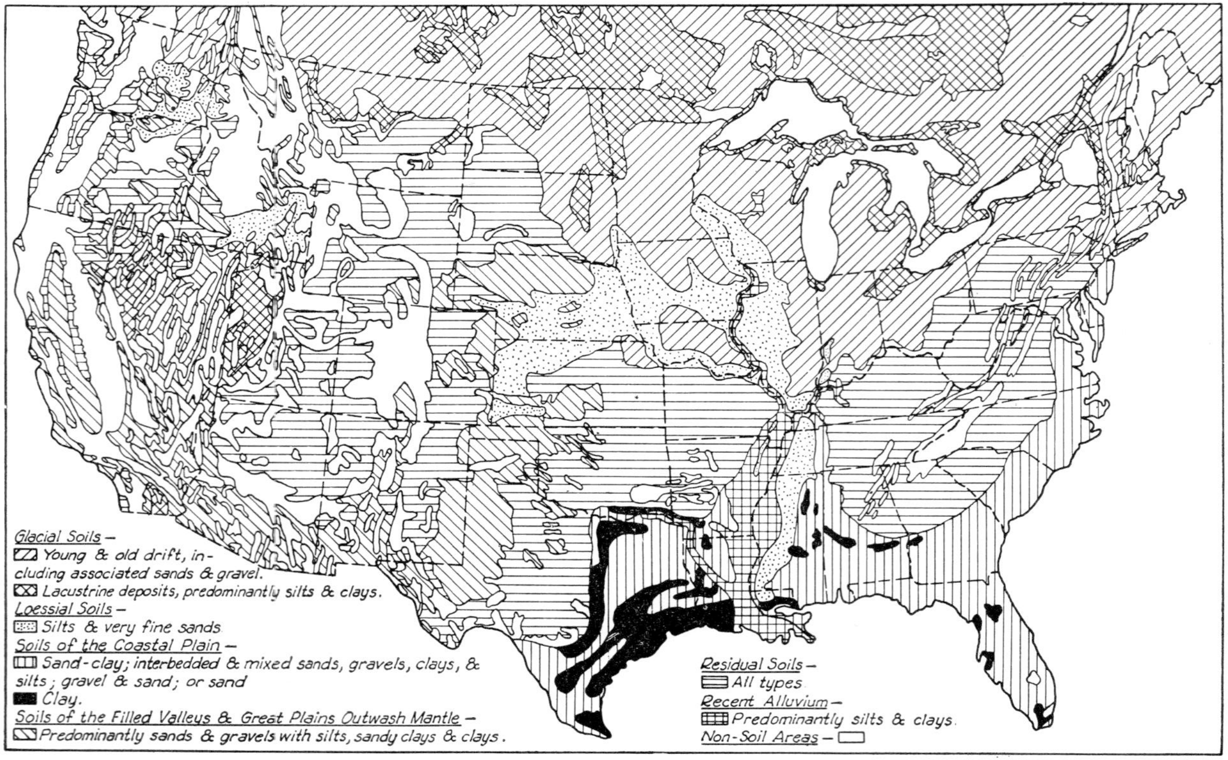

Soil Deposits of the United States and Canada (after Flint, Belcher, et al.).

PLATE 4.

CHAPTER 4

Character of Natural Deposits

4.1. Origin of Natural Deposits

Most of the rock deposits encountered by the foundation engineer were once sediments deposited beneath shallow seas. The seas covered vast areas, and the conditions of sedimentation were fairly uniform. Although many of these deposits have since been altered by folding and faulting, some have been metamorphosed, and some transformed into igneous rocks by magmatic action; yet many of the formations, even those of great extent, have remained essentially uniform. Since the bearing capacity of all but the softest intact rocks is very great, the foundation engineer is concerned primarily with defects and discontinuities in the rock formations, such as bedding planes, joints, solution channels, and zones of alteration by chemical or physical weathering.

Man appears to have come upon the earth at a time of remarkably great tectonic activity. In contrast to conditions throughout much of the earth's history, the land masses today are unusually large and high. Processes of erosion and weathering are active, there are relatively few shallow seas of sedimentation, and many sediments are being deposited on the continents themselves. Moreover, the present times follow closely after, or possibly within, the glacial age, one of the most unique periods in the earth's entire history. In many parts of the world, the existing topography is the direct result of the action of glaciers which plowed up old soils, ground up rocks, and deposited the materials in random fashion partly on land and partly under water. The influence of glaciation was also important in other regions. During the glacial epoch the climate was colder and the rainfall greater. As a consequence, large inland lakes developed and the flow of rivers was abnormally great.

Hence, the deposits of soil with which the engineer deals present an extraordinary variety in contrast to the relative uniformity of the rocks. Many of the problems of the foundation engineer are associated with soft or loose deposits not older than those of the glacial age. The engineer must learn the character of the deposit at each site in order to forecast the most unfavorable conditions he may encounter. *He dares not assume uniformity.*

In connection with any project, the program of subsurface exploration must be adequate to disclose the essential character of the deposit and particularly its possible variations from point to point. Yet, economy dictates that no greater expenditure should be made than necessary to produce the desired results. This end cannot be achieved if the engineer does not have at least a rudimentary knowledge of the structure of various kinds of deposits. Such knowledge will assist him in interpreting information as it is obtained from the field and laboratory, and will help him to recognize the stage at which further information will not be worth the added cost.

In the following articles, some of the more common types of natural deposits will be described, and typical examples will be presented of the results of exploratory programs to determine their characteristics. A highly generalized map showing the locations in the United States and southern Canada where the various types of deposits are encountered is shown in Plate 4.

4.2. Deposits Associated with Glaciation

Introduction. These deposits are probably the most variable of all that have been made by nature. The great continental glaciers covered much of the surface of the land north of the 40th parallel. Elsewhere most mountains were buried or partly covered by Alpine glaciers, even in the tropics. The ice excavated loose rocks and soils, transported and deposited them. All materials laid down by glaciers are known as *drift.* Those deposited directly out of the ice are called *till.* The meltwater flowing away from the ice also carried debris and deposited it in broad sheets known as *outwash.* The concentration of meltwater into torrential streams of variable velocity depending on the seasons gave rise to *glaciofluvial deposits.* In some instances, the meltwater was dammed between high ground and the glacier itself. Thus glacial lakes were formed in which sediments known as *glacial lake deposits* were laid down.

Moraines. The ice of every active glacier continually flows toward the outer edges carrying drift. Near the edges, melting takes place and the drift is concentrated at the bottom of the ice where part of it becomes fixed to the frozen ground. This part constitutes *ground moraine.* It consists primarily of till of erratic composition. A small amount of ground moraine is deposited when an ice sheet is growing, a larger amount when it is shrinking. Where there has been alternate growth and shrinkage of a glacier near its extremity, there may be several distinct sheets of till. Figure 4.1 represents a cross section through a part of downtown Chicago, where there are at least three successive ground moraines.

If the edge of the ice remains stationary for a few years or more, the drift accumulates in a ridge at the face of the glacier. Such accumu-

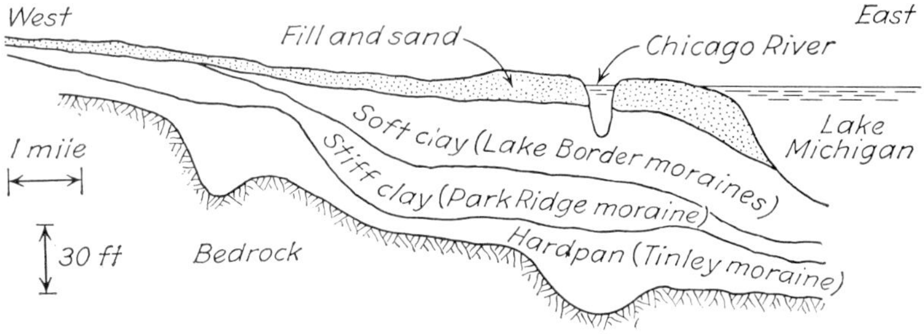

FIG. 4.1. Simplified cross section through glacial deposits beneath Lake Street in Chicago showing successive ground moraines.

lations are known as *terminal moraines.* They consist largely of till, but they may be stratified in places once occupied by pools of meltwater, and they may contain deposits of outwash having irregular shapes. Commonly, there is considerable outwash of sand, gravel, or silt on the side of the moraine that slopes away from the ice. Figure 4.2 shows a cross section through such a moraine.

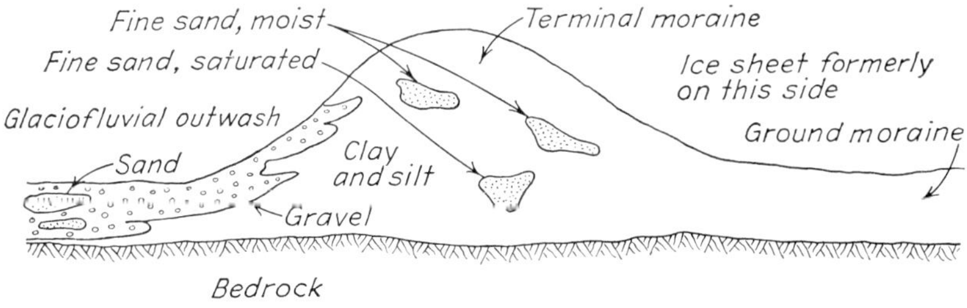

FIG. 4.2. Idealized cross section through typical terminal moraine.

In the central part of the United States are vast areas of ground moraine with fairly level surfaces. These are known as *till plains.* The different till sheets beneath such plains may be separated by old topsoils and by buried channels.

All types of moraines are likely to contain some waterlaid clays and silts formed in temporary ponds. Moreover, there may be various deposits of uniform sands and gravels laid down in channels and tubes in the ice. These appear as irregularly shaped inclusions such as those shown in Fig. 4.3. It has been noted that some moraines are composed of clay till having exceptionally uniform water content whereas others formed under nearly similar conditions may exhibit extremely erratic changes in water content. The contrast between two moraine deposits of this type is shown in Fig. 4.4

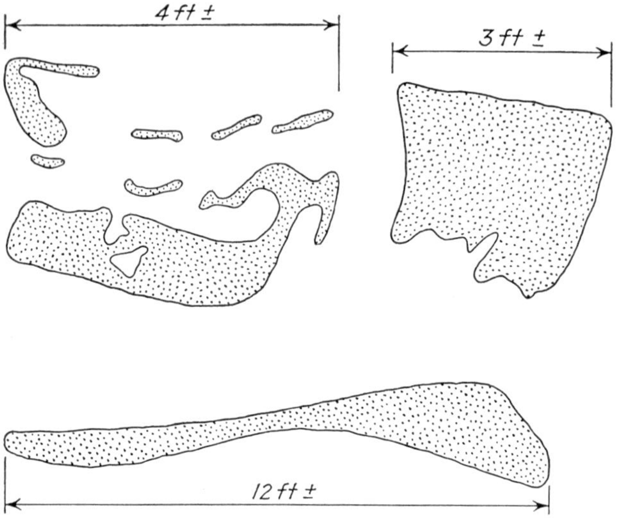

FIG. 4.3. Sketches of coarse sand inclusions encountered during tunneling operations in ground moraine consisting of soft clay.

Morainic areas are likely to be poorly drained, especially if the till was deposited during one of the more recent glacial advances. In the poorly drained pockets, deep beds of peat are often encountered.

Glaciofluvial Deposits. Tremendous quantities of water flowed from the faces of the continental glaciers during the warm seasons carrying coarse material for short distances and transporting sands, silts, and clays for greater distances. Temporary channels quickly became choked with debris, and new channels formed elsewhere. The resulting deposits, especially if they were formed close to the glaciers, consist of lenses of coarse and fine materials, some loose and some dense. They are among the most variable natural coarse-grained sediments. Figure 4.5 is a photograph of a road cut through a glaciofluvial deposit near Denver. A record of the penetration of steel rails driven by a pile driver into glaciofluvial gravel at Port Alberni, B. C., is shown in Fig. 4.6. Wide variations in relative density and in grain size are apparent, in spite of the fact that the explorations were made within relatively small distances of each other.

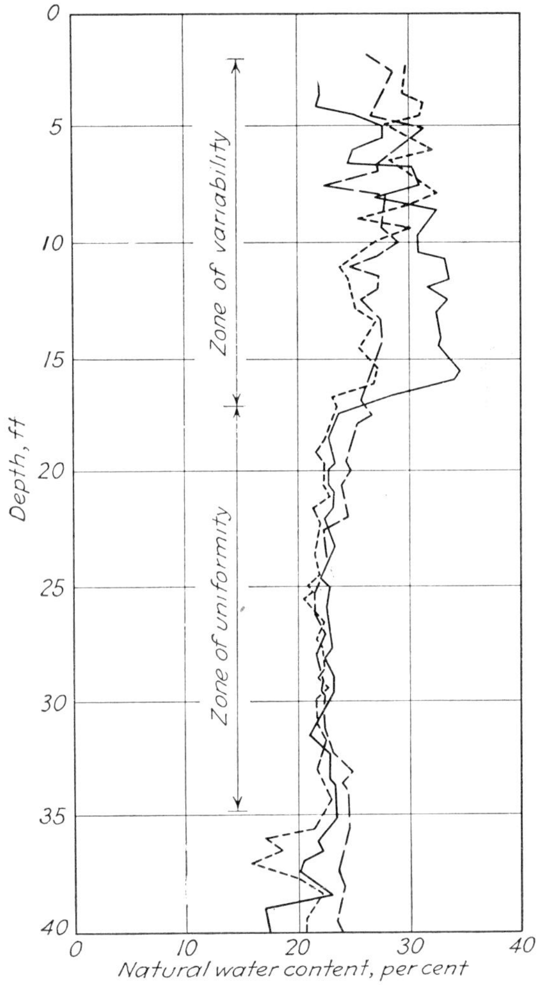

FIG. 4.4. Variations in water content along vertical lines extending from variable moraine above to uniform moraine below surface of contact. The two moraines are indistinguishable to the eye.

Glacial Lake Deposits. In contrast, the deposits formed in the relatively quiet waters of glacial lakes have a high degree of regularity, especially in horizontal directions. However, they are commonly laminated or varved. The inflowing streams carried silt and

FIG. 4.5. Road cut through glaciofluvial deposit near Denver, Colo.

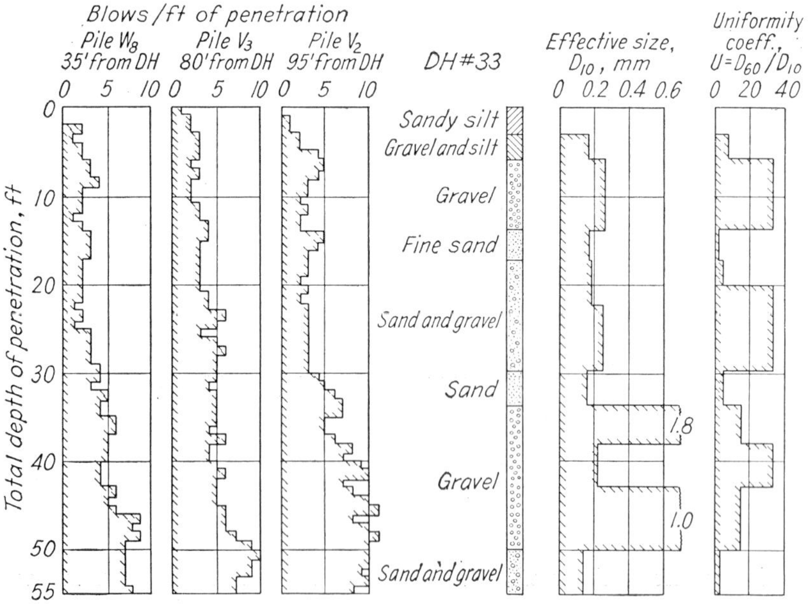

FIG. 4.6. Penetration resistance of glaciofluvial gravel deposit at Port Alberni, B.C. (courtesy of K. Terzaghi).

clay into the lake during the summer, and a major portion of both constituents settled to the bottom. In the winter the melting of the glaciers ceased, and no new sediments were brought to the lake. The suspended clay continued to settle throughout the winter, but all the silt reached the bottom within a few days. As a consequence, the deposit developed a banded character consisting of layers of silty clay or silt and thinner layers of clay. Figure 4.7*a* is a photograph

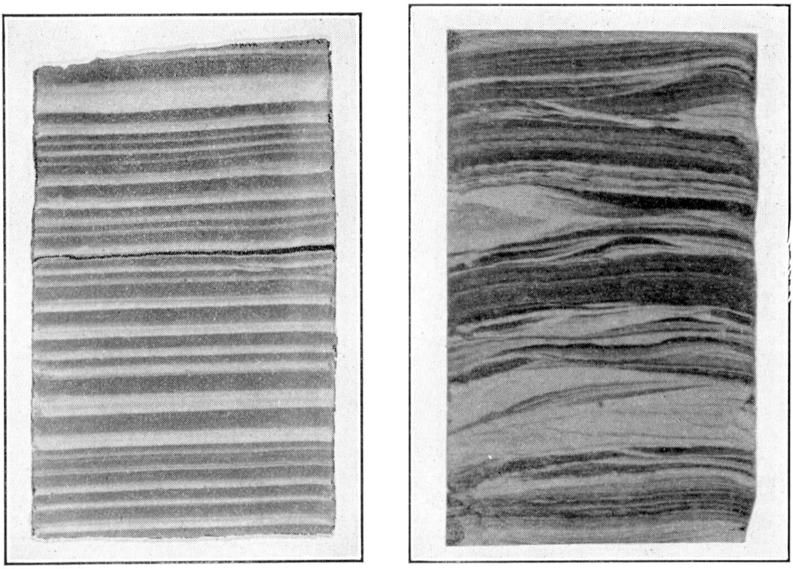

(*a*) (*b*)

FIG. 4.7. (*a*) Typical varved clay (after Hvorslev); (*b*) laminated clay from bed of glacial Lake Agassiz.

of a typical varved clay. In some instances where the bottom was shallow enough to be influenced by currents the details of banding may be very intricate. The photograph, Fig. 4.7*b*, is an example from the bed of old glacial Lake Agassiz.

Many glacial lakes of large extent were formed around the present Great Lakes and in the valley of the Red River of the North. Similar deposits, although not so strikingly laminated, were made in other lakes formed in the glacial period on account of the increase in precipitation. The valley floor near Salt Lake, Utah, once occupied by Lake Bonneville, is an example. The basin in which the City of Mexico is located was similarly filled, but the principal constituents of the sediment were of volcanic origin.

Buried Valleys. For various reasons, the levels of the glacial lakes fluctuated widely from time to time. While the water was at a low level, tributary streams eroded valleys in the surrounding lands. As the water slowly rose in the lakes, the streams became filled with

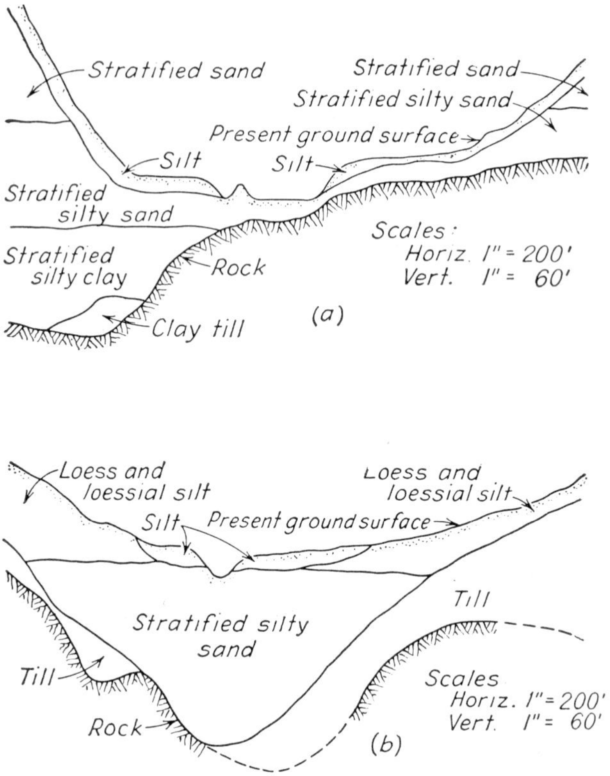

FIG. 4.8. Cross section through two silt-filled buried valleys in Illinois. In (*a*) the entire valley was covered by the sand in which the present stream channel is located; in (*b*) the valley was covered by till which was subsequently eroded and replaced by alluvium, and then covered by loess.

sediments, often mixed with the remains of plant life. The entire topography may then have been covered by lake sediments or even by till in later glacial advances, and the channels with soft filling may have been completely buried. The present streams do not necessarily flow in the old valleys, and the subsurface relationships are often very complex. A cross section through a buried valley in Jackson County, Ill., in shown in Fig. 4.8*a*. It may be observed that the old fill is now covered by a high bluff and that the present river valley is at a considerable distance from that of its predecessor. The valley shown in Fig. 4.8*b* was originally eroded in the bedrock, was enlarged by glacial ice and covered by till, then reexcavated in the till, filled with sediments, and covered by loess. The existing stream is very small and is eroding the loess.

Such buried channels are very common near the Great Lakes. They are also found on the seaboard because the level of the ocean was probably as much as 200 ft lower than at present during several periods in the glacial epoch. A rational interpretation of test borings in the vicinity of buried valleys requires at least a rudimentary knowledge of the geological history of the region.

4.3. River Deposits

Stream-channel Deposits. In its lower reaches, a stream overloaded with sediments deposits the material in its bed and the bed of the stream is raised. The water then spills over into an adjacent depression and establishes a new channel. Such a stream is said

FIG. 4.9. Typical braided stream bed.

to be *braided* because it consists of a series of rivulets continually joining each other and then separating. The position of the rivulets constantly shifts. The fill deposited in each of the channels has a relative density and a grain size dependent on the velocity of flow and on the character of the sediments being furnished to the river. Hence, the cross section of such deposits consists of an aggregate of lenticular elements in each of which conditions are fairly uniform but each of which differs from its neighbor in unpredictable fashion. Figure 4.9 is a photograph of a typical braided stream.

The upper reaches of streams that rise in mountainous regions often rest directly upon the bedrock. The rock surface is likely to be covered by large boulders and somewhat smaller rock fragments.

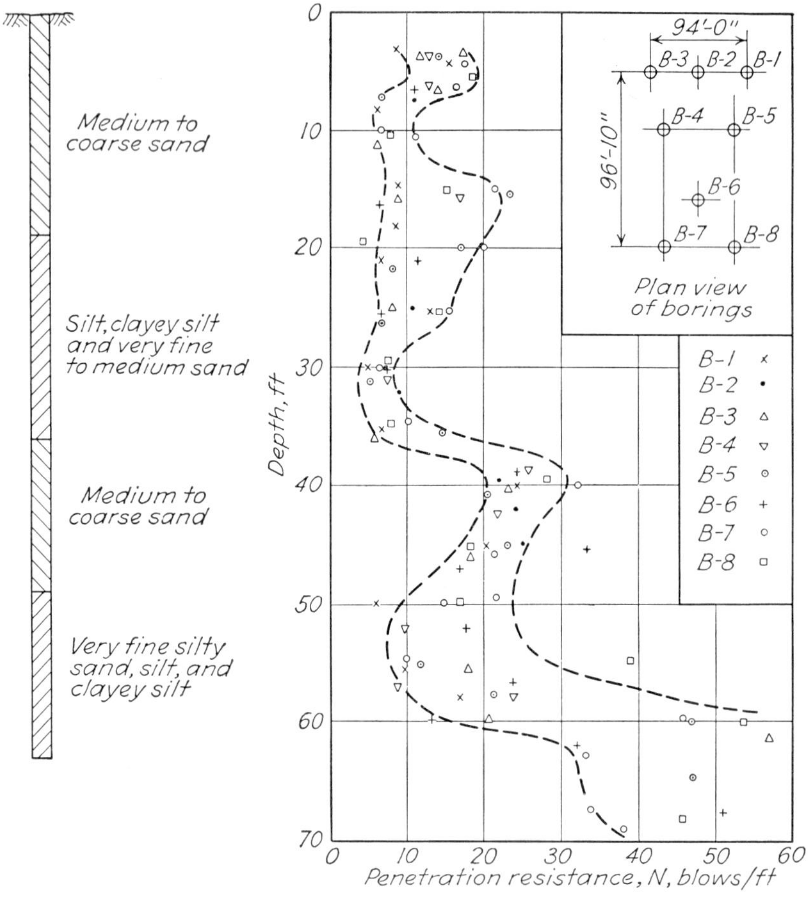

FIG. 4.10. Variations in penetration resistance of sands and silts in the valley of the Cedar River, Cedar Rapids, Ia.

Changes in elevation of different parts of the earth's surface, together with changes in climatic conditions, alter the character of the material deposited by a stream at any given point. As a consequence, the deposits below the bed of many streams exhibit major variations in character due to these causes as well as the minor variations described above. Variations in the penetration resistance of silts and sands in the valley of the Cedar River in Iowa, once an outlet for glacial meltwater, are shown in Fig. 4.10 as an example.

Flood-plain Deposits. Many rivers deposit little sediment in their channels, but during floods silt-laden water spreads over the adjacent lowlands. The coarser materials settle rapidly whereas the finer silts and clays settle more slowly and become segregated. The resulting deposits have a laminated character. Commonly they consist predominantly of fine sand and silt with thin partings of sand or clay. Although they are quite uniform in horizontal directions, they change vertically in an erratic fashion. The permeability in horizontal directions is very great compared to that in the vertical direction. A photograph of a sample of such a deposit from the flood plain of the Mississippi River near Baton Rouge is shown in Fig. 4.11.

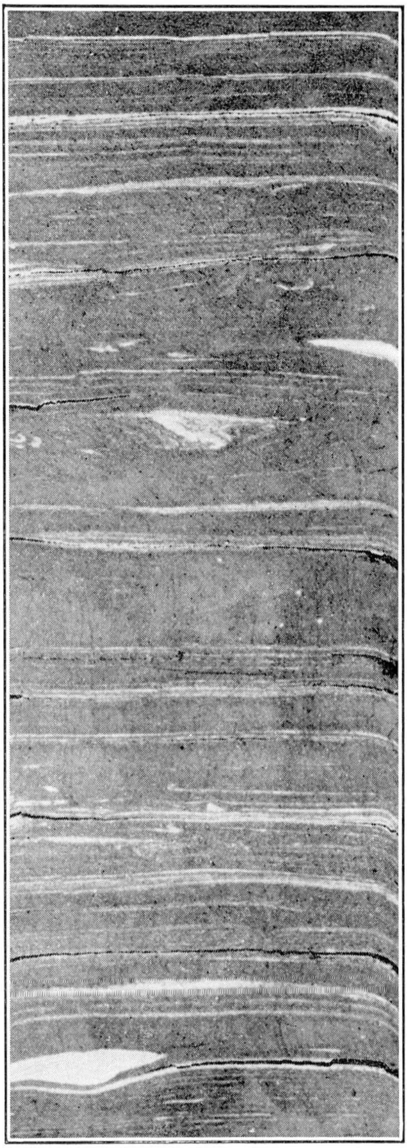

FIG. 4.11. Two-inch tube sample from flood-plain deposit of Mississippi River near Baton Rouge, La.

Delta Deposits. When the water of a sediment-laden stream enters a lake or sea, it experiences a sudden check in velocity, and its sediments drop. The structure of a delta formed under these conditions is simple but there are likely to be many local variations because the burden of the stream depends upon the climate and because shore currents continually shift. The variations in permeability of a delta deposit formed during the glacial epoch near Chicopee, Mass., are shown in Fig. 4.12.

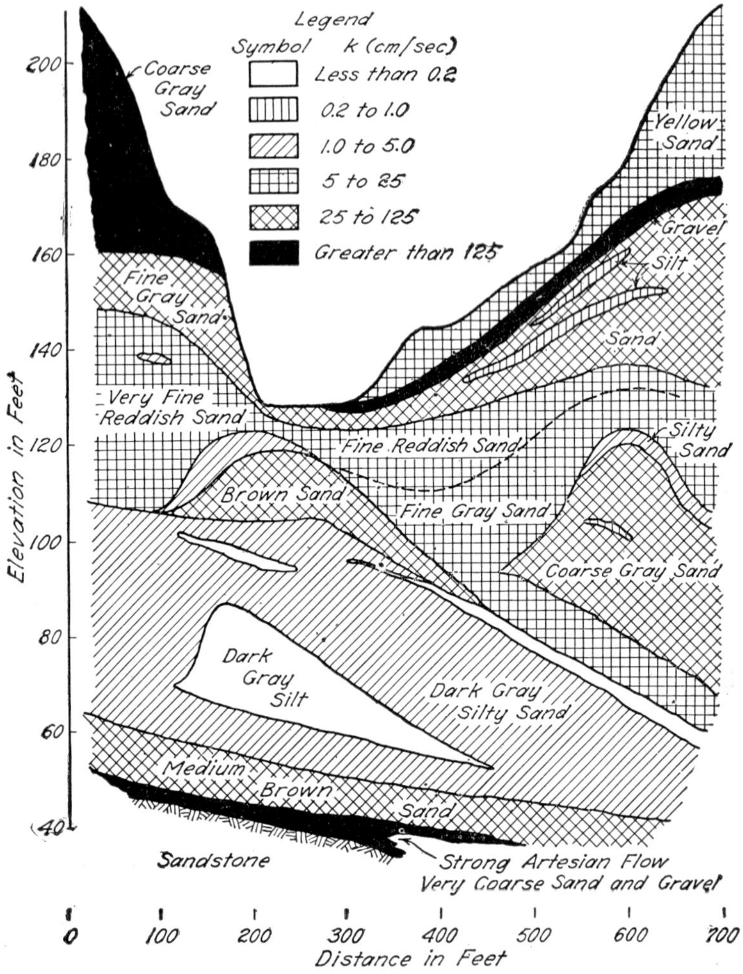

FIG. 4.12. Permeability profile of glacial delta deposit near Chicopee, Mass. (after K. Terzaghi).

4.4. Windblown Deposits

Dune Sand. The sweep of the wind across large sand-covered areas, such as some desert regions, flood plains of broad rivers, or even beaches, rolls grains of fine sand over each other and piles them up in the form of dunes. The coarser particles do not participate in the movement whereas the silt-size grains are blown away. By this

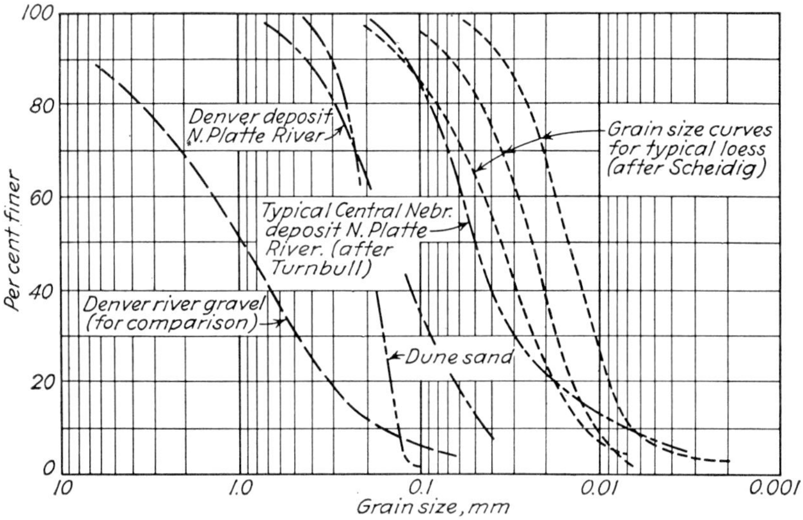

FIG. 4.13. Typical grain-size curves for various wind-blown deposits.

process of selection, the grains of dune sand become very uniform in size. Indeed, dune sands are among the most uniform of natural formations. With increasing distance from the source, the sand becomes finer, and within a few miles it may grade into loess having

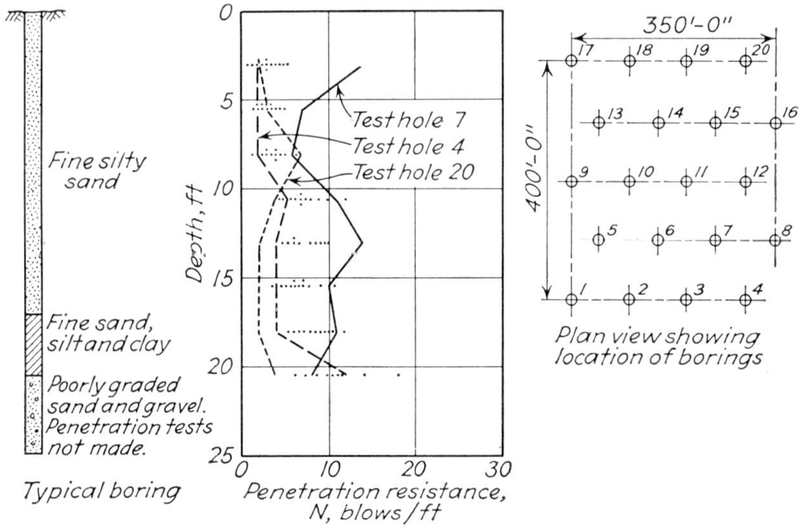

FIG. 4.14. Results of standard penetration tests in 20 borings on one building site above extremely loose, fine, partly wind-blown sand in Denver, Colo.

grains of silt size. Grain-size curves for several wind-blown soils are shown in Fig. 4.13. In the transition zone of very fine sand, extremely loose deposits have been encountered. The results of standard penetration tests at one such site in Denver are shown in Fig. 4.14. The corresponding grain-size curve is plotted in Fig. 4.13.

Loess. The grain size of the windblown silts that form a loess deposit decreases in a general way with increasing distance from the source. Nevertheless, loess deposits are usually quite uniform. Their cohesion is due in part to the precipitation of chemicals or colloids leached by rainwater from the upper weathered zone. Hence, the cohesion may vary considerably from place to place. Deep weathering or saturation may lead to destruction of the characteristic root-hole structure and vertical cleavage. These processes may occur locally and introduce variations into the deposit. Loess is widespread in the United States, especially in the lower Mississippi Valley and in the Central States from Kansas and Nebraska to Illinois. It is often associated with glaciation because of the broad expanses of unvegetated outwash plains and rivers that existed close to the glaciers.

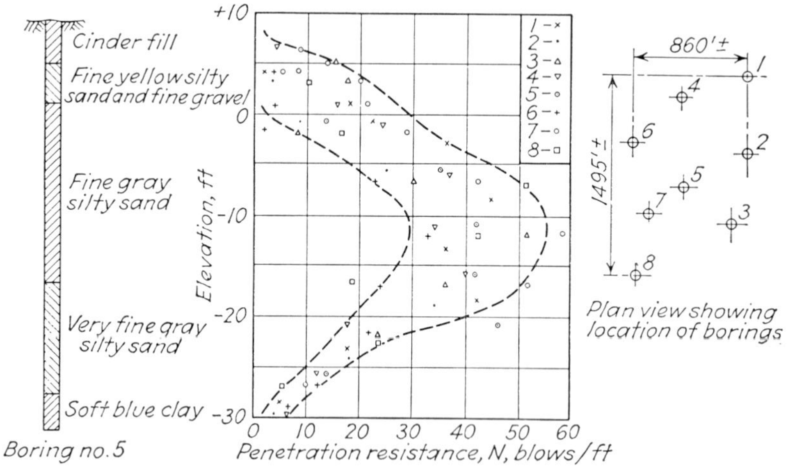

FIG. 4.15. Results of standard penetration tests in sandy shore deposits at southern end of Lake Michigan.

4.5. Shore Deposits

The action of waves and shore currents in lakes and oceans builds up beach and shore deposits primarily of sand and gravel. These deposits may be relatively uniform and of moderate to high density.

Results of standard penetration tests in a sandy shore deposit at the southern end of Lake Michigan are shown in Fig. 4.15.

On the other hand, if the shoreline has fluctuated on account of changes in the water level, the deposits of sand may alternate in an

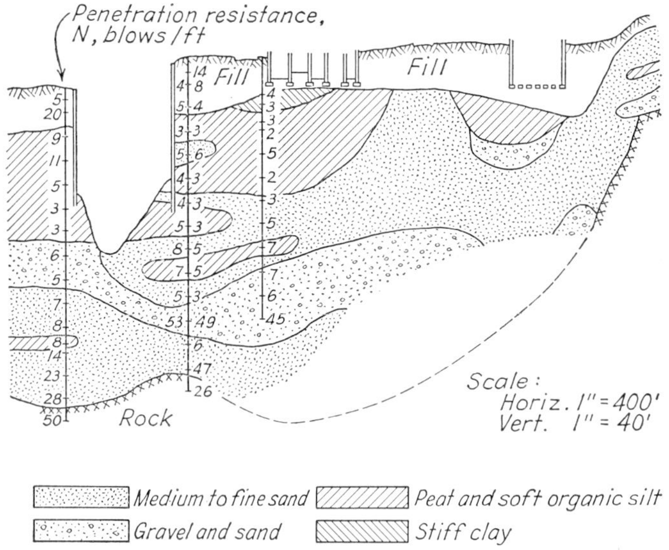

FIG. 4.16. Cross section through composite shore deposit near mouth of Milwaukee River, Wis.

erratic manner with organic silts and peats. Such formations are known as *composite shore deposits.* The one illustrated by Fig. 4.16 is located near the mouth of the Milwaukee River in Wisconsin.

4.6. Bedrock and Residual Soils

Defects in Unweathered Bedrock. The foundation engineer is commonly not seriously troubled by structural defects in the bedrock. However, certain types of rock deserve careful scrutiny.

Limestone and, to a less extent, dolomite are soluble in groundwater. Strata of these rocks may contain large channels, often filled with clay, or even extensive caverns. The absence of surface drainage or the presence of sink holes in a limestone region is indicative of solution cavities. Such cavities may endanger the stability of foundations and must be located.

Shales are often relatively compressible and commonly contain partings along which horizontal movements readily occur. Special precautions may be required to provide for lateral forces exerted by the foundations of retaining walls and bridge abutments. The distinction between stiff clay and shale is somewhat arbitrary, and swelling clays may be mistaken for soft shales. Under some conditions, laboratory tests may be needed for proper identification.

Occasionally bridge and building foundations must be located in fault zones where excavations encounter strongly upturned or jointed strata. It is always desirable to obtain at least a rough conception of the general geology in the vicinity and of the special geological features at the site if rock excavation is contemplated or if foundations are to rest on rock.

In mineralized regions, mining operations or production by solution from wells may have led to removal or weakening of some bedrock strata. This possibility requires investigation, and particular attention should be given to the detection of long-abandoned workings.

Processes of Weathering. The oldest rocks exposed on the face of the earth appear to be metamorphosed sediments. They have been extensively deformed by heat and pressure, and they have in places recrystallized into granite and other typical igneous rocks. Yet, their sedimentary origin bespeaks the effectiveness of the agents of weathering throughout geological time, inasmuch as these ancient rocks are composed of the products of weathering of still older rocks.

Some of the agents of weathering are purely physical. One of the most important of these is the *differential expansion* and *contraction* that arises on account of heating and cooling masses of rock containing minerals of different thermal properties. Others are the *abrasive action* of wind and water, especially when these agents are transporting sediments that aid the grinding action; the *expansive force of freezing water;* and in many localities, especially in the past, the *plucking and grinding* action of glaciers. Important as these physical agents are, however, they are not capable of reducing the size of individual fragments below about 0.01 mm. More complete distintegration can be accomplished only by chemical processes.

The principal agents of chemical weathering are *oxidation, hydration, carbonation,* and *solution.* Almost all materials combine with oxygen, especially in humid climates, but iron compounds are particularly susceptible. The combination with oxygen is usually associated with increased volume, whereupon the oxidized portion is likely to split away from the unaltered material. Similarly, the increase in volume accompanying the formation of new minerals by hydration is responsible for disintegration of the original minerals. Falling rain dissolves small quantities of carbon dioxide from the atmosphere and thus becomes a weak solution of carbonic acid. The chemical combination of certain minerals with carbonic acid leads to the formation of softer and weaker minerals; more important, however, is the fact that weak carbonic acid is a much more effective solvent than pure water. It attacks not only materials containing calcium carbonate but also those containing silicates.

As a rule, physical weathering predominates in frigid or dry climates, whereas chemical weathering is more important in warm humid climates. Plant and animal life contribute to both processes. However, rocks may also be subjected to chemical decomposition when far below the surface, because of the circulation of magmatic waters charged with chemicals. Granites and gneisses thus may locally have been transformed into soft rocks of claylike character, and hard quartzites into cohesionless sand.

When the processes of weathering have eliminated the cohesive bonds between most of the constituents of the rock, the material becomes soil. Yet, even after this state has been reached, further weathering occurs. Ultimately, the disintegration may be so complete that only certain substances, the clay minerals, are present.

Weathered Rock. The processes of weathering rarely attack all parts of a rock mass uniformly; the rate and extent of the disintegration depend on many factors including the variations in mineralogical composition of the original rock, and the type and pattern of such discontinuities as bedding planes, joints, and fault zones. Hence detailed studies are often necessary to locate and judge the strength and compressibility of weathered zones in a rock formation.

In the tropics it is not uncommon to find formations that have the appearance of sound rock, but that have been chemically altered for depths of several hundred feet to such an extent that the material may be less capable than soft clays of supporting heavy foundations.

In humid tropical regions, the rapid weathering of solid rocks into weak clays leads to many slides on the hillsides. The subsurface conditions may be very complex because the subsoil may consist of residual soils and slide material of similar origin in a jumbled mass. Similar phenomena occur on a smaller scale in southeastern United States.

When the bedrock consists of shale or schist, especially if the latter contains chlorite, mica, or talc, the partly disintegrated material is known as *detritus.* The presence of detritus is also a common cause of slide topography, especially in West Virginia and parts of Ohio and Pennsylvania.

Where there is a gradual transition from residual soil to sound rock, the problem of the engineer is to decide upon the most appropriate depth of foundation. This may require careful investigation of the physical properties of the materials found at various depths.

Residual Soil. If the products of rock weathering accumulate without being eroded, residual soils are formed. In the tropics, where the depth and rate of weathering are very great, the deposits consist primarily of clays. In southern United States deposits 50 ft thick are occasionally encountered, and thicknesses of 10 to 20 ft are not

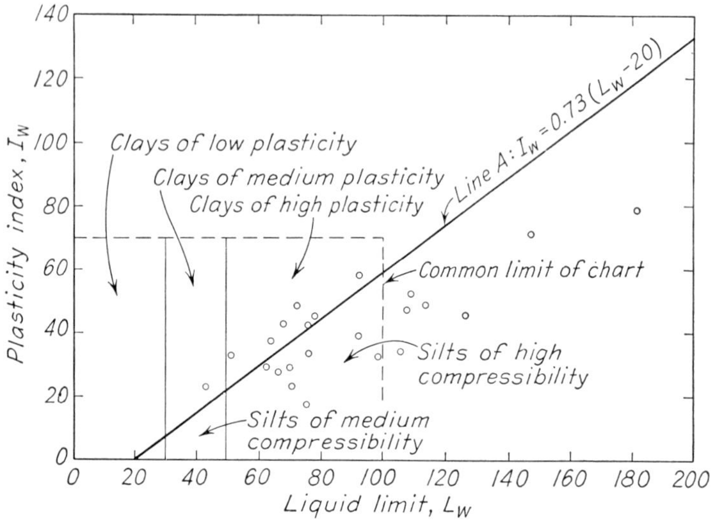

FIG. 4.17. Plasticity chart showing variations in characteristics of residual clay developed from sandstone near Bogota, Colombia.

uncommon. In arctic regions, freezing and thawing predominate, and the products of weathering usually consist of rock fragments and sand. The depth of such residual soils is usually not more than 5 ft.

Deposits of residual soils, especially those formed primarily by chemical weathering, may possess properties as variable as those of transported soils. Seemingly unimportant differences in the parent material, in the topography, in the character of the groundwater, and in climatic conditions lead to marked differences in the physical characteristics of the soils. Residual clays may be soft saturated soils that exhibit extreme variations in plasticity and other properties even within a limited region. Figure 4.17 is a plasticity chart indicating the characteristics of a series of samples taken from a depth of 10 ft beneath the gently sloping surface of one hillside near Bogota, Colombia. Variations in all of the other physical properties were found to be equally striking.

Boring Rig with Rock-drilling Attachment

Portable boring rig widely used for subsurface investigations. In soil, the hole is advanced by the wash-boring process and may be protected by 2½-in. casing. Samples may be taken with a split spoon in connection with the standard penetration test, or with thin-walled tubes of 2-in. outside diameter. Rock or large boulders may be drilled with the aid of the attachment shown in the photograph.

PLATE 5.

CHAPTER 5

Program of Subsurface Exploration

5.1. Development of Exploratory Program

Introduction. The preceding chapter demonstrates that very few natural soil deposits are even approximately uniform, and many are extremely erratic. It is obvious that in an erratic deposit no program of subsurface exploration can lead to more than a rough idea of the average values for the physical constants of the subsurface material and of the probable variations from these values.

The nature of the deposit is an important factor in determining the method of soil exploration that will yield the greatest amount of useful information. If, for example, the foundation of an important structure is to be established above a fairly homogeneous layer of clay, a considerable amount of testing of undisturbed samples may be justified because the test results permit a relatively accurate forecast of both the amount and the rate of settlement. On the other hand, if the same structure is to be located above a deposit composed of pockets and lenses of sand, silt, and clay, a comprehensive testing program would not be justified because it would provide little more information than could be obtained merely by determining the index properties of representative samples. Much more useful information could be obtained at less cost by making an adequate number of penetrometer measurements that would disclose the pattern of the various soft and stiff elements in the subsoil. If the soil profile is complex, an elaborate program of soil testing is likely to be out of place.

The magnitude and character of the exploratory program should also be chosen in consideration of the importance of the project under construction. If the job involves only a small expenditure, extensive programs of soil exploration cannot be justified economically. It is cheaper to take advantage of whatever information may already be available and to use a liberal factor of safety in the design.

Preliminary Exploration. The program of subsurface exploration should be preceded by a fact-finding survey. In such a survey the engineer responsible for the exploration should prepare a digest of all available information on soil conditions near the site and on the behavior of other structures located in the vicinity. In highly developed regions with unfavorable subsurface conditions, useful information is likely to be available in technical journals and published reports, but in most rural areas or in areas being newly developed for industry, information concerning structural behavior may be lacking. However, the engineer should not overlook the maps and publications of state and federal geological surveys, or reports of soil surveys prepared in connection with agriculture or highway construction. The type of information to be obtained from these sources has been discussed in Art. 1.8 under the geological and pedological systems of classification.

On the basis of the information obtained by the fact-finding survey, the type of preliminary exploration is selected. In most instances, unless the underlying material is bedrock, auger borings or wash borings supplemented by the results of the standard penetration test are suitable. Moreover, for many projects, no further subsurface exploration is necessary. This is likely to be the case if the loads on the subsoil will be small and a large factor of safety can be used without excessive cost, if the structure can be founded on rock or strata of high bearing capacity, or if an ordinary structure is to be erected in an area where much practical experience has been summarized in the form of reliable empirical rules or building codes.

Detailed Exploration. When the preliminary exploratory program does not provide sufficient information for design or construction, further investigations are required. The methods should be selected to obtain the most pertinent information at the least cost. As a rule, the properties of fairly uniform deposits of soft clay and silt can be investigated most economically by obtaining continuous samples in 2-in. steel tubes and performing appropriate laboratory tests (see Chapter 15). Erratic deposits of soft silt and clay can be examined by means of penetration tests combined with enough 2-in. tube borings to permit interpretation of the penetrometer data. Standard penetration tests or dynamic cone penetrometer tests are appropriate for sands.

On some jobs involving structures of great importance or foundation conditions of exceptional difficulty, additional information may still be required. It may be advisable to obtain large-diameter undisturbed samples from critical strata, to conduct load tests, to make field pumping tests, or to conduct other special tests. Since such studies are always expensive, they should be undertaken only to investigate specific questions that the cheaper procedures are inadequate to answer.

Number and Depth of Borings. The program of soil exploration should not be determined in final form beforehand, but should be developed step by step as information accumulates. By this procedure, the maximum amount of information can be obtained for a given expenditure and the program can be stopped as soon as adequate data have been collected. Hence, no definite rules can be established for the exploratory program.

For buildings or structures of ordinary size, it is customary and satisfactory to anticipate making four borings, one at each corner of the structure. Unless bedrock is encountered, the first boring should ordinarily extend to the maximum depth within which the stress caused by the structure could conceivably produce excessive settlement. This depth may have to be established on the basis of approximate stress and settlement computations, as indicated in Chapter 15. For a heavily loaded structure the first boring should ordinarily extend to a depth equal to twice the least width of the structure. Beneath lightly loaded structures with widely spaced columns, the depth of the first boring should not be less than twice the probable width of the largest footing.

The second boring will serve to indicate whether soil conditions are likely to be relatively uniform or erratic. If the results are similar to those for the first boring, the second boring may ordinarily be discontinued when it has penetrated all the soft or compressible strata encountered in the first.

If the borings encounter rock and the conditions are such that the structure may be founded upon rock, cores should be obtained for a depth of 5 to 10 ft to make sure that bedrock, rather than boulders, has been reached. If there is evidence of solution channels or deep weathering, the cores should be continued until sound rock is encountered.

As the exploratory program develops, the engineer should study the implications of all new information. If additional borings, penetration tests, or special investigations are needed, they should be so located and planned that each addition to the program will provide the maximum increase of knowledge at the current stage of the investigation.

Conclusion. No matter how complete the program of soil exploration and testing may be, there always remains a large margin of uncertainty concerning the exact nature of subsurface conditions at a given site. This fact is of outstanding practical importance. It makes foundation design fundamentally different in its basic concepts from all other branches of structural design. The engineer cannot proceed as he would with materials having well-defined properties, such as steel, concrete, or timber. Although the latter materials are not perfectly uniform, yet they can almost always be considered so in design. The foundation engineer must often wait to obtain his final data concerning soil conditions until he can observe what happens in the field, and he should always make use of every fragment of evidence. Soil tests performed on a few samples taken from an erratic deposit do not provide a satisfactory basis for design because the engineer is interested in the behavior of the deposit as a whole rather than of a few specimens taken from it.

Throughout Part C of this book, particular attention will be given to the uncertainties inevitably involved in the information concerning subsurface conditions. Methods will be developed to permit design on a rational and economical basis, in spite of the handicap arising from these uncertainties.

Part B

TYPES OF FOUNDATIONS AND METHODS OF CONSTRUCTION

One of the requisites for the successful practice of foundation engineering is an adequate knowledge of the engineering properties of soils and rocks, the natural materials upon which the engineer's structures depend for support. Part A has dealt with this aspect of the engineer's background.

Part B is concerned with a second equally important aspect, the commonly used types of foundations and methods of construction. Here tradition, experience, and industrial competition have played a part more important than that of science, and partly for this reason foundation engineering is properly considered an art.

By a sort of evolutionary process, strongly influenced by the relative costs of labor and materials, certain types of foundations and certain construction procedures have come into general use in America. These are briefly described in the following pages. Thus, Part B consists of a summary of current practice, to which the engineer should adhere unless highly unusual conditions arise.

James C. Meem (1866–1936)

Chief engineer for the contracting firm that constructed a large part of the early subway system in New York. He was keenly interested in the pressures exerted by earth against temporary supports, and was one of the first to observe that the pressure acting against the upper part of the bracing in an open cut is greater than that acting against the lower part. This observation was directly contrary to the theoretical concepts then current. These concepts were based upon assumed and unrealistic stress-deformation characteristics of soils. Later, when soil mechanics demonstrated the importance of the actual stress-deformation characteristics, new theories were developed that substantiated Meem's conclusions.

PLATE 6.

CHAPTER 6

Excavating and Bracing

6.1. Introduction

The foundations for most structures are established below the surface of the ground. Therefore, they cannot be constructed until the soil or rock above the base level of the foundations has been excavated.

Ordinarily it is not the function of the foundation engineer to select the equipment for excavating at a given site, or to design the bracing if any is required. These are considered to be within the province of the contractor. It is generally the engineer's duty, however, to approve or disapprove the construction procedure proposed by the builder and to check the design of the bracing. In exercising these functions, the engineer is properly concerned only with the ability of the proposed procedures to permit satisfactory construction of the structure as designed; he should not ordinarily presume to dictate the manner in which the construction should be carried out.

On large or complicated jobs, it may be impossible to prepare a design without also deciding upon the method of construction. When the design and the construction procedures are related so closely that they must be considered as a unit, the engineer is obliged to specify the construction method and possibly to design the bracing.

In pervious soils, excavation below the water table usually requires drainage of the site either before or during construction. In this chapter we shall assume that the water table is normally below the bottom of the excavation or that it has been temporarily lowered. The techniques for drainage will be discussed in the next chapter.

Moreover, in this chapter we shall discuss only the general aspects of excavating and of providing support for the sides of the pits or cuts. Details or modifications appropriate in connection with specific foundation conditions or types of structures will be discussed in Part C.

6.2. Open Excavations with Unsupported Slopes

Shallow excavations can be made without the provision of support for the surrounding material if there is adequate space to establish slopes at which the material can stand. The steepness of the slopes is a function of the type and character of the soil or rock, the climatic and weather conditions, the depth of the excavation, and the length of time the excavation must remain open. As a rule, construction slopes are made as steep as the material will permit because the occurrence of a few small slides is generally not serious. The cost of removing the material affected by the slides may be considerably less than that of the additional excavation required to provide flatter slopes.

The steepest slopes that can be used in a given locality are best determined by experience. Most sands contain a small amount of cementing material or display a small amount of apparent cohesion on account of the moisture they contain. This cementation or cohesion cannot safely be depended upon for the stability of permanently exposed slopes, but it is ordinarily utilized for the periods during which excavations stand open. Although permanent slopes of about 1 vertical to 1½ horizontal are usually considered standard for sandy soils, construction slopes as steep as 1 vertical to ½ horizontal are not uncommon.

The maximum slope at which a clay soil can stand is a function of the depth of the cut and of the shearing resistance of the clay. If the clay is soft below the base level of the excavation, flat slopes may be required to avoid a rise of the bottom. Furthermore, stiff or hard clays commonly possess or develop cracks near the ground surface. If these cracks become filled with water, the hydrostatic pressure greatly reduces the factor of safety and may cause slope failures. The water in the cracks also softens the clay progressively so that the safety of the slope is likely to decrease with time. For these reasons, bracing is often used to support the sides of excavations in clay, in spite of the fact that the clay would stand briefly to the necessary height without lateral support.

6.3. Sheeting and Bracing for Shallow Excavations

Many building sites extend to the edges of the property lines or are adjacent to other sites upon which structures already exist. Under these circumstances, the sides of the excavation must be made vertical and must usually be supported. Several methods of bracing are in common use.

If the hole is not to extend to a depth greater than about 12 ft, it is common practice to drive vertical planks known as *sheeting* around the boundary of the proposed excavation. The depth to which the sheeting is driven is usually kept near that of the bottom of the hole as excavation progresses. The sheeting is held in place by means of horizontal beams called *wales* that in turn are commonly supported by horizontal *struts* extending from side to side of the excavation. The struts are usually of timber, but, if the excavation is not more than about 5 ft wide, extensible metal pipes called *trench braces* are

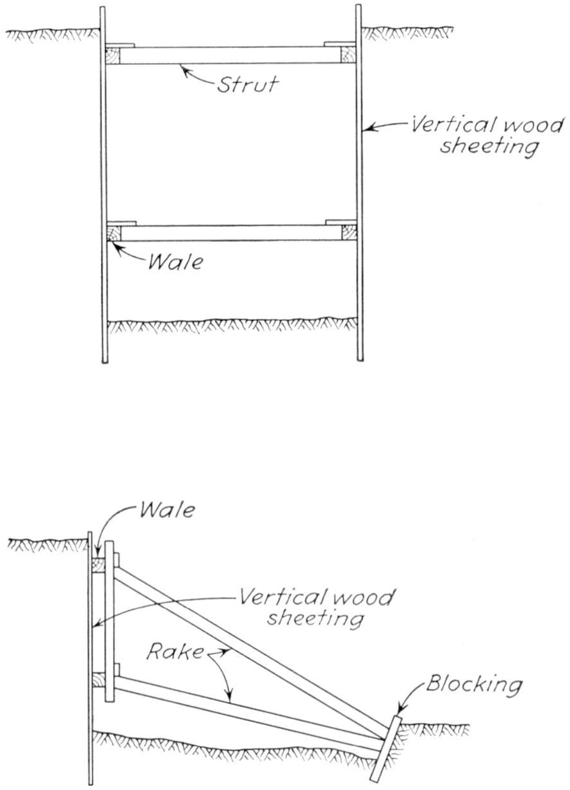

FIG. 6.1 Common methods for bracing sides of shallow excavations.

commonly used. If the excavation is too wide for the use of struts extending across the entire width, the wales may be supported by inclined struts known as *rakes*. The use of rakes requires that the soil in the base of the excavation be firm enough to provide adequate support for the inclined members.

Two typical arrangements for shallow bracing are shown in Fig. 6.1.

6.4. Sheeting and Bracing for Deep Excavations

When the depth of excavation exceeds 15 or 20 ft, the use of vertical timber sheeting generally becomes uneconomical, and other methods of sheeting and bracing are commonly employed. According to one procedure, steel sheet piles are driven around the boundary of the excavation. As the soil is removed from the enclosure, wales and struts are inserted.

The types of sheet piles commonly used for this purpose are shown in Fig. 6.2. The strength and stiffness of the *arch web* type *b* exceeds that of the *flat web a*, whereas that of the *Z piling c* is the greatest for all rolled sheet-pile sections. Consequently, types *a* and *b* are used for shallow excavations and type *c* for the deepest excavations or those where the heaviest pressures are expected.

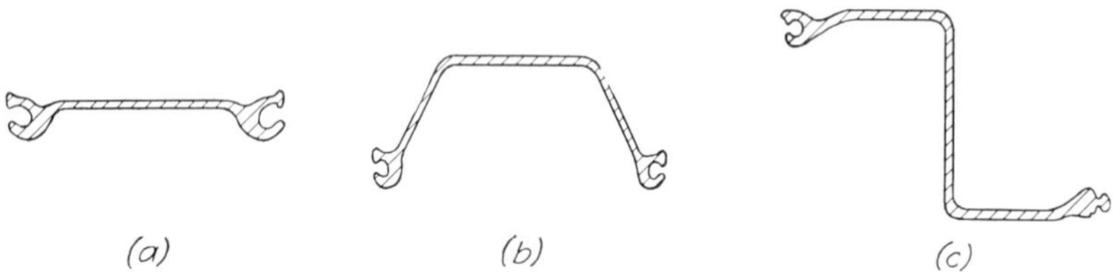

FIG. 6.2. Types of steel sheet piles commonly used for bracing sides of deep excavations. (*a*) Flat web; (*b*) arch web; (*c*) Z piling.

As soon as excavation has proceeded for a few feet, wales and struts are inserted as shown in Fig. 6.3. The wales are commonly of steel, and the struts may be of steel or wood. Excavation then proceeds to a lower level, and another set of wales and struts is installed. This process continues until the excavation is complete. In most soils it is advisable to drive the sheet piles several feet below the bottom of the excavation to prevent local heaves. In some instances the embedment may eliminate the necessity for installing a strut at the bottom of the cut. It is important to provide vertical support for the bracing. This may be done by keeping posts beneath the system of bracing to transfer its weight to the underlying soil or by suspending the bracing from beams extending across the top of the cut.

In most types of soil, it is possible to expose a vertical face of several square feet without danger of collapse of the ground. It may then be possible to eliminate the sheet piles and to replace them with a series of H piles spaced at 4 to 8 ft. The H piles are driven with their flanges parallel to the sides of the excavation as shown in Fig. 6.3*b*. As the soil next to the piles is removed, horizontal boards known as *lagging* are introduced as shown in the figure and are wedged against the soil outside the cut. As the general depth of excavation advances from one level to another, wales and struts are inserted in the same manner as for steel sheeting.

If the width of a deep excavation is too great to permit the use of struts across the entire excavation, inclined bracing may be used,

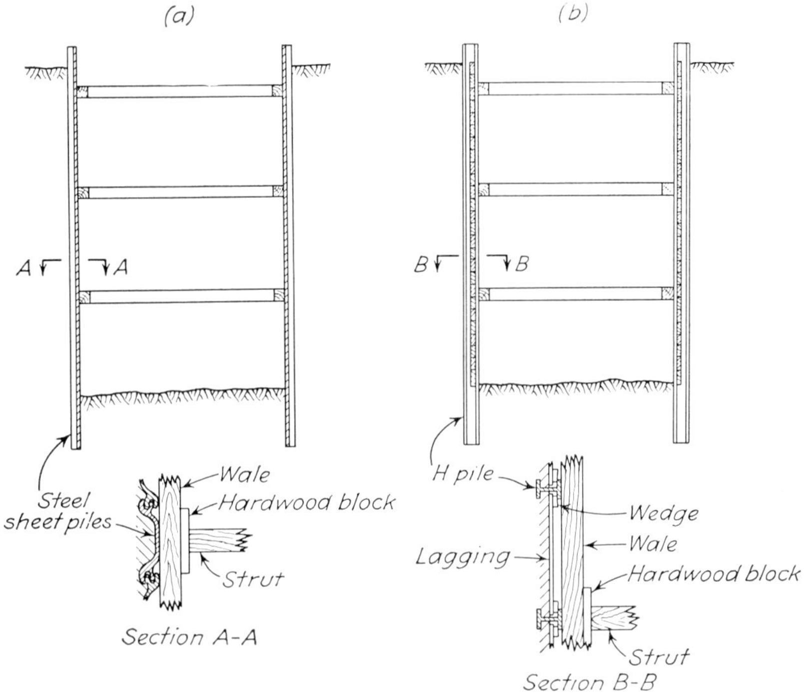

FIG. 6.3. Cross sections through typical bracing in deep excavation. (*a*) Sides retained by steel sheet piles; (*b*) sides retained by H piles and lagging.

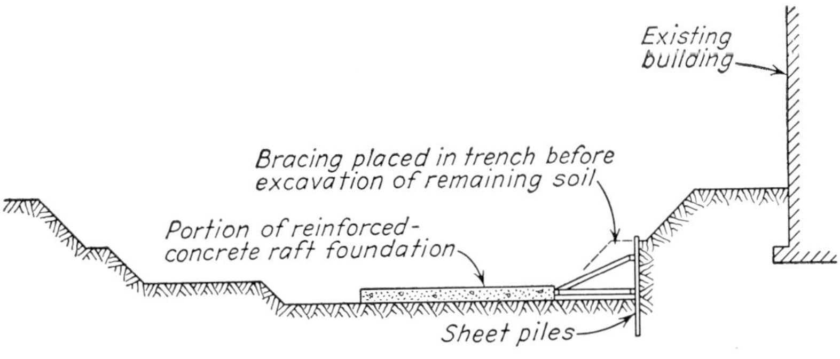

FIG. 6.4. Typical bracing in deep, wide excavation.

provided there is adequate support for the inclined members. In some instances, it is necessary to excavate the central portion of the site to its maximum depth and to place part of the permanent foundation. The completed part of the foundation then serves as support for the inclined bracing required when the remainder of the site is excavated. This procedure is shown in Fig. 6.4.

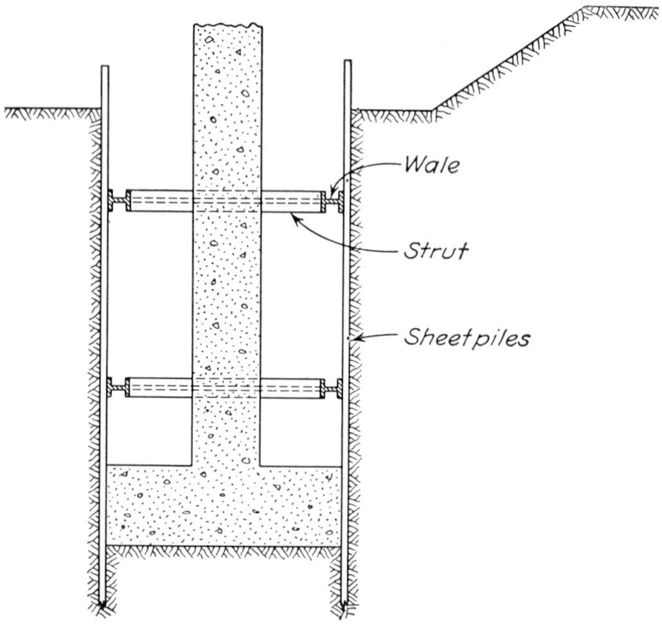

FIG. 6.5. Method of constructing basement walls in trenches before general excavation.

Sometimes it is preferable to complete the exterior walls of a structure before removing the material in the space occupied by the basements. The walls are constructed in narrow braced trenches as shown in Fig. 6.5. After the walls have been completed and the floor system established at the top, the block of soil between the walls may be excavated. The floor provides the bracing for the upper part of the walls and additional bracing may be inserted as needed while excavation progresses.

6.5. Movements Associated with Excavation

Excavation always involves removal of material and, as a consequence, a change in the state of stress in the rock or soil beneath and beside the excavated space. This is true whether or not the sides of the cut are supported by bracing. Inasmuch as no material can experience a change in stress without corresponding deformations, excavation is always associated with movements of the adjacent ground surface. Such movements usually have the character of settlements, but under certain rather rare conditions the ground surface may rise.

On the other hand, properly designed and carefully placed bracing can materially reduce the change in lateral pressure in the material

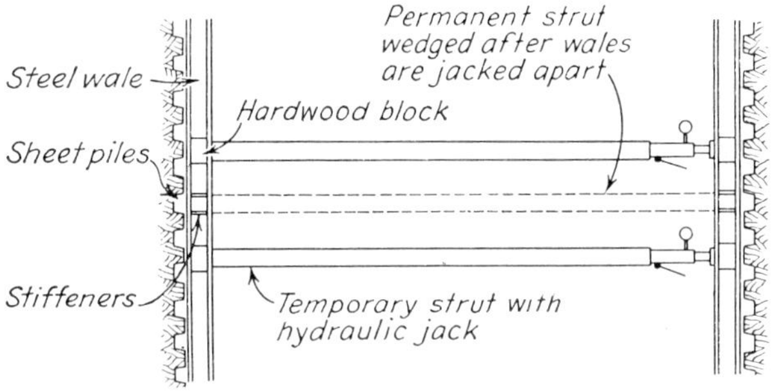

FIG. 6.6. Method of prestressing struts for bracing sides of open cuts.

adjacent to the excavation and is thus capable of reducing the settlements to a value that may be regarded as the practicable minimum for a particular job. Where the settlements might damage adjacent structures it is one of the duties of the foundation engineer to make sure that the method proposed for bracing the sides is capable of reducing the settlement to a tolerable value. If this condition is satisfied and the settlements become excessive, the damage may properly be ascribed to lack of good workmanship on the part of the construction forces. These matters are discussed in greater detail in Chapter 18.

A common and desirable precaution for reducing movements of the adjacent ground to the practicable minimum is to prestress each strut as it is inserted. One method is illustrated in Fig. 6.6. Before a strut is inserted, two auxiliary struts equipped with hydraulic jacks are used to increase the distance between the wales against which the strut is to bear. The strut is then inserted and wedged tightly, so that a stress of several tons remains in the strut when the jacks are released.

Robert Stephenson (1803–1859)

Engineer of construction for some of the earliest English railroads, including the Birmingham and London Railway, first to enter London (1838). This new form of transportation demanded the construction of deep cuts and tunnels, some of which had to be driven through cohesionless sand below the groundwater table. Stephenson successfully met these unprecedented problems and developed many techniques in the art of drainage. He was a pioneer in structural engineering as well, serving as engineer for such famous early structures as the Menai Straits tubular girder and the first railroad bridge across the St. Lawrence River at Montreal.

Plate 7.

CHAPTER 7

Drainage and Stabilization

7.1. Introduction

When the depth of excavation is greater than the distance to the free water surface in a pervious soil having a coefficient of permeability greater than about 10^{-3} cm/sec, the soil must be drained to permit construction of foundations in the dry. If the coefficient of permeability of the soil is within the range 10^{-3} to 10^{-5} cm/sec, the quantity of water that seeps into the excavation may be inconsequential but drainage may still be required to maintain the stability of the sides and bottom of the excavation. If the coefficient of permeability is smaller than about 10^{-7} cm/sec, the soil is likely to possess sufficient cohesion to overcome the influence of the seepage forces, and drainage may not be required, even if the excavation extends for a considerable depth below the water table.

After completion of structures with basements, it is often necessary to maintain the water level in a lowered position. This requires the installation of permanent drains.

The physical processes of drainage have been outlined in Art. 2.5. The possibility of inducing settlements by drainage is considered in Art. 18.3. Specific procedures required for the drainage of various types of foundations in different subsurface formations will be described in Part C. The present chapter is concerned primarily with the general techniques of drainage and the equipment necessary to accomplish it.

7.2. Ditches and Sumps

Where space permits, ditches may be used to lower the water table in sand or in other materials made pervious by cracks or joints. In silty or fine sands, the side slopes must ordinarily be relatively flat on account of the seepage pressures exerted by the entering water.

The relatively flat slopes required for open ditches in sand generally preclude the use of ditches for lowering the water table more than a very few feet. However, open ditches are commonly used in the bottom of an excavation to collect the water that seeps into the hole. Such ditches lead to sumps from which the water is pumped.

A *sump* is a pit with its bottom below the level of the ditches that enter it. Considerable care must often be exercised to prevent sand and silt beside and beneath the sump from washing in and being pumped out with the water. To reduce the loss of sand by pumpage and to prevent the consequent instability, it is often desirable to line the sides of the sump and to cover the bottom with coarse-grained material that acts as a filter. Such a sump for an open cut in sand is shown in Fig. 7.1.

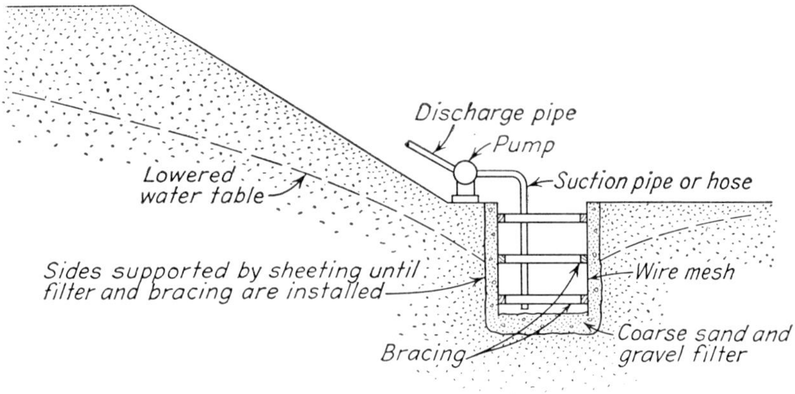

Fig. 7.1. Filter-protected sump for open cut in sand.

Drainage for either temporary or permanent construction may also be accomplished by excavating trenches rather than ditches, placing drain tiles or perforated pipes in the trenches, and filling over the drains with pervious material. To prevent washing the fine material out of the backfill or surrounding soil, it may be necessary to surround the drains with granular material satisfying the requirements for a filter. The width of the openings in the drain pipe should preferably be equal to about the 60 per cent size (D_{60}) of the surrounding material.

7.3. Well Points

The water table in granular materials may be lowered for a depth of 15 to 18 ft by means of well points. A *well point* is a perforated pipe about 3 ft long and 1½ inches in diameter, covered by a cylindrical screen to prevent the entrance of fine particles. It is attached to the bottom of a 1½- or 2-in. riser pipe inserted vertically in the ground. The point can be jetted into the ground without driving.

On the job, a line of well points spaced at 2 to 5 ft is connected to a 6-, 8-, or 10-in. header pipe laid on the surface of the ground. The header, in turn, is connected to a suction pump. The various parts of the assembly are shown in Fig. 7.2.

If the depth of excavation is greater than about 15 ft below the water table, several stages of well points are required. The first

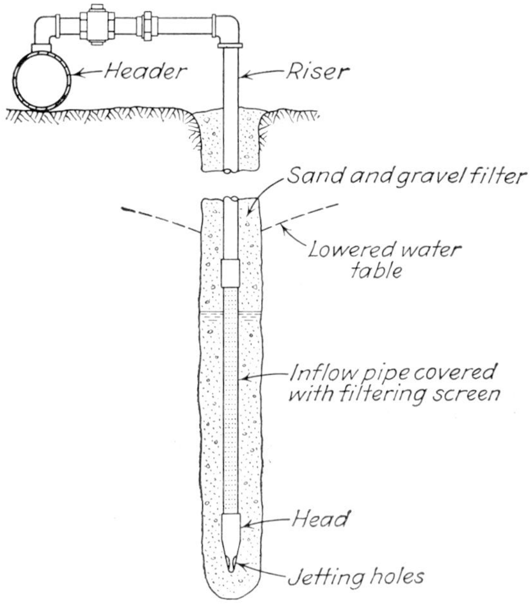

FIG. 7.2. Details of well-point assembly.

excavation is made to a depth of about 15 ft, whereupon additional well points are jetted into the ground before the next 15 ft are excavated. The points are generally arranged in such a manner that the sides of the excavation consist of slopes connected by flat berms containing drainage ditches. This arrangement, known as a *multiple-stage setup,* is shown in Fig. 7.3.

If the permeability is less than about 10^{-4} cm/sec, drainage cannot be accomplished simply by pumping from well points because the capillary forces prevent the flow of water from the pores of the soil. However, drainage can be accomplished by consolidation. This may be done by means of the *vacuum method* of well-point operation, Fig. 7.4. In this method, the well points are jetted into the ground with water under sufficient pressure to produce a hole about 8 inches in diameter around the entire well point. A filter of medium to coarse sand is formed around the point and pipe by shoveling the sand into the hole as rapidly as possible. In the uppermost 2 or 3 ft, an impervious material such as clay is tamped to form a seal.

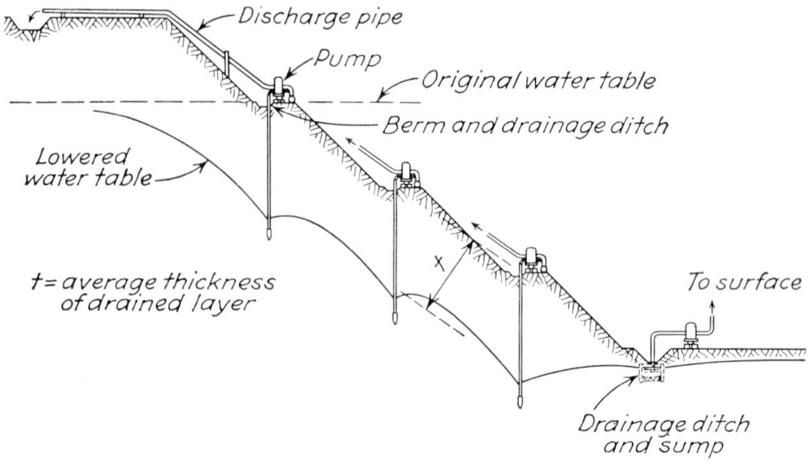

FIG. 7.3. Multiple-stage well-point setup.

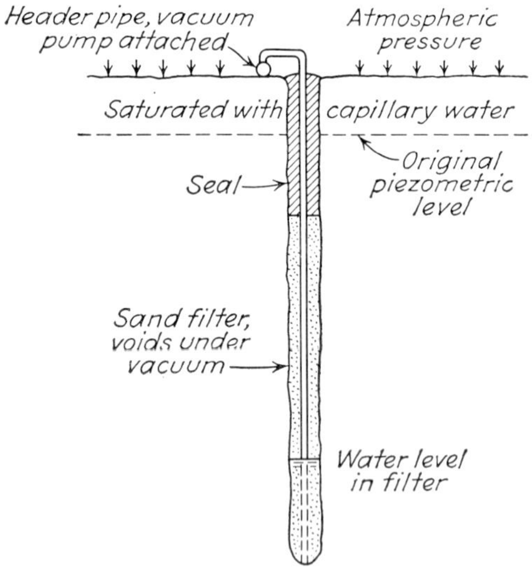

FIG. 7.4. Vacuum well-point installation.

Pumping is then done with equipment capable of maintaining a vacuum in the well points and the surrounding filter. The pressure around the well points is thereby reduced to a small fraction of atmospheric pressure whereas the surface of the ground is acted upon by the weight of the atmosphere. Thus, the soil becomes consolidated under a pressure of about 1 ton/sq ft.

The vacuum process is highly effective in silts and organic silts, but the time required to achieve consolidation and stability is likely to be several weeks.

7.4. Deep-well Pumps

For very deep excavations the multiple-stage well-point setup has the disadvantage that the water level is pulled down rather abruptly at the edges of the excavation. As a consequence, the hydraulic

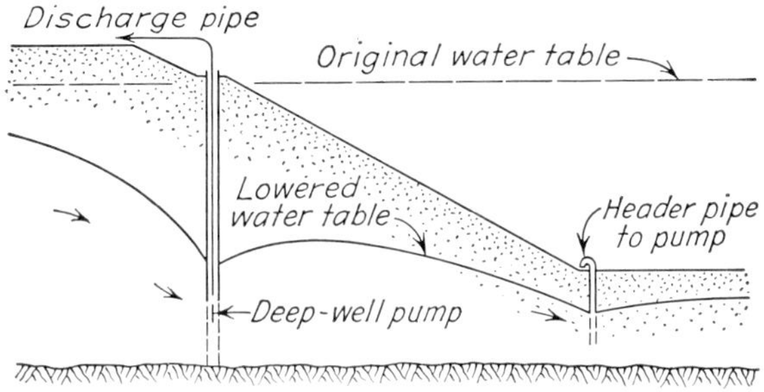

FIG. 7.5. Drainage by means of deep-well pumps.

gradient near the excavation is quite large and the resulting seepage pressures may lead to instability of the side slopes. Under these circumstances, it is safer and sometimes more economical to install large-diameter drainage wells equipped with deep-well pumps. A typical arrangement for such wells is shown in Fig. 7.5.

The drainage wells for such an installation consist of cased holes with diameters commonly ranging from 6 to 24 in. The casing is perforated in the pervious zones. In some installations the pumping unit has a multistage arrangement of turbine pumps and is mounted near the bottom of the casing on a rotating shaft. The motor is mounted at the top of the casing. A 6-in. pump of this type is capable of discharging 120 to 200 gpm against a head of 100 ft and requires a 10- to 15-hp motor. In recent years submersible motors, attached directly to the pumps, have found increasing use. Since deep-well pumps are quite expensive, they are commonly used only on large jobs.

7.5. Consolidation by Sand Piles

In many instances, it is necessary to build a structure or to construct an embankment on fine-grained soils with low shearing resistance. The initial strength of the soils may be too low to support the weight of the structure without failure. However, if the weak soils can be drained rapidly enough to let consolidation occur at nearly the same rate as the load is applied, the strength of the material may increase sufficiently to permit safe construction.

For the purpose of accelerating drainage in relatively impervious deposits, vertical drains may be installed. In the United States, these drains commonly consist of columns of sand about 2 ft in diameter arranged in a pattern of squares or triangles at a spacing of 10 to 15 ft. The ground surface at the top of the drains is cov-

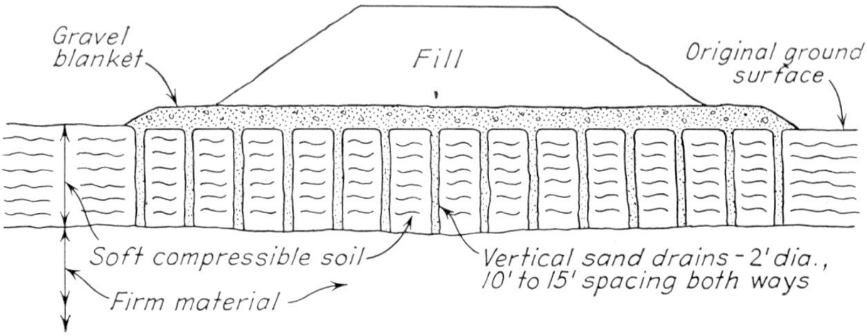

FIG. 7.6. Sand-drain installation.

ered by a pervious blanket, and the structure or embankment is constructed on top of the blanket, Fig. 7.6. As the weight increases, the water is squeezed out of the subsoil into the drains from which it escapes through the drainage blanket to ditches. The rate at which consolidation will occur can be controlled by varying the spacing and diameter of the drains.

7.6. Miscellaneous Methods of Drainage and Stabilization

The stabilization of sands and silts by drainage is not always practicable. Hence, various other methods have been devised, most of which involve the injection of slurries or solutions into the voids of the soil. These materials harden to varying degrees and impart cohesion to the injected soil. Since they partly fill the voids, they also reduce the permeability.

The injection of *cement grout* has been attempted on many occasions. Experience has demonstrated that the method may lead to very satisfactory results, but only if the soil is relatively homogeneous and unstratified, and if the grain size is not too small. The grout will not penetrate the voids of a loose soil with an effective size D_{10} less than about 0.5 mm or a dense soil with an effective size less than about 1.5 mm. Thus, cement grouting is not appropriate for soils much finer than coarse sands.

The grain size of cement particles limits the fineness of a sand that is suitable for cement grouting. Clay suspensions of any desired fineness can be obtained, however, by removing the coarser fractions from natural clays. This has led to attempts to grout soils with *clay slurries*. In practice, it is found that the penetration of the slurry is impeded by the formation of a filter skin that seals the voids. The development of the filter skin appears to be strongly influenced by the electrolytes present in the groundwater; this fact introduces an element of considerable uncertainty regarding the effectiveness of the procedure. As a matter of fact, it appears that the materials that can be successfully injected with clay slurries have roughly the same characteristics as those suitable for cement grouting. Moreover, although clay grouting may greatly reduce the permeability of a sand, it does not significantly increase the strength.

Solidification of soils by *injection of chemicals* has been widely practiced. The most common procedure consists of the successive injection of solutions of water glass and calcium chloride which react in the soil to form a cohesive binder. In other methods a single solution is injected containing a buffer that delays and controls the time of setting. The chemical injection methods are highly successful in clean relatively homogeneous sands having an effective size greater than 0.1 mm, but the efficacy of the procedures decreases rapidly as the grain size of the sand decreases. Furthermore, it depends greatly on the chemical composition of the groundwater.

All the injection procedures are expensive and, even under favorable conditions, are uncertain. Although numerous successful applications have been made, many other attempts have resulted in disappointing failures. Hence, stabilization of this type should usually be considered only under exceptional circumstances where the risk of an unsuccessful attempt is worth taking in view of the possible benefits from a successful application. In any event, injection should not be considered a routine matter and should not be undertaken without the advice of competent and experienced specialists.

If the coefficient of permeability of the soil lies between about 10^{-4} and 10^{-6} cm/sec, no injection procedure is satisfactory. On a few jobs in Europe, stabilization has been accomplished by *electro-osmosis*, an electrical method for creating seepage pressures and producing consolidation. The method is very expensive, and in the United States similar soils are usually stabilized by the vacuum well-point method.

In several instances soils have been rendered impervious and stable by *freezing* the water in the voids. The method has most frequently been used in sinking shafts or in tunneling, but it has also been successfully employed for the temporary stabilization of at least one large landslide in silt. Freezing is accomplished by circulating a refrigerant through a series of double pipes inserted in the soil to be stabilized. Each double pipe consists of an outer casing into which the cold liquid is pumped and an inner pipe through which it returns. Several weeks or months may be required to freeze a block of soil with a volume of a few hundred cubic yards. On account of this fact as well as the cost of the refrigerating equipment, the method is expensive.

John Rennie (1761–1821)

One of the most famous of civil engineers. His works include the drainage of the Fens, the construction of Waterloo and London Bridges, and the building of such waterfront structures as the London Dock, the East India Dock, and the Plymouth Breakwater. For the latter structures he devised a hollow-wall construction to spread the load over a large base in order not to overstress the soft soils. In 1785 he designed the foundation for the Albion Mills, on very soft ground near the bank of the Thames. The foundation, which consisted of a masonry raft, was established at such a depth that the weight of the structure was equal to that of the excavated soil. Thus, in this design, Rennie anticipated one of the techniques of modern soil mechanics by more than a century.

PLATE 8.

CHAPTER 8

Footing and Raft Foundations

8.1. Types of Footings

A footing is an enlargement of the base of a column or wall for the purpose of transmitting the load to the subsoil at a pressure suited to the properties of the soil. A footing that supports a single column is known as an *individual column footing*, an *isolated footing* or a *spread footing*. The footing beneath a wall is known as a *wall footing* or a *continuous footing*. If a footing supports several columns, it is called a *combined footing*. A particular form of combined footing commonly used if one of the columns supports an exterior wall is a *cantilever footing*. The various types are illustrated in Fig. 8.1.

8.2. Historical Development

Footings undoubtedly represent the oldest form of foundation. Until the middle of the nineteenth century, most footings consisted of masonry. If they were constructed of stone cut and dressed to specific sizes, they were known as *dimension-stone footings*. In contrast, *rubble-stone footings* were constructed of pieces of random size joined by mortar. Masonry footings were adequate for most structures until the development of tall buildings with heavy column loads. Such loads required large and heavy footings that occupied valuable basement space.

In the earliest attempts to enlarge the areas of footings without corresponding increase in weight, timber grillages were constructed and conventional masonry footings built on them. The grillage usually consisted of two layers of oak timbers 12 in. square. The timbers in the two layers were laid in directions at right angles to each other and were allowed to project about 3 ft beyond the edges of the overlying stone. The use of timber grillages was limited to locations beneath the permanent water table. Relatively few buildings were founded upon such footings. One of the earliest, however, is still standing (Pullman Building, Chicago, 1884).

In 1891 a grillage consisting of steel railroad rails embedded in concrete was devised as an improvement over the timber grillage (John Wellborn Root, Montauk Block, Chicago). The rail grillage was an important forward step because it saved much weight and

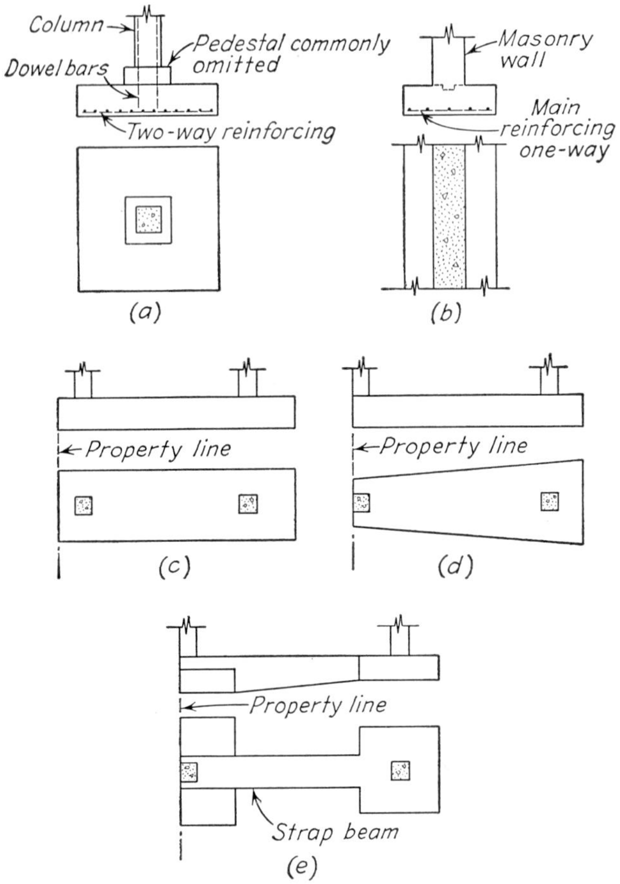

FIG. 8.1. Types of footings. (*a*) Individual column footing; (*b*) wall footing; (*c*) and (*d*) combined footings; (*e*) cantilever footing.

increased space in the basement. Within the following decade, railroad rails were superseded by steel I beams which occupied slightly more space but which were appreciably more economical of steel. Typical grillage foundations of timber, railroad rails, and steel I beams are shown in Fig. 8.2.

The steel I beam proved admirably suited for the construction of cantilever footings. These were introduced in 1887 almost simulta-

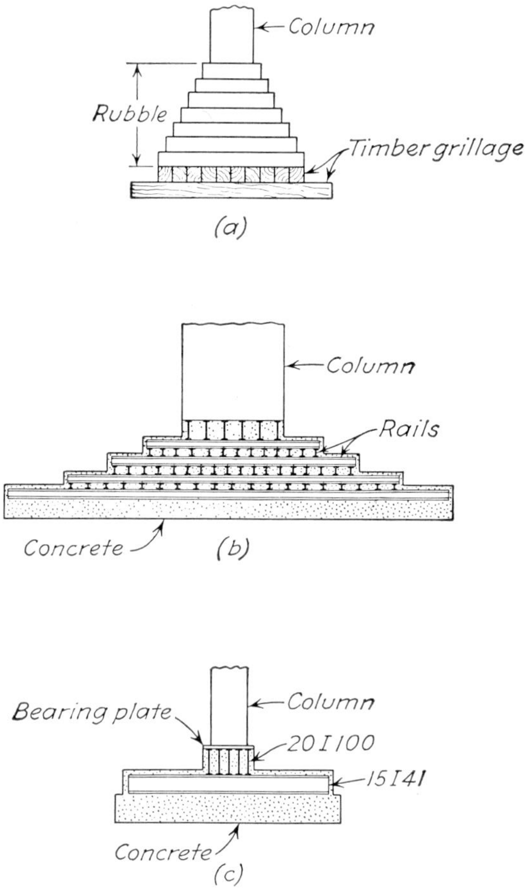

FIG. 8.2. Grillage foundations of (a) timber, (b) railroad rails, (c) steel I beams.

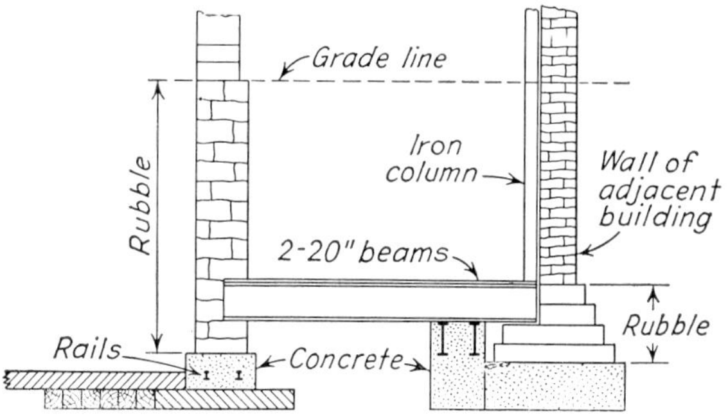

FIG. 8.3. Cantilever footing supporting exterior column of Auditorium Building, Chicago, 1887.

neously in two buildings in Chicago. One of these early footings is illustrated in Fig. 8.3.

With the advent of reinforced concrete shortly after 1900, grillage footings were almost entirely superseded by reinforced-concrete footings which are still the dominant type.

8.3. General Considerations

In temperate latitudes footings are commonly located at a depth not less than that of normal frost penetration. In warmer climates, and especially in semiarid regions, the minimum depth of footings may be governed by the greatest depth at which seasonal changes in moisture cause appreciable shrinkage and swelling of the soil.

The elevation at which a footing is established depends upon the character of the subsoil, the load to be supported, and the cost of the foundation. Ordinarily the footing is located at the highest level where adequate supporting material may be found. In some instances, if an especially firm layer is encountered at greater depth, it may be more economical to establish the footing at a lower elevation because the area required for the footing is smaller.

The excavation for a reinforced-concrete footing should be kept dry so that the reinforcement can be set and held in its proper position while the concrete is being placed. To do this in waterbearing soil it may be necessary to pump either from sumps or from a previously installed system of drains. Forms are usually required around the sides of the footing. The necessity for pumping and for supporting the sides of the excavations in which the footings are placed may add appreciably to the cost of a footing foundation.

8.4. Allowable Soil Pressure

In the earliest days of foundation engineering the area of a footing was selected according to the judgment of the engineer on the basis of his experience. In most localities simple rules of thumb developed. For example, in some parts of the United States, the width of a continuous footing in feet was made equal to the number of stories in the structure. No attempt was made to provide larger footings for the support of heavier loads.

In the early 1870's the proportioning of footings was placed on a more rational basis. Progressive engineers of that day recommended that the areas of footings on a given site be made proportional to the loads on the footings and that the center of gravity of the load on each footing be made to coincide with the centroid of the footing. It was believed that the settlements of all footings beneath a structure would be equal and that no footing would tilt if these recommendations were conscientiously followed.

Furthermore, it was believed that for each soil there existed a specific pressure under which the settlements of the various footings would not exceed reasonable values. This pressure was called the *allowable soil pressure* and was generally specified in the building codes or ordinances of various cities.

It is now known that the allowable pressure for a given site depends upon several factors other than the character of the subsoil. Furthermore, it is known that different footings beneath the same structure are not likely to settle the same amount even under the same soil pressure. Nevertheless, the concept of the allowable soil pressure has the great advantage of simplicity and it is sufficiently reliable for designing the footings of most structures. Suitable modifications of the method as indicated by experience and the contributions of soil mechanics will be incorporated in the procedures for selecting the allowable soil pressure, Part C.

8.5. Conventional Procedure of Proportioning Footings for Equal Settlement

In its simplest form, the method of proportioning footings according to an allowable soil pressure is as follows: The load acting at the base of each column is determined. The weight of the footing is then estimated and added to the column load. The total load is divided by the allowable soil pressure to determine the area required for the footing. After the size of the footing has been determined, its weight is calculated and the value assumed in the computation is revised if necessary. After the dimensions of the footing are established, the footing is designed.

Except for the choice of the allowable soil pressure, the most difficult step in the procedure is the determination of the load for which the footing should be proportioned. It has been generally believed that the settlement of a footing is caused primarily by the dead load plus only the amount of live load that acts on the footing for an extensive period of time. Under these circumstances, equal settlement would be achieved by choosing the areas of the footings in proportion to the dead load plus a fraction of the design live load. However, it is also generally believed that the allowable soil pressure should not be exceeded beneath any footing, even if the maximum probable wind and live loads should act upon the footing for a short time. These two requirements lead to the following procedure which represents conventional practice at the present time:

1. Determine the dead load for each column, including the estimated weight of the footing.
2. Determine the maximum live load, including wind and earthquake load, that may act on the footing. This value is usually established by the building code.
3. Determine the ratio of maximum live load to dead load for each footing.
4. Select the footing for which this ratio is the largest, and determine the area of this footing by dividing the sum of the dead load and maximum live load by the allowable soil pressure.
5. To the dead load of this same footing add the live load that will actually be present often enough or long enough to govern the settlement. This live load is termed the *reduced live load.*
6. Divide the sum of the dead load and the reduced live load on this footing by the area of the footing to obtain the *reduced allowable soil pressure.*
7. Use the reduced allowable soil pressure for determining the area of all other footings, considering the dead load and reduced live load for these footings.

If this procedure is used, the soil pressure will be the same beneath all footings for dead load plus the reduced live load. According to the concepts on which the procedure is based, this should lead to equal settlement of the footings. Furthermore, the allowable soil pressure will not be exceeded, even if the wind load and the maximum live load specified by the building code should act on any footing, because the reduced allowable soil pressure is chosen for the footing having the maximum ratio of wind and live load to dead load.

Certain modifications of the conventional procedure are also in common use. For example, it is sometimes specified that the allowable soil pressure may be increased by an arbitrary percentage when the dead load, wind load, and maximum live load are assumed to act simultaneously. Subsequently, it will be shown that the pressure that may be considered allowable for settlement is unrelated to the pressure at which the soil beneath the footing may rupture; corresponding modifications of the conventional method will be made in Part C. Yet the essence of the procedure will not be altered.

Illustrative Design. DP B-1. Proportioning of Footings

The conventional method of proportioning footings for uniform settlement is illustrated in DP B-1. In general, the design computations follow the procedure discussed in the preceding paragraphs.

DP B-1
Footings
Sh.1 of 2

Footing Design. Required Areas
Conventional Method for Uniform Settlement

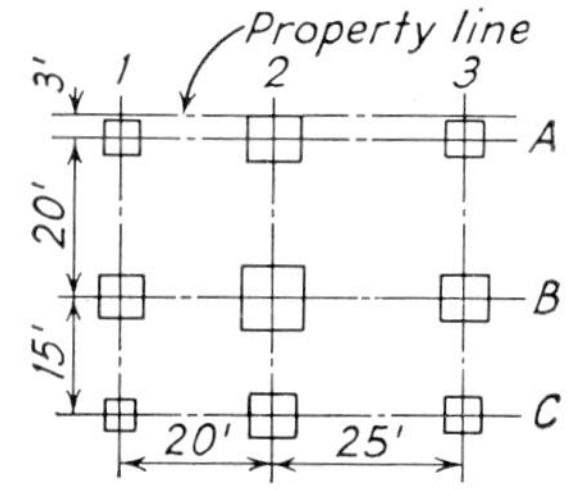

Footing Plan

Column Loads to Footings

Footing	DL	Max.LL	Max.LL / DL	Red.LL	DL + Max.LL	DL + Red.LL
A-1	42^k	35^k	0.83	17.5^k	77^k	59.5^k
A-2	78	80	1.03	40.0	158	118.0
A-3	51	45	0.88	22.5	96	73.5
B-1	65	62	0.95	31.0	127	96.0
B-2	120	142	1.18	71.0	262	191.0
B-3	75	76	1.01	38.0	151	113.0
C-1	34	26	0.77	13.0	60	47.0
C-2	68	62	0.91	31.0	130	99.0
C-3	42	33	0.79	16.5	75	58.5

Footing B-2 governs design.

For total DL + max. LL allow $4.00^k/\square'$

Assume footing depth = 2'-0" $\times 0.15 = 0.30$

$3.70^k/\square'$ = allowable soil pressure corrected for footing weight

Required area $= \frac{262}{3.70} = 71.0\square'$

Use 8'-6" x 8'-6". $A = 72.2\square'$

Column DL + red. LL = 191.0^k

$\div 72.2 = 2.64^k/\square'$

Add for weight of footing $= 0.30$

Reduced allowable soil pressure $= 2.94^k/\square'$ = soil pressure desired under all footings for DL + reduced LL

DP B-1
Footings
Sh. 2 of 2

Correct Reduced Allowable Soil Pressure for Weights of Other Footings

Footings A-1, A-3, C-1, C-3 assume depth = 1'-3" 2.94 k/□'
x 0.15 = 0.19
Reduced allowable soil pr. corr. for footing wt. = 2.75 k/□'

Footings A-2, B-1, B-3, C-2 assume depth = 1'-6" 2.94 k/□'
x 0.15 = 0.23
Reduced allowable soil pr. corr. for footing wt. = 2.71 k/□'

Summary of Footing Sizes

Footing	Column DL+Red.LL	Corr. Reduced Soil Pressure	Required Area	Footing Size	Furnished Area
A-1	59.5 k	2.75 k/□'	21.6 □'	4'-9"sq	22.5 □'
A-2	118.0	2.71	43.6	6'-0"x7'-3"	43.5
A-3	73.5	2.75	26.7	5'-3"sq	27.6
B-1	96.0	2.71	35.3	6'-0" "	36.0
B-2	191.0	2.64	72.4	8'-6" "	72.2
B-3	113.0	2.71	41.8	6'-6" "	42.2
C-1	47.0	2.75	17.1	4'-3" "	18.0
C-2	99.0	2.71	36.4	6'-3" "	39.0
C-3	58.5	2.75	21.3	4'-9" "	22.5

However, it should be noted that an estimate is not made of the total weight of the footing which must be added to the column load before division by the soil pressure. Instead, the column load is divided by a soil pressure that has been corrected for the weight of the footing. This procedure requires fewer computations; moreover, at the outset of a design it is easier to estimate the depth of a footing than its volume. At least a portion of the structural design must be completed before the exact depth of the footing can be established. However, since the plan dimensions are commonly determined to the nearest 3 in., a difference of several inches in depth does not change the required area of the footing.

8.6. Combined Footings

If the loads from several columns are transmitted to the same footing, the footing should be proportioned so that its centroid coincides with the center of gravity of the column loads under normal conditions, and so that the maximum pressure beneath the footing does not exceed the allowable soil pressure under the most severe loading. Combined footings are customarily used along the walls of buildings at property lines where the footing for a wall column cannot extend outside the limits of the structure, Fig. 8.1 *c*, *d*, and *e*. Under these circumstances, the wall footing is usually combined with an interior footing by one of the three methods shown.

8.7. Raft Foundations

A *raft* or *mat* foundation is a combined footing that covers the entire area beneath a structure and supports all the walls and columns. Wherever the building loads are so heavy or the allowable soil pressure so small that individual footings would cover more than half the building area, a raft foundation is likely to be more economical than footings.

Ordinarily, rafts are designed as reinforced-concrete flat slabs. The downward loads on the raft are the loads from the individual columns or walls. If the center of gravity of the loads coincides with the centroid of the raft, the upward load is regarded as a uniform pressure equal to the sum of the downward loads divided by the area of the raft. The weight of the raft is not considered in the structural design because it is assumed to be carried directly by the subsoil. Since this method of analysis does not take into account the moments and shears caused by differential settlement, it is customary to reinforce the raft more heavily than required according to the analysis.

Raft foundations are also used to reduce the settlement of structures located above highly compressible deposits. Under these conditions, the depth at which the raft is established is sometimes made so great that the weight of the structure plus that of the raft is wholly compensated by the weight of the excavated soil. The settlement of the structure is then likely to be insignificant. Where complete compensation is impracticable, a shallower raft may be acceptable if the net increase in load is small enough to lead to tolerable settlements.

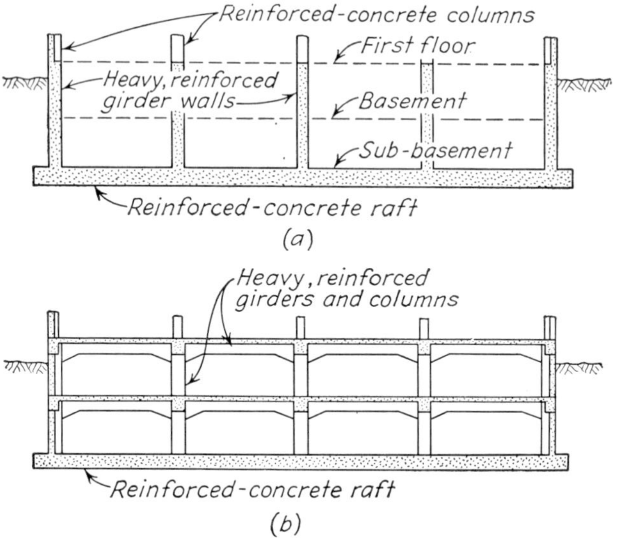

FIG. 8.4. Methods of stiffening large raft foundations. (*a*) Use of ribs or walls as T beams; (*b*) rigid-frame construction.

If the column loads are not more or less uniformly distributed or if the subsoil is such that large differential settlements would tend to develop, large rafts must be stiffened to prevent excessive deformations. This stiffening has been accomplished by using partitions as the stems of T beams connected to the raft (Fig. 8.4*a*), by constructing a cellular or rigid-frame foundation (Fig. 8.4*b*), or in some instances by utilizing the stiffness of a reinforced-concrete superstructure. The larger the foundation, the more expensive these procedures become; often pile or pier foundations are preferable.

8.8. Soil-supported Floors

Many modern industrial buildings are one-story structures covering large areas. The principal function of the columns is to support the roof and, in some latitudes, the weight of snow. All other loads are applied directly to the floor slab. Thus the footings are likely to be lightly loaded whereas the floors may be required to support heavy loads, often intermittently. Moreover, although it is common practice to establish the footings upon the original soil, the floors are frequently established on compacted fills several feet in thickness. Thus the live load on the floor, the dead load of the floor slab, and the weight of the fill may combine to produce stresses in the subsoil that require more serious consideration than those due to the footings themselves.

8.9. Drainage, Waterproofing and Damp-proofing

When a basement is established below normal groundwater level, special precautions must be taken to prevent seepage into the structure. Two general methods are in common use: *drainage,* whereby the water is prevented from reaching the exterior of the structure, and *waterproofing,* whereby entrance of the water adjacent to the structure is prevented by some sort of impervious barrier. The two methods are often combined.

Drainage may be suitable where the seepage is small enough to permit removal of the water at little expense, usually by gravity flow into sewers or ditches. The most common installations for this purpose are footing drains, Fig. 8.5*a* and *b*, and floor drains, Fig. 8.5*c*. The footing drains consist of open-jointed tile lines laid in trenches backfilled to within about a foot of the ground surface with free-draining material. The upper foot of backfill is preferably of less pervious soil that will prevent inflow of surface waters. Where the presence of water in the drains might reduce the strength of the soil beneath the footings, type *b* is more suitable than type *a*. If the soil being drained is likely to clog the backfill, the backfill material should satisfy the requirements for a filter, Art. 2.5.

Floor drains, Fig. 8.5*c,* are not usually necessary if the footing drains are effective. In some instances, however, there may be slow seepage from beneath the structure and the drainage may be advisable.

If the quantity of water that would be collected in a system of drains would be excessive, it may be necessary to waterproof the basement and permit the structure to be subjected to the full water pressure. The most positive procedure is the *membrane method.* In this method, a membrane consisting of alternate layers of fabric and bituminous material is constructed on or near the exterior of the building. The bituminous material is applied hot. Such a membrane is sufficiently flexible and extensible to maintain its integrity even if a moderate amount of cracking occurs in the walls or floors.

Membrane waterproofing to be fully effective must be continuous over the entire surface of the structure that extends below water level. This requires the placement of a concrete subfloor upon which the

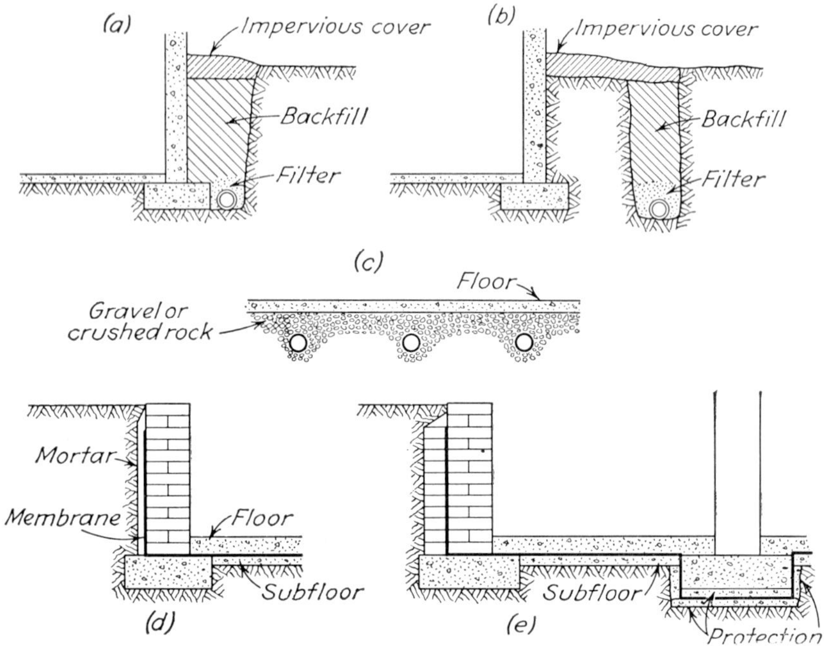

FIG. 8.5. Methods of draining and waterproofing basements. (*a*) Footing drain; (*b*) intercepting drain; (*c*) floor drain; (*d*) membrane waterproofing where exterior wall is accessible from outside; (*e*) membrane waterproofing where outside of wall is not accessible (after W. C. Huntington).

membrane is built before the structural floor is constructed, and requires special details at walls and footings, Fig. 8.5 *d* and *e*. Figure 8.5*e* also shows the arrangement necessary if the exterior wall is not accessible for covering with the membrane.

Since the walls and floors inside the membrane are acted upon by full water pressure, they must be designed for this force as well as for earth pressure.

The seepage of capillary moisture through basement walls and floors can be reduced by placing pervious material outside the walls and beneath the floors. Ordinarily, the exterior of the wall is painted with a bituminous material, and the floor is cast upon a layer of impregnated building felt resting on the pervious material. These measures commonly prevent entrance of moisture into the wall or floor. Various admixtures are available for decreasing the permeability of the concrete itself and for reducing the flow of moisture by application on the inside of the structure. The effectiveness of these substances is extremely variable.

Batter-Pile Foundation (1951)

Foundation piles for the spillway structure of the Morganza Floodway, lower Mississippi Valley. The piles vary from 80 to 120 ft in length, are of precast reinforced concrete, and are driven on a batter of 1 horizontal to 2 vertical. When the photograph was taken, the piles in the rear had not yet been cut off at the specified level. The leads of the pile driver, rising 136 ft high, are among the tallest yet built.

PLATE 9.

CHAPTER 9

Pile Foundations

9.1. Types of Piles

Timber Piles. The origin of pile foundations is lost in antiquity. In very early times, if the subsoil was obviously too weak for the support of a structure, rows of stakes were driven into the ground to provide a more firm bearing. Such foundations were used by the Aztecs in North America before the coming of the white man. They were used at a much earlier date in Europe. The art of driving piles was well established in Roman times and the details of such foundations were recorded by Vitruvius in 59 A.D. The use of timber piles for consolidating loose strata or for penetrating soft layers to reach underlying firmer material became common in the United States in the middle of the nineteenth century. Today, piles of timber are successfully used in many parts of the world and are undoubtedly more common than any other type. Under many circumstances, they provide dependable, economical foundations.

However, timber piles cannot withstand the stresses due to hard driving that may be required to penetrate highly resistant layers. Damage to the points may be reduced by the use of steel shoes, but the danger of breaking the piles can be reduced only by limiting the energy and the number of the blows of the pile-driving hammer. Hence, it is customary to limit the working load on timber piles to about 25 tons and in many localities to smaller values.

Although wood piles last indefinitely when located below permanent water table, they are subject to decay above water. Nevertheless, a considerable degree of permanence can be achieved by treating wood piles with creosote under pressure. Trestle structures are in existence that have been supported on such piles for as long as 40 years without deterioration, but experience with creosoted piles beneath permanent buildings does not exceed about 25 years (Cincinnati Union Terminal, 1929).

Concrete Piles. Just before 1900, several types of concrete piles were devised. Since that time, numerous variations have been introduced, and today a variety of piles is available from which the engineer may choose those best adapted to a particular project. Con-

Cast-in-Place Concrete Piles

This chart furnishes general information concerning the more common types of cast-in-place concrete piles used in the United States and Canada. Values given for maximum lengths should be regarded as approximate. Greater lengths may be obtained for some types of piles, but generally at considerable added expense. Other variations in dimensions and details of installation can sometimes be arranged; for unusual conditions the manufacturers or pile contractors should be consulted.

a. Raymond Standard Pile

Shell : Thin sheet steel, corrugated and reinforced with spiral wire. Closed at bottom with 8" dia. boot.

Taper : 1" in 2'-6".

Length : Up to 38 ft. Not available in greater lengths.

Driving : Steel mandrel inserted in shell. Shell and mandrel driven into ground. Mandrel removed. Shell filled with concrete.

b. Raymond Step-Taper Pile

Shell : Thin sheet steel, corrugated, closed at bottom with steel boot. Shell sections 8 ft long, screw connected.

Taper : None within each 8-ft section. Steps result from successive dia. increases of 1" per section from point to butt. Point diameters available from $8\frac{1}{2}$ to $13\frac{1}{2}$ in.

Length : Up to 60 ft not uncommon. Max. about 90 ft.

Driving : Steel mandrel, stepped to fit shell, inserted and driven with shell into ground. Mandrel removed. Shell filled with concrete.

c. Union Metal Pile or Monotube

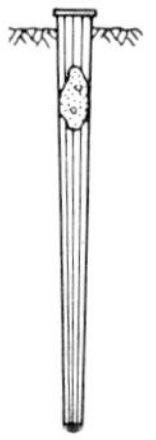

Shell : Fluted sheet steel, thickness $\frac{1}{8}$" to $\frac{5}{16}$". Closed at bottom with 8" dia. steel point.

Taper : Standard types, 1" in 2'-6" up to 25 ft, 1" in 4'-0" up to 40 ft, 1" in 7'-0" up 75 ft. Upper part of longer piles made by practically cylindrical extensions (taper $\frac{5}{8}$" in 20'-0").

Length : Up to about 60 ft not uncommon. Max. about 125 ft.

Driving : Steel shell driven without mandrel. Shell filled with concrete.

d. MacArthur Compressed Uncased Pile, Western Compressed Uncased Pile } *With or without Pedestal*

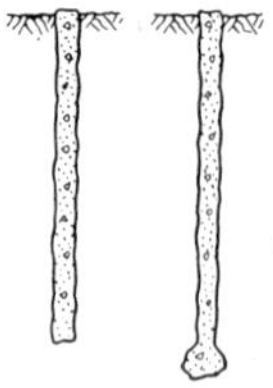

Shell : None.

Taper : None.

Length : Up to 60 ft not uncommon.

Driving : Heavy 14-in. steel drive pipe and steel core driven together. Core removed and concrete deposited in pipe. Core and hammer positioned on concrete to press it into contact with soil as drive pipe is withdrawn. Pedestal may be formed by ramming concrete after drive pipe is withdrawn a few ft.

FIG. 9.1. Principal varieties of cast-in-place concrete piles.

e. MacArthur Cased Pile / Western Cased Pile } *With or without Pedestal*

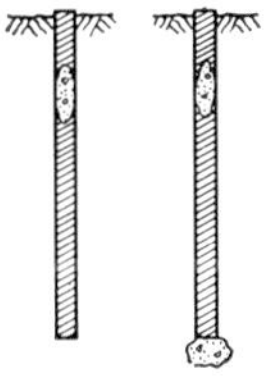

Shell : *Thin sheet steel, corrugated.*

Taper : *None.*

Length : *Up to 80 ft not uncommon.*

Driving : *Heavy steel drive pipe and steel core driven together. Core removed and 12-in. shell dropped inside drive pipe. Shell filled with concrete and drive pipe withdrawn. Pedestal may be formed by ramming charge of concrete in the drive pipe before shell is dropped inside.*

f. Western Button Bottom Pile. With or without Pedestal

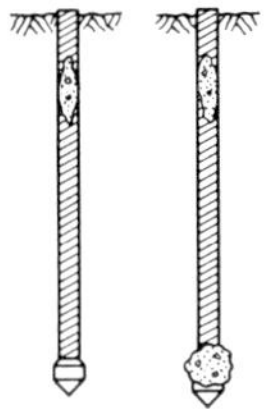

Shell : *Thin sheet steel, corrugated.*

Taper : *None*

Length : *Up to about 110 ft.*

Driving : *Heavy steel drive pipe with 17" dia. precast concrete button driven into ground without use of core. Shell dropped inside drive pipe making contact with button. Shell filled with concrete and drive pipe withdrawn. Pedestal may be formed around button by ramming charge of concrete in drive pipe before shell is dropped inside.*

g. Steel Pipe Pile, Concrete Filled

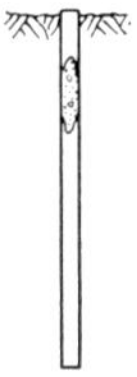

Shell : *Steel pipe, 10" to 20" dia., thickness $\frac{3}{16}$" to $\frac{1}{2}$". Commonly closed at bottom with steel boot.*

Taper : *None.*

Length : *Up to 150 ft not uncommon.*

Driving : *Steel pipe driven without core. For long piles, sections joined by internal sleeves or by welding. Open-ended pipe commonly cleaned by air jet or by combination of air and water. Pipe filled with concrete.*

h. Franki Standard Pile

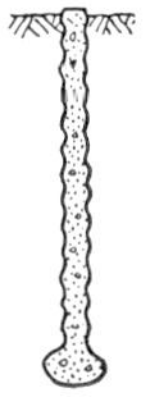

Shell : *None.*

Taper : *None. Rough bulbous shaft, 20" dia. min.*

Length : *Up to about 120 ft*

Driving : *Heavy steel pipe plugged at bottom with relatively dry concrete. Pipe dragged into ground by hammer falling directly on plug. Pedestal formed by ramming plug and additional small charges of concrete into soil at end of pipe. Rough shaft formed by ramming successive quantities of concrete while progressively raising steel pipe.*

FIG. 9.1. (*Continued*)

crete piles may be divided into two principal categories, *cast-in-place* and *precast*.

The principal varieties of cast-in-place concrete piles used in North America are illustrated and described in Fig. 9.1.

Precast piles can be constructed in a variety of forms, two of which are illustrated in Fig. 9.2. Such piles must be reinforced to

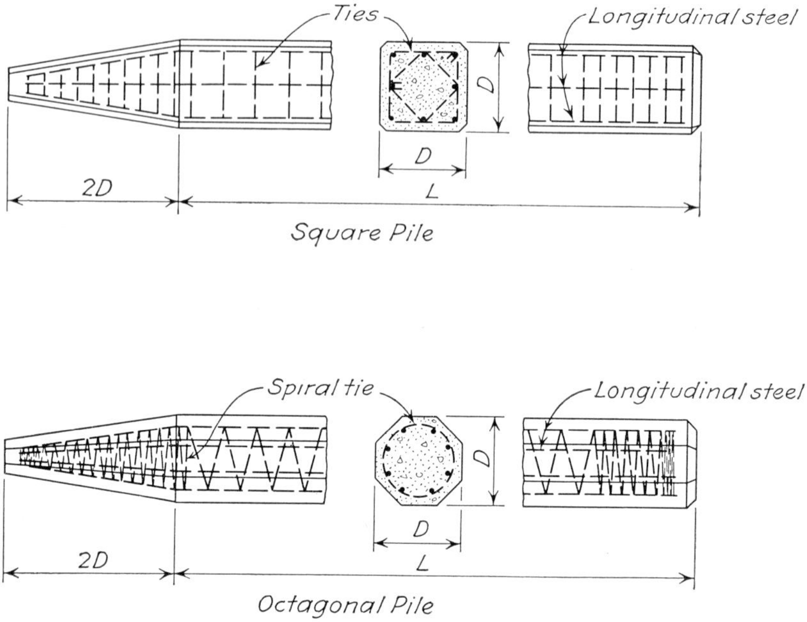

FIG. 9.2. Varieties of precast concrete piles.

withstand handling until they are ready to be driven and must also be reinforced to resist the stresses caused by driving.

Most varieties of concrete piles can be driven to high resistance without damage. As a consequence, the allowable load on such piles is commonly greater than that on timber piles. Under most conditions, they are not subject to deterioration, and they can be used above the water table. In sea water, however, they require protection at and above the water line. Precast piles are commonly used in the approach spans of bridges where they are permitted to extend above the ground surface as columns that serve to support the deck of the bridge.

Where the upper part of a pile is to project above the water table, a *composite* pile may be suitable. Such a pile consists of a lower portion of untreated timber and an upper portion of concrete. The two parts are connected by a strong joint, and the pile is driven so that the connection is located permanently below the water table.

Pipe Piles and H Piles. In recent years, the need has grown for extremely long piles with high bearing capacity. To satisfy these requirements, *concrete-filled steel pipes* and *steel H piles* have been used. The pipes range in diameter from 10 to 30 in. and have a wall thickness of 0.2 in. or more. They may be driven with the lower end open, or closed with a steel plate to keep out soil and water. The open-ended piles must be cleaned out before they are filled with concrete. For a power plant in Muskegon, Mich., 24-in. piles of this type were driven in one piece to a depth of 200 ft.

Piles of these types are rarely used unless they reach a stratum of exceptionally high supporting capacity. Their cost is relatively great, but the bearing capacity of an individual unit is relatively large.

Shorter pipe piles of smaller supporting capacity are also used. In these instances the wall thickness is sometimes reduced to $\frac{3}{16}$ in. or even to $\frac{1}{8}$ in.

Steel H piles have been used for the foundations of bridges and buildings for a period of about 30 years. Such piles penetrate into the ground more readily than other types because they displace relatively little material. As a consequence, they are likely to go deeper into the subsoil than many other types.

Batter Piles. Piles driven vertically into the ground are capable of supporting vertical loads but are not generally considered adequate to support appreciable lateral loads. When provision must be made to resist horizontal as well as vertical forces, *batter piles* are commonly used. These are piles driven at an inclination to the vertical. The inclination commonly lies between the limits of 1 horizontal to 6 vertical and about 5 horizontal to 12 vertical.

9.2. Pile Driving

Pile Drivers. Piles are commonly driven by means of a hammer supported by a crane or by a special device known as a pile driver, Fig. 9.3. The pile driver consists of a pair of vertical guides known as *leads* within which the hammer is held. The leads are carried on a frame in such a way that they can be supported in a vertical position or in an inclined position. The framework also contains the hoisting engine, which is usually operated by steam. The assembly may be mounted on a base suitable for operation on land, on a barge for operation on the water, or on a railroad car.

Hammers. Originally, pile drivers were equipped with hammers that fell from the top of the leads to the top of the pile. Such

Fig. 9.3. Common type of pile driver.

devices are known as drop hammers. At the present time, however, a hammer generally consists of a unit containing a ram lifted by steam pressure and allowed to fall by gravity with or without the assistance of steam pressure. If the fall is due to gravity alone, the hammer is called *single acting.* If steam pressure adds to the downward energy, the hammer is called *double acting.* The rams of the most common single-acting hammers in current use weigh 3000 to 5000 lb and fall 3 ft. The most common double-acting hammers deliver an equivalent amount of energy. The lighter hammers are used for driving wood piles and the heavier ones for concrete and steel piles. The rams of the heaviest single-acting steam hammers used thus far weigh 14,000 lb and fall 3 ft.

Driving Resistance. Piles are ordinarily driven to a resistance measured by the number of blows required for the last inch of penetration. For wood piles driven by hammers delivering 15,000 ft-lb of energy, the final number of blows per inch should not be more than about 4 to reduce the danger of breaking or of brooming the piles. Resistances of 6 to 8 blows per inch are commonly specified for concrete and steel piles. Resistances of more than 10 blows per inch usually constitute excessively severe requirements.

Jetting and Spudding. If piles must penetrate through dense layers of sand or gravel in order to continue through underlying soft deposits, the sand or gravel may be loosened by jetting. In this procedure, a stream of water is discharged near the point or along the sides of the pile through a pipe 2 to 3 inches in diameter. The jetting pump must be capable of delivering about 500 gpm at a pressure of 150 to 200 lb/sq in. The water loosens the sand and creates a quick condition so that the pile drops through the sand to the underlying material.

Relatively thin deposits of stiff clays or soft rock may sometimes be penetrated with spuds consisting of hard metal points driven into the ground and withdrawn before the piles are inserted. Jetting is a relatively common practice, but spudding is not commonly utilized.

Pre-excavation. When piles are driven through saturated plastic clay, they displace a volume of soil that may be as great as that of the piles. This displacement usually takes the form of a heave of the ground which may lift adjacent structures or piles already driven. If the piles depend for their support on a firm layer beneath the plastic clay, the heave may cause loss of bearing capacity. In many instances the heaved piles may be redriven to firm bearing, but occasionally they cannot be broken loose from the skin of consolidated clay that may develop around each pile within a few days after driving. Moreover, the displacements may cause not only heave but also undesirable lateral movements, especially if there are adjacent excavations extending below the level from which the piles are driven. If the piles are of the composite type, the heave may cause separation at the joint.

Where the displacements are likely to be objectionable, they can be considerably reduced by removing part of the soil in the space to be occupied by each pile. This may sometimes be done by driving an open-ended pipe at the location of the pile until it becomes plugged with clay, removing the pipe and cleaning out the plug, and repeating the operation until sufficient clay has been removed. It may also be done by the use of a rotary cutting tool combined with water jets which transform the clay where the pile is to be located into a heavy slurry. In some instances earth augers can be used to form the holes. These procedures are known as *pre-excavating* or *coring*. Usually pre-excavation appreciably increases the cost of the pile foundation.

9.3. Action of Piles under Downward Loads

Piles are commonly divided into two kinds, *point-bearing* and *friction*. The point-bearing pile derives practically all its support from the rock or soil near the point, and very little from the soil surrounding the upper part. On the other hand, a friction pile derives its support principally from the surrounding soil through the development of shearing resistance between the soil and the pile. A very small percentage of the load is carried by the soil near the lower end of the pile.

In contrast to the relatively simple conditions of support implied by the classification of piles into these two categories, the manner in which piles are actually supported is by no means simple. For example, under many conditions encountered in nature, the stiffness or relative density of the subsoil increases in a general way with depth. Piles may be driven through the soft upper layers and through progressively stiffer layers until they are believed to be adequate. Such piles derive an appreciable part of their support from the surrounding material, especially in the lower portion of their length, but they may also derive considerable support by direct bearing at their points.

One of the most important engineering decisions in connection with any pile job is the choice of the most appropriate type of pile to suit the particular circumstances. Many factors enter into the final decision, and it appears unlikely that simple, definite rules can be formulated for the sure guidance of the inexperienced engineer.

Nevertheless, a sound conception of the manner in which piles of different characteristics behave during driving and of how they subsequently transmit their load to the soil is an invaluable asset and can serve as a basis for the development of good judgment as the engineer's experience accumulates.

A point-bearing pile surrounded by soil is sometimes erroneously regarded as a structural member that transfers its load like a column from the top of the pile to the bottom where it is delivered to the underlying rock or soil. This concept leads to the conclusion that the stresses in the pile should not exceed those that would be considered tolerable in a column of the same dimensions and materials. However, experience has amply demonstrated that structural failures of driven piles are so rare that the eventuality need seldom be considered in design. During load tests on piles, if structural failure of the pile itself occurs, it usually takes place at or above the ground surface where the projecting part of the pile is not surrounded by soil. Furthermore, both experience and theory have demonstrated that there is no danger that a point-bearing pile may buckle on account of inadequate lateral support, provided it is surrounded by even the very softest soils.

These observations lead to the conclusion that the capacity of a point-bearing pile depends almost entirely on the capacity of the material upon which the point finds its bearing, and on the degree to which the point of the pile has a satisfactory seat on the bearing material. It is obvious that the ultimate bearing capacity of the pile increases with increasing bearing area, whence it may be concluded that the capacity of piles with large point diameters is greater than that of piles with small point diameters. On the other hand, if the bearing stratum is at considerable depth or is overlain by moderately resistant material, it may not be possible to drive a large-diameter pile to a firm seat on the bearing stratum, whereas a more slender pile that displaces less soil may successfully reach the firm material and may have a higher capacity.

However, the volume of soil displaced by a pile is not the only significant factor that determines the pile's ability to penetrate deeply enough, for a given expenditure of energy, to reach the bearing stratum. Probably a more important factor in this connection is the capacity of the pile to transmit the energy of the hammer from the butt to the point. This ability depends primarily upon the material of which the pile is made. The stresses are most easily transmitted to the point through a steel pile or, in the case of some types of cast-in-place concrete piles, through steel mandrels that extend to the point.

The driving stresses are transmitted somewhat less efficiently through precast concrete piles and still less efficiently through wood piles. Also important is the loss of energy due to the plastic deformation of certain parts of the pile or to elastic vibrations set up under the impact of the hammer blow. To prevent these energy losses, the cross-sectional area of the pile or the mandrel should be adequate. It has been demonstrated, for example, that two cylindrical steel piles of the same external diameter driven with the same hammer will penetrate different distances into a firm stratum, depending upon the thickness of the steel tube. The thicker tube penetrates more deeply.

If the supporting stratum does not consist of extremely resistant material such as sound rock or firmly cemented hardpan but, in contrast, is a fairly thick but only relatively firm deposit, two different types of piles deserve consideration. A pile of small point diameter, but of a type capable of transmitting the driving stresses to the point without excessive loss of energy, is likely to penetrate for several feet into the firm stratum and develop high capacity by a combination of direct bearing at the point and intense skin friction within the zone of embedment in the firm stratum. The contribution of the skin friction may be markedly increased if the lower part of the pile has even a small taper.

The other alternative if the bearing stratum is not exceptionally firm is the use of a pile with a very large point. The point may consist of a plate or precast concrete button, or it may have the form of an enlargement or pedestal made by expanding fresh concrete into a soft deposit directly above the bearing stratum. In any event, the capacity of the pile is not enhanced by any wedging action due to penetration of the pile into the bearing stratum, but on the contrary is governed entirely by the capacity and compressibility of the soil upon which the bottom of the pile rests. Piles with enlarged points are also appropriate if the bearing stratum is very firm but is so thin that piles of small diameter could punch through it.

The capacity of friction piles depends upon the characteristics of the material surrounding the pile. Hence, as a general rule the structural strength of the pile itself is of secondary importance. If friction piles have parallel sides, the load is transferred from the pile to the soil exclusively through shear. If the pile has a taper, a portion of the load is transmitted by direct bearing, but still the greater part is transferred by shear. If the piles are driven in soft clay, for which the frictional resistance is negligible, the difference in supporting capacity between parallel-sided and tapered piles is relatively small. However, for soils possessing appreciable frictional resistance, such as sands, silts, and clays containing air, the wedge action of a tapered pile increases the lateral pressure and increases the shearing resistance correspondingly. Hence, a tapered pile is likely to be very advantageous under such circumstances, and, as a rule, the heavier the taper the shorter the pile required to attain a given capacity.

The foregoing discussion serves as a basis for a general understanding of the manner in which piles support their loads and gives some indication concerning the most appropriate types of piles for given conditions.

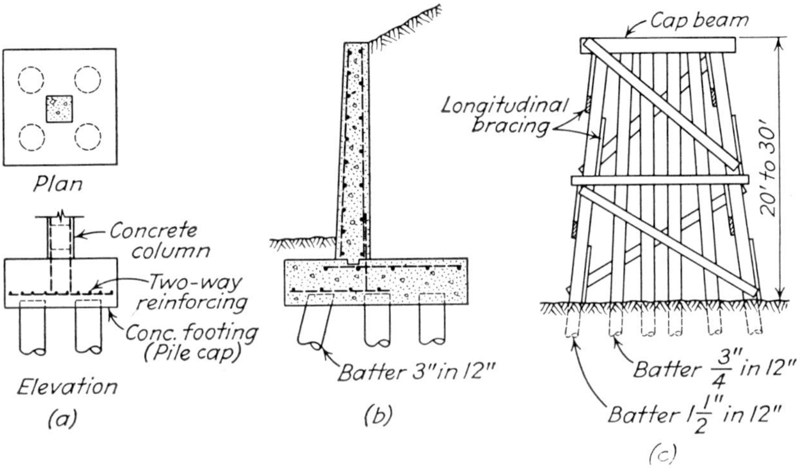

FIG. 9.4. Typical pile foundations for various kinds of structures. (*a*) Simple pile footing; (*b*) retaining wall; (*c*) bridge trestle bent.

9.4. Pile Foundations

The loads exerted by a structure are usually transferred to groups of piles through *pile footings* or *pile caps*, Fig. 9.4*a*. If less than three piles are provided beneath a single column, the eccentricity caused by unavoidable deviations of the piles from their proper position is likely to produce bending in the piles. Hence, unless provision is made to resist the bending by means of girders or other structural framing, it is not considered good practice to use less than three piles in a group. Similarly, it is inadvisable to support a wall by a single row of piles; the piles are customarily driven in pairs or in a staggered line.

In groups, piles are customarily driven at a spacing of two to three times the butt diameter. Under some conditions, however, variations from this spacing are often advisable. These will be discussed in

Part C. The behavior of a group of piles connected to the base of a structure or to a pile cap may not be directly related to the behavior of single piles subjected to the same load per pile in the same deposit. The essential differences between the action of single piles and pile groups or pile foundations are also discussed in Part C.

Many types of pile-supported structures are subjected to lateral loads applied at an elevation considerably higher than that of the base of the foundation. Therefore, the foundation must resist not only lateral forces but also moment. Beneath such structures as lock walls, retaining walls, and ordinary buildings, the downward pile loads due to the weight of the structure are usually greater than the upward loads due to the moment caused by the lateral forces, and none of the piles must resist uplift. On the other hand, the piles beneath the windward side of tall steel towers or tall piston-type gas holders must often be counted upon to provide the reaction against uplift forces. Finally piles may be required to resist the uplift due to the buoyancy of tanks and similar structures established below the groundwater level.

When a pile foundation will be subjected to lateral loads, one of the principal decisions to be made by the designer is whether or not batter piles will be required. This decision cannot be made without an estimate of the ability of vertical piles to withstand horizontal loads. Unfortunately, there is little information on which to base such an estimate. If the soil immediately below the pile caps consists of sand, irrespective of its relative density, or of silt or clay having an N value greater than about 5, it is reasonable to allow a horizontal load at the top of each pile of at least 1500 lb. In denser or stiffer soils greater loads can be resisted, but no decision to use higher values should be made without careful study of the subsurface conditions and structural requirements. In particular, it is important to judge whether there may be a tendency for the soil in which the piles are embedded to move laterally for any reason whatsoever. If such a tendency may exist, the vertical piles should not be counted upon to resist the movement, but should be expected to move together with the surrounding soil. Attention must then be turned to the causes of the lateral instability of the soil itself.

When the horizontal load per pile exceeds the value suitable for vertical piles, batter piles must be used in combination with vertical piles. Batter piles are common beneath retaining walls, Fig. 9.4*b*, bridge piers, and abutments. They are also used to provide lateral stability in the transverse rows of piles, known as *bents,* that constitute the vertical supports for trestles, Fig. 9.4*c*. If both the batter and vertical piles beneath a structure are point-bearing and are all driven to the same stratum, the axial capacity of all the piles is generally assumed to be equal. The horizontal load per batter pile is then considered to be the horizontal component of the axial load. If the piles depend only on skin friction for their support and are all of equal length, the same assumption is usually made.

The uplift resistance of piles depends upon many factors, including the type and dimensions of the piles and the subsoil conditions. It is obvious that only skin friction can be effective in resisting the upward loads. The uplift resistance for various conditions will be discussed in Part C.

The final choice of the type of pile for any one job is dictated not only by the probable performance of the foundation but also by economy. Economic comparisons should be based on the cost of the entire foundation rather than on the cost of the piles alone. For example, the cost of twelve 20-ton wood piles might be less than that of four 60-ton concrete piles, but the larger pile cap required to transfer the column load to the wood piles might increase the cost of the wood-pile foundation above that of the concrete-pile foundation.

Daniel E. Moran (1864–1937)

One of the foremost foundation builders and engineers of recent times. His responsibilities included the substructures of the Philadelphia-Camden, Ambassador, George Washington, and San Francisco-Oakland Bay bridges, and of many of the tallest buildings in the financial district of New York. In 1894 he invented improvements of the air lock that greatly facilitated the use of the pneumatic method of pier construction. In 1936, he devised techniques for controlling the sinking of the large open-dredged caisson that formed the center pier or anchorage of the San Francisco-Oakland Bay bridge. This pier was 92 by 197 ft in plan and was established at a depth of 220 ft below the water level.

PLATE 10.

CHAPTER 10

Pier Foundations

10.1. Definitions

In foundation engineering the term *pier* has two different meanings. According to one usage, a pier is an underground structural member that serves the same purpose as a footing, namely, to transmit the load to a stratum capable of supporting it without danger of failure or excessive settlement. In contrast to a footing, however, the ratio of base width to depth of foundation for piers is usually less than about ¼, whereas for footings this ratio commonly exceeds unity.

According to the second usage, a pier is the support, usually of concrete or masonry, for the superstructure of a bridge. The pier usually rises above the ground surface, and very commonly also extends through a body of water to a level above maximum high water. According to this definition a pier may be regarded as a structure in itself, which must in turn be supported on an adequate foundation. To avoid confusion, the term *pier shaft* will be used with reference to the part above the foundation. The base of a pier shaft may rest directly on a firm stratum, or it may be supported on piles, on timber grillages or cribs, or on several piers as defined in the preceding paragraph. A pier shaft located at the end of a bridge and subjected to lateral earth pressure is known as an *abutment.*

No sharp distinction can always be drawn between piers and piles. Large-diameter steel pipes, driven open-ended and subsequently cleaned out and filled with concrete, may properly be considered either as piers or piles. The pipes themselves might be regarded as the linings of shafts or as caissons. The terminology in these respects differs considerably in different localities.

10.2. Methods of Construction

General. The methods of constructing piers are divided into two principal groups. In one, a hole is excavated to the base level of the foundation and the pier built inside the hole. Usually, the sides of the excavation must be lined and braced to prevent them from caving.

Such a hole is known as a *sheeted pit* or a *cased hole,* depending upon whether the lining consists of individual boards or sheet piles, or of cylindrical metal shells. If the ground surface is below water, the area to be occupied by the pier may be surrounded by a relatively water-tight enclosure known as a *cofferdam.* Within the protection of the cofferdam the excavation is carried to the desired level and the pier constructed.

The other principal method for constructing piers is the caisson method. A *caisson* is a hollow shaft or box that is sunk into position and becomes the outer part of the finished pier. To facilitate sinking, the lower part of the caisson has a cutting edge that is sometimes made larger than the upper part to reduce the friction between the caisson and the soil. The material inside the cutting edge is removed by dredging through openings in the top, or by hand excavation. The lower part of the caisson may be sealed from the atmosphere and filled with air under pressure to keep a space free of water in which men can work. This procedure, known as the *compressed-air method,* permits removal of obstructions beneath the cutting edge, and facilitates cleaning the bottom of the excavation.

Piers in Sheeted Pits and Cased Holes. Piers constructed in sheeted pits are commonly used for the support of buildings founded on a resistant material overlain by soft or loose deposits. The deepest excavations yet made for piers to support a building are of this type (Cleveland Terminal Tower, 224 ft to bedrock). The most common procedure for constructing such a pier originated in Chicago in 1892 (General Wm. SooySmith, Stock Exchange), and is known as the *Chicago method.* It is particularly appropriate for clays in which waterbearing inclusions are not prevalent.

In the Chicago method a circular hole at least 3.5 ft in diameter is excavated by hand for a depth varying from 2 to 6 ft, depending on the consistency of the clay. The walls of the hole are then lined with vertical boards, known as *lagging.* The lagging is held in place by two circular steel rings, Fig. 10.1*a*. Excavation is then continued until another set of lagging and rings can be installed. When the hole reaches the stratum on which the foundation is to be supported, the bottom may be enlarged or *belled out* to increase the bearing area. The rings and lagging are left in place as the hole is filled with concrete.

If a few waterbearing seams or pockets are encountered in the use of the Chicago method, various construction expedients can be used to cut off the water. However, if the subsoil contains numerous thin but wet layers or if it consists of saturated soft silts, the *Gow method,* Fig. 10.1*b*, may be more suitable. According to this method, circular steel shells about 6 ft long are driven successively into the ground, and the soil within each shell is excavated by hand before the next lower shell is driven. Each shell has a diameter 2 in. less than that of the shell above. If the soil at the base of the hole can stand without support, a bell can be formed as in the Chicago method.

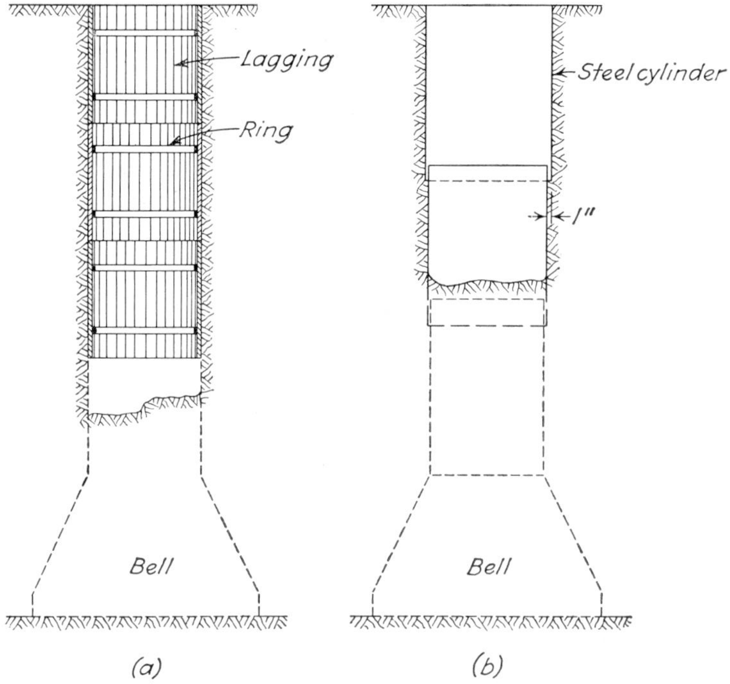

FIG. 10.1. Procedures for excavating shafts for piers (*a*) Chicago method; (*b*) Gow method.

In firm dry ground, piers are often established in unlined holes excavated by machines equipped with rotating augers or buckets provided with cutting vanes. Under favorable conditions such machines can rapidly and cheaply excavate shafts ranging in diameter from 12 to 48 in. to depths of 40 ft, and at a slower rate to even greater depths. The buckets may possess reamers capable of forming a bell at the bottom of the shaft. The efficiency of the method is greatly decreased if water is encountered even in relatively small quantities, and any appreciable flow that may induce sloughing of the soil at the boundaries of the hole is likely to render the procedure impracticable, principally on account of the impossibility of inspecting or correcting conditions at the bottom of the hole at the time concrete is to be placed.

On the other hand, if subsurface conditions are too unfavorable to permit installation of foundations by any of the methods described in the preceding paragraphs, large-diameter steel pipes may be installed by a procedure that combines the methods of pile driving and open excavation. The pipes are driven open-ended, a few feet at a time, and are cleaned out by means of air jets, water jets, or cable tools similar to those used for percussion drilling (Fig. 3.3). Such shafts are almost always carried to rock. They are commonly seated in the rock, whereupon the hole is continued for several feet

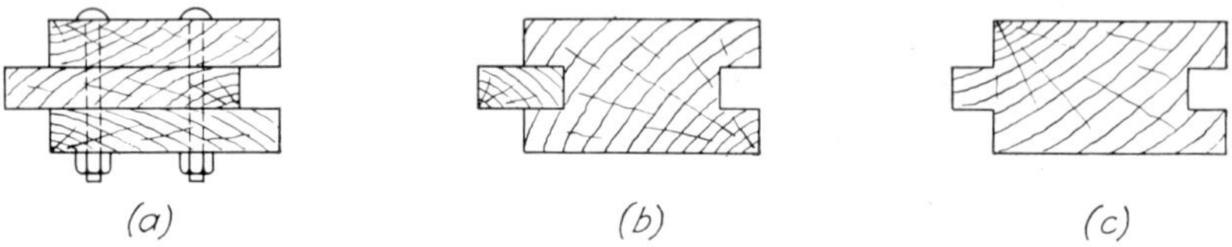

FIG. 10.2. Types of wood sheet piles used for cofferdams in shallow water. (*a*) Wakefield; (*b*) splined; (*c*) tongue and groove.

by percussion or rotary drilling before the shaft is finally cleaned and concreted. Piers formed by this method are expensive but they can usually be assigned very high loads, and they can be installed under almost any subsurface conditions. The development of the large-diameter pipe pier has greatly reduced the circumstances under which compressed-air methods must be employed.

Cofferdams. Where piers are to be constructed in water and the depth of water does not exceed 8 or 10 ft, they are commonly built in cofferdams made of wood sheet piles. The piles may have one of the various forms shown in Fig. 10.2. They are driven around the area to be unwatered and are braced close to the water line by wales and struts. The lower part of the sheet piles is supported by the soil in which the piles are embedded.

For greater depths wood sheet piles are inadequate, and the enclosure is usually made of steel sheet piles. One of the simplest types of cofferdams consists of a sheet-pile enclosure with internal bracing, Fig. 10.3. The sheet piles are driven until their lower ends are embedded and sealed in the underlying soil. Ordinarily, they extend for the full depth of the proposed pier. Before the cofferdam is pumped out, one set of bracing is installed just above the water line. The water level is then lowered to the elevation at which another set of bracing must be installed. Alternate lowering of the water level and installation of bracing continue until the cofferdam is pumped out, whereupon the rest of the excavation is completed in the dry. One of the principal difficulties with this procedure is the rather great leakage through the interlocks of the sheet piling, especially at the beginning of dewatering. Cinders, manure, and various other materials are often dumped outside the cofferdam to plug the leaks. As the water level is lowered, the sheet piles bend inward and the interlocks become tighter. If the depth of the cofferdam is more than about 50 ft, it may be impracticable to lower the water table

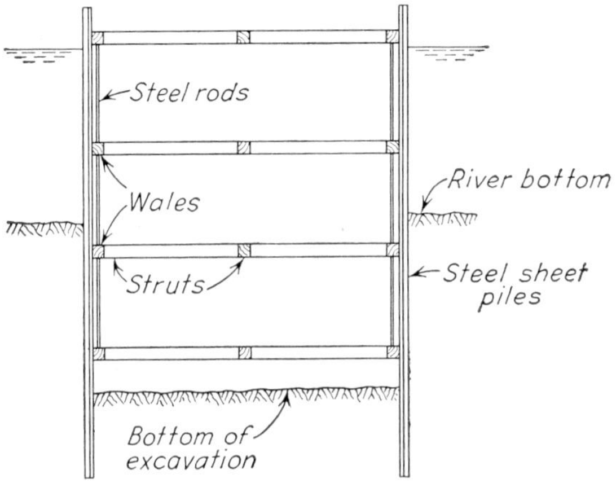

FIG. 10.3. Braced single-wall sheet-pile cofferdam.

far enough to expose bottom. Under these circumstances, the excavation can be made by dredging under water. A concrete seal is then placed on the bottom before the cofferdam is dewatered.

For larger excavations in not over about 20 ft of water, a single wall of sheet piles is often driven and braced with berms of earth, Fig. 10.4*a*. A cofferdam of this type is easily attacked by currents in the water and must usually be protected on the outside by riprap. The double-wall sheet-pile cofferdam, Fig. 10.4*b*, is suitable for somewhat greater depths of water. It consists of two lines of sheet piles connected by tie rods. The space between the piles is filled with rock or soil.

Where the depth of water is very great, cellular sheet-pile cofferdams are used. The two principal arrangements are shown in Fig. 10.5. Each cell is filled with rock or gravel. The circular type has the advantage that each cell is independently stable. Hence, during construction the cofferdam is quite invulnerable to sudden floods or storms. For small bridge piers, single cells are sometimes made

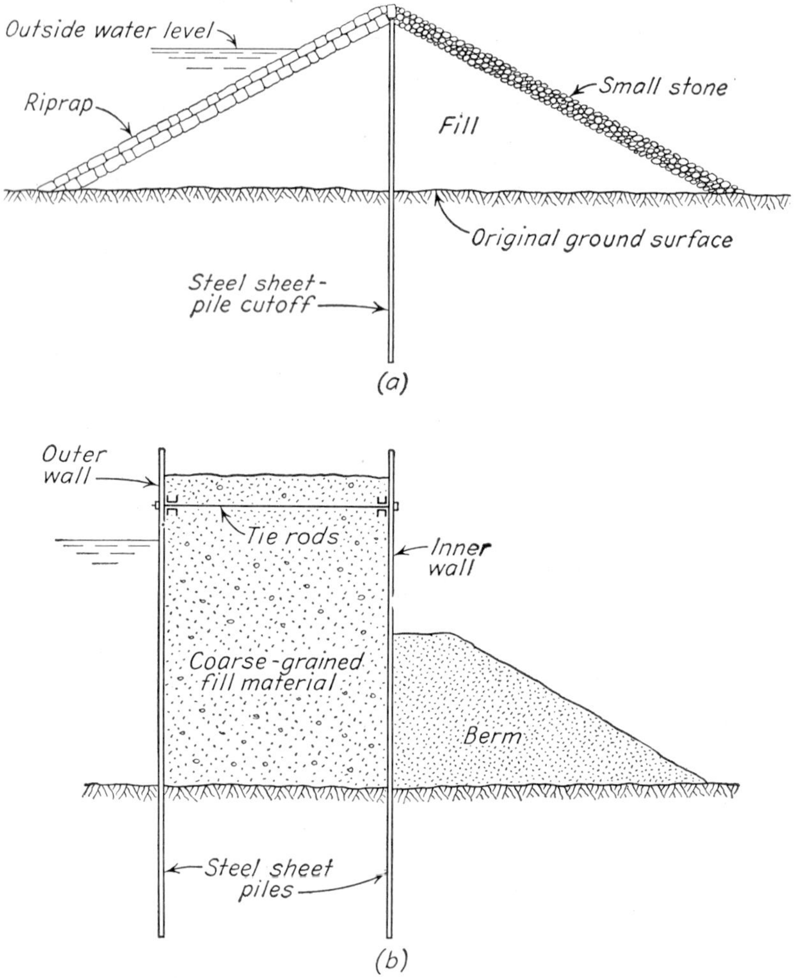

FIG. 10.4. (*a*) Single-wall sheet-pile cofferdam protected by embankments; (*b*) double-wall sheet-pile cofferdam.

large enough to surround the entire foundation. Since such cells cannot be filled, the sheet piles must be supported by circular rings.

Caissons. If the ground surface is above water, construction of a caisson may be started on the ground directly above the area where its base will be located. If the ground level is below water, the lower part of the caisson may be constructed elsewhere, floated into place, and sunk, Fig. 10.6*a*. As an alternative, a ring of sheet piles may be driven to form an enclosure in which an island of sand is con-

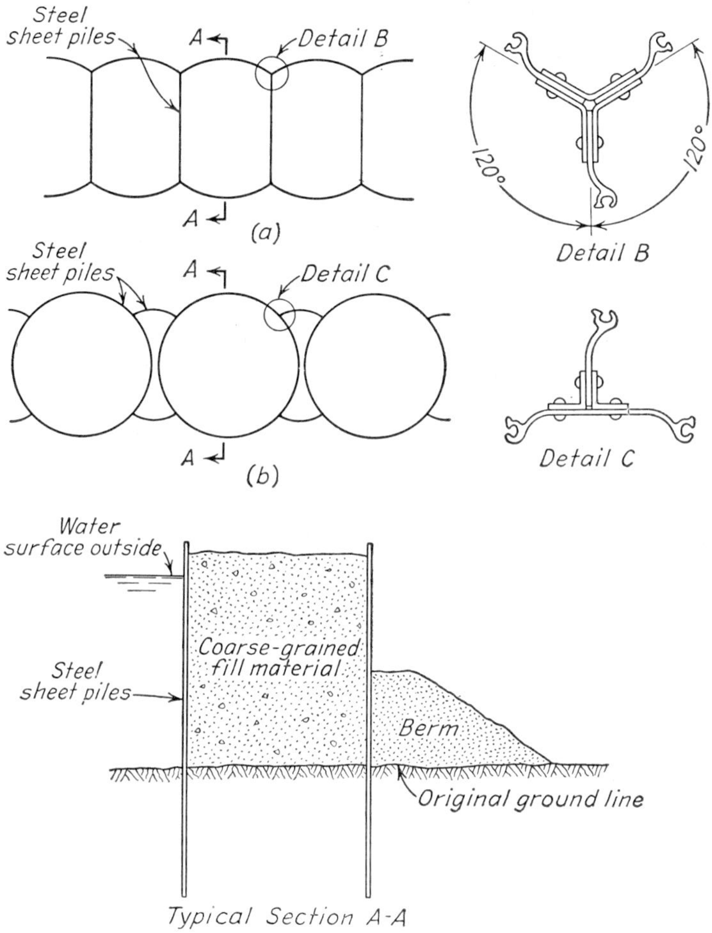

FIG. 10.5. Cellular cofferdams. (*a*) Diaphragm type; (*b*) circular type.

structed. The caisson is then started on the sand and sunk through it as if the ground surface were above water table, Fig. 10.6*b*.

Open caissons are usually sunk by dredging. Hence, they must be provided with a number of wells extending from top to bottom through which the dredging can be done. The wells should be large enough to provide easy passage of the excavating buckets but small

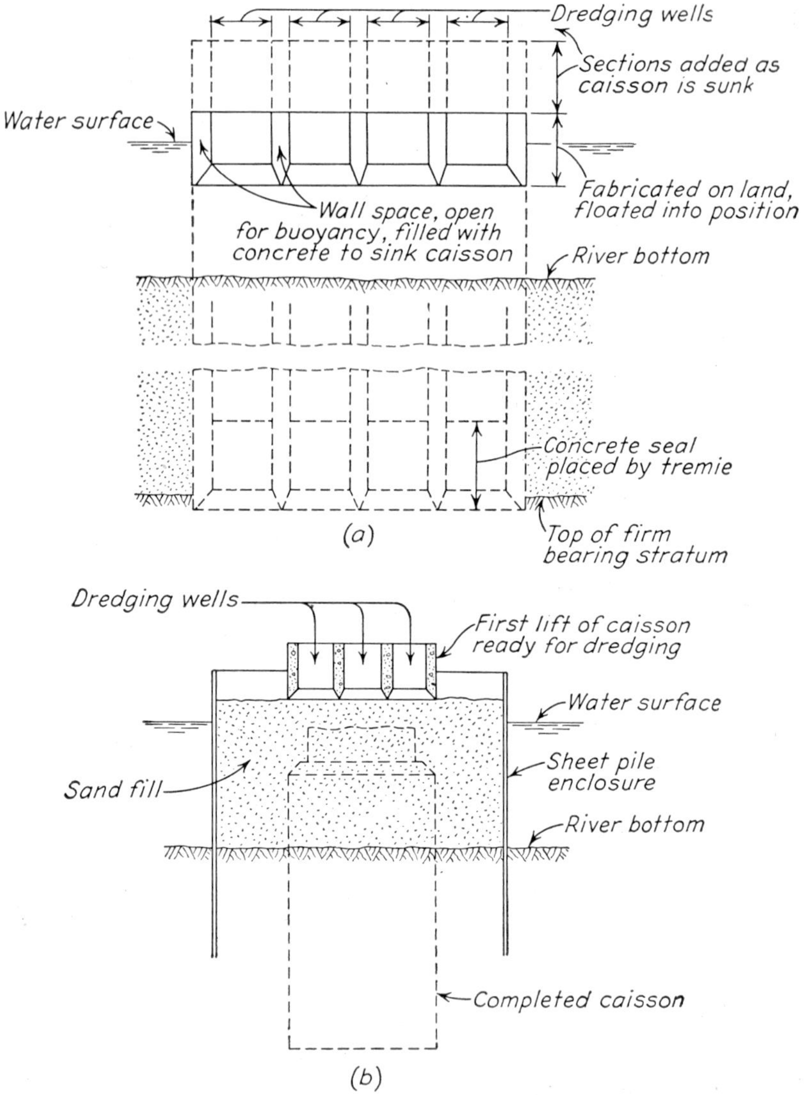

FIG. 10.6. Cross section through typical open caissons. (*a*) Floating type; (*b*) sand-island construction.

enough to leave substantial walls to provide some of the weight necessary for sinking the caisson. In small caissons a single central well is provided. In very large caissons several may be used. One of the piers for the San Francisco-Oakland Bridge extends to the record depth of 242 ft below water level. It was constructed by means of

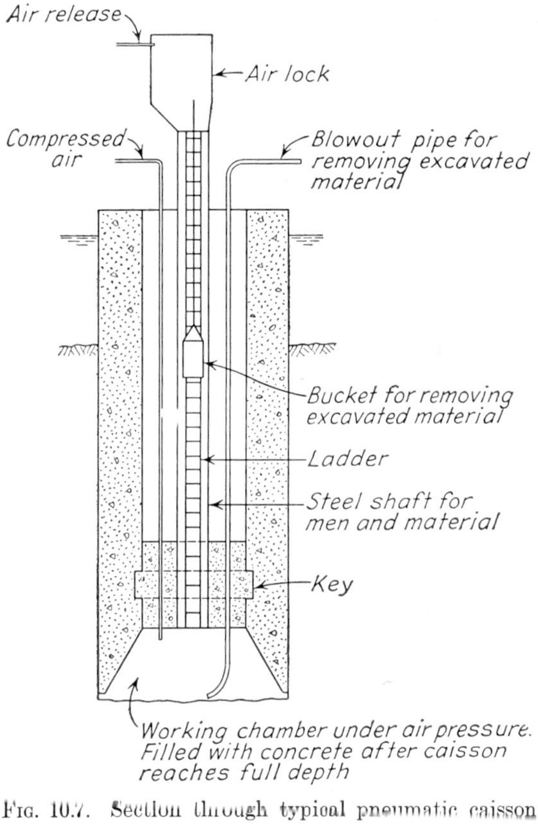

FIG. 10.7. Section through typical pneumatic caisson.

an open-dredged caisson which contained 28 wells. The weight of a caisson must at all times be sufficient to overcome the side friction against the structure. In some instances, loads must be added and jetting facilities used to force the caisson down. After the caisson has reached its final position, the bottom is sealed with concrete deposited under water by means of a tremie.

The pneumatic caisson, Fig. 10.7, is used for depths between about 40 and 110 ft. Compressed air prevents water and mud from entering the working chamber. The working chamber is usually about

6 ft high and is provided at the top with openings for passage of men and materials. The openings are protected by air locks. The use of pneumatic caissons permits the removal of sunken logs, large boulders, etc., which may be encountered by the cutting edge. The

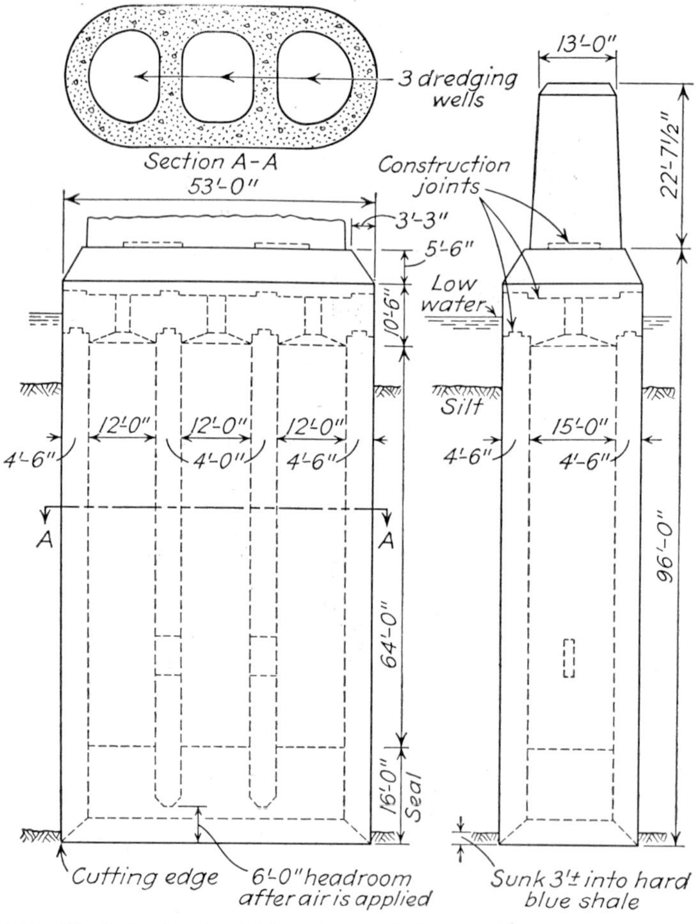

FIG. 10.8. Typical pier for Milwaukee and Rock Island Railroad Bridge at Kansas City (courtesy Howard, Needles, Tammen and Bergendoff).

foundation bed can be prepared carefully and the concrete placed in the dry. On the other hand, the cost of construction is relatively great compared to open dredging. Furthermore, the depth is limited by the air pressure under which men can work.

Figure 10.8 shows a typical pier for the Milwaukee and Rock

Island Railroad Bridge at Kansas City, Mo. The foundation is a caisson 96 ft high. The caisson was sunk by open dredging to the elevation of the shale, whereupon it was converted to the pneumatic type. It was then sunk into the shale to provide a good seat for the bottom. This procedure combined the economy of open dredging with the reliability and certainty of the pneumatic process.

Pier Shafts and Abutments (1950)

Construction photograph of grade separation structure on the New Jersey Turnpike. In the foreground is the rear of one abutment before backfilling. The wing walls are of the cantilever type and are built monolithically with the front wall. In the background are several pier shafts. The substructure is ready to receive the steel girder spans.

PLATE 11.

CHAPTER 11

Pier Shafts, Retaining Walls, and Abutments

11.1. Pier Shafts

The dimensions of the top of a pier shaft for a bridge are determined by practical considerations such as the magnitude of the bridge-shoe reactions, the distance required to provide for expansion of the

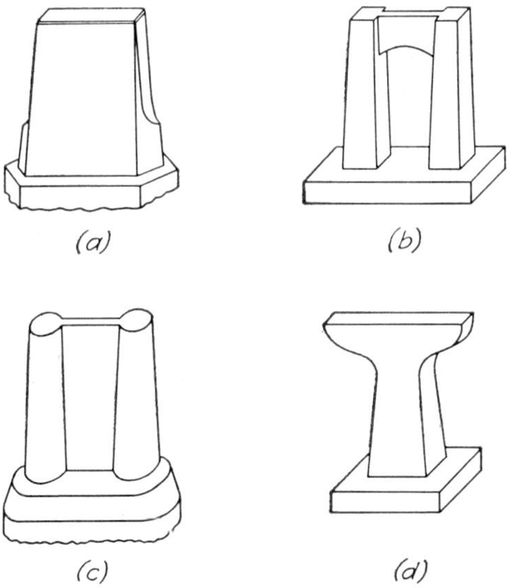

FIG. 11.1. Typical pier shafts for railway and highway bridges. (*a*) Solid shaft; (*b*) and (*c*) double shafts; (*d*) mushroom shaft.

superstructure, and the distance between trusses or girders. Frequently, the top of the shaft is provided with a coping or overhang which extends about 6 in. beyond the edges of the top of the shaft proper. If the shaft extends through a body of water, its shape may be streamlined below high water to prevent eddy currents and scour.

In northern latitudes the upstream edge may be provided with an inclined cutting edge to lift and break cakes of ice. For the sake of appearance, a batter of about 1 in./ft is commonly given to the shaft as a whole.

Solid shafts, Fig. 11.1*a*, are commonly used for railroad bridges. The double shaft, Figs. 11.1*b* and *c*, is often adopted for highway bridges, although type *b* is also common for railroads. The mushroom pier, Fig. 11.1*d*, is one of several types occasionally used to avoid skew spans in passing over existing railroad tracks or highways.

Most modern bridge piers are of reinforced concrete. For protection against the elements, stone facing is sometimes used, especially near the water line.

Although pier shafts are commonly regarded as part of the substructure of a bridge, they are not part of the foundation in the sense that their design requires consideration of the properties of the subsurface materials. Therefore, they will not be considered further in this text.

11.2. Retaining Walls

A retaining wall is a structure that provides lateral support for a mass of soil and that owes its stability primarily to its own weight and to the weight of any soil located directly above its base. Retaining walls constitute inherent parts of many foundations and their design is one of the functions of the foundation engineer.

Before about 1900 retaining walls were usually constructed of stone masonry. Since that time concrete, either plain or reinforced, has been the predominant material. The most common types in current use are gravity, semigravity, cantilever, counterfort, and crib walls.

The *gravity wall*, Fig. 11.2*a*, depends for its stability entirely upon the weight of the stone or concrete masonry and of any soil resting on the masonry. No reinforcement is provided except in concrete walls, where a nominal amount of steel is placed near the exposed faces to prevent surface cracking due to temperature changes.

The *semigravity wall*, Fig. 11.2*b*, is somewhat more slender than a gravity wall and requires reinforcement consisting of vertical bars along the inner face and dowels continuing into the footing. It likewise is provided with temperature steel near the exposed face.

The *cantilever wall*, Fig. 11.2*c*, consists of a concrete stem and a concrete base slab, both relatively thin and fully reinforced to resist the moments and shears to which they are subjected.

The *counterfort wall*, Fig. 11.2*d*, consists of a thin concrete face slab, usually vertical, supported at intervals on the inner side by vertical slabs or counterforts that meet the face slab at right angles. Both the face slab and the counterforts are connected to a base slab, and the space above the base slab and between the counterforts is backfilled with soil. All the slabs are fully reinforced.

The preceding four types are known as *monolithic walls,* in contrast to *crib walls,* Fig. 11.2*e*, which consist of individual structural units

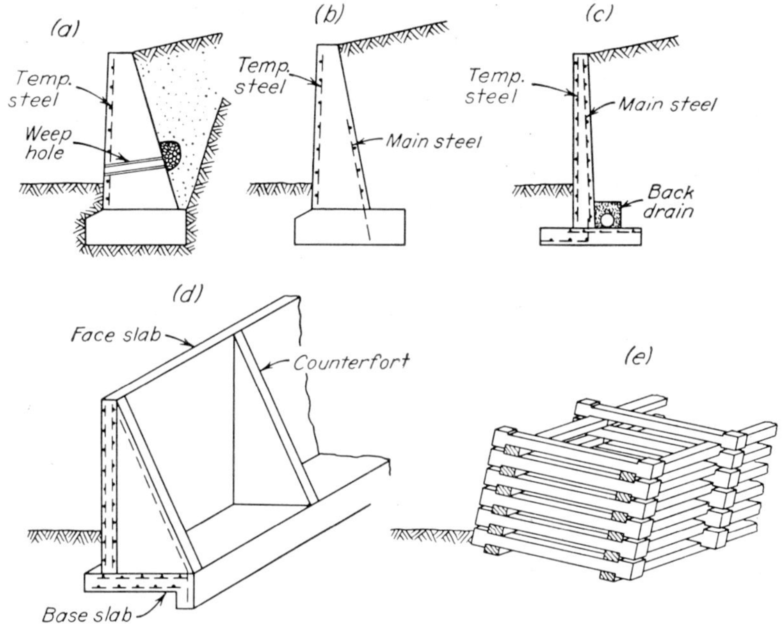

FIG. 11.2. Types of retaining walls. (*a*) Gravity section; (*b*) semigravity section; (*c*) cantilever section; (*d*) counterfort wall; (*e*) crib wall.

assembled at the site into a series of hollow bottomless cells known as cribs. The cribs are filled with soil, and their stability depends not only on the weight of the units and their filling, but also on the strength of the soil used for the filling. The units themselves may consist of reinforced concrete, fabricated metal, or timber.

Of the monolithic types, those most commonly constructed today are the cantilever and the semigravity. Cantilever walls generally have the advantage of lowest first cost and are widely used in connection with buildings and highways. However, because of the relatively small thickness of the concrete sections they may be vulnerable to the effects of freezing and thawing, expansion and contraction, and concrete deterioration. Therefore, where permanence and low maintenance costs are primary considerations, as for railroad structures, the thicker semigravity walls are sometimes considered preferable.

All retaining walls are expected to withstand the pressure of the earth that they support, but they are not usually designed to resist water pressure in addition to the earth pressure. Therefore, well-designed retaining walls are provided with means for draining the water that would otherwise accumulate in the backfill. The drains commonly consist of pipes known as weepholes, Fig. 11.2*a*, which have a diameter of 6 or 8 in. and which extend through the stem of the wall and are protected against clogging by pockets of gravel in the backfill. The drains should be spaced at about 10 ft both vertically and horizontally; in counterfort walls there should be at least one drain in each pocket between adjacent counterforts.

Weepholes are not highly efficient in draining semipervious backfills. Unless the pockets of gravel satisfy the requirements for a filter (Art. 2.5), they are likely to become clogged. In freezing weather the outlets may become obstructed with ice. For these reasons, a *continuous back drain,* Fig. 11.2*c*, is considered preferable when the physical conditions at the site permit its use. The drain consists of a perforated pipe with a diameter not less than 10 in. The pipe, which must be surrounded by a filter, usually terminates at an open ditch where it should be accessible for cleaning.

The character of the material used for backfill has an important influence on the forces acting against the inner face of a retaining wall. Clean sands and gravels are considered superior to all other soils because they are free draining, are not susceptible to frost action, and do not become less stable with the passing of time. Silty sands, silts, or granular soils containing a small percentage of clay are less desirable because they cannot be drained readily, are likely to be subject to frost action, and may experience a decrease of shearing strength with accumulation of moisture. Clays are undesirable as backfill because they can hardly be drained, are likely to experience alternate swelling and shrinking with the seasons, and may lose much of their shearing strength if moisture accumulates. If shrinkage cracks in a clay backfill become filled with rainwater, the wall may be subjected to full hydrostatic pressure as well as earth pressure, even if drains have been provided. Wherever possible, it is considered good practice to insert a wedge-shaped body of free-draining material between the wall and a clay backfill, as shown in Fig. 11.2*a*.

11.3. Abutments

An abutment serves two principal functions. It supports the end of a bridge span, and it provides at least some lateral support for the soil or rock upon which the roadway rests immediately adjacent to the bridge. Hence, an abutment combines the functions of a pier shaft and a retaining wall.

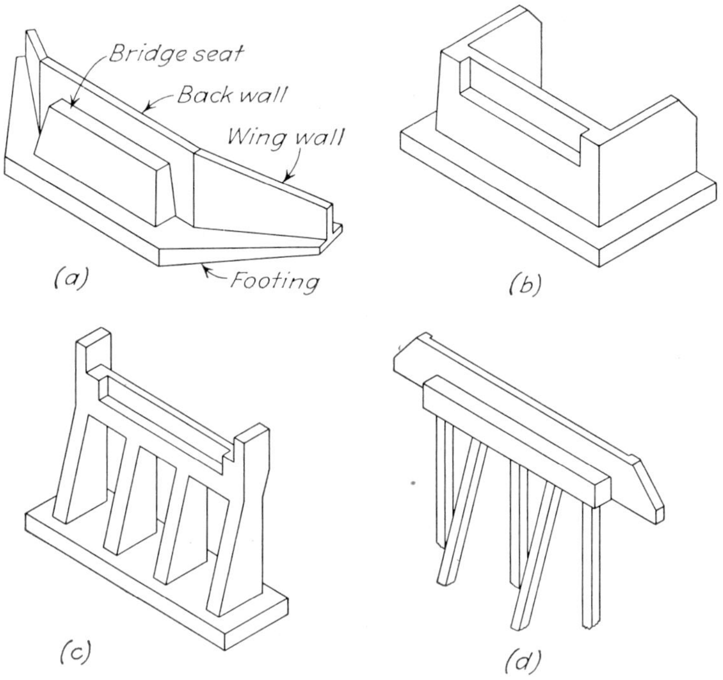

FIG. 11.3. Types of abutments. (*a*) Typical gravity abutment with wing walls; (*b*) U abutment; (*c*) spill-through abutment; (*d*) pile-bent abutment with stub wings.

One of the most common types of abutment is shown in Fig. 11.3*a*. It consists of a central pier supporting the bridge seat, and two wing walls to retain the fill. All three elements rest on a single footing. If the wing walls are at right angles to the pier the structure is known as a U abutment, Fig. 11.3*b*. The wing walls of a U abutment are sometimes tied together to reduce their tendency to overturn.

The spill-through or open abutment, Fig. 11.3*c*, is also widely used. It consists of two or more vertical columns carrying a beam that supports the bridge seat. The fill extends on its natural slope from the bottom of the beam through the openings between the columns. In its extreme form a spill-through abutment is no more than a row of piles driven through the fill and supporting a bridge seat, Fig. 11.3*d*. The bridge seat is usually provided with small wings to keep the bridge shoes free of soil. Another common variation is a simple pier with small wings near the top. The fill in this case spills around the abutment.

Lazurus White (1874–1953)

Foundation constructor especially noted for his skill in executing difficult shoring and underpinning operations. He was one of the first engineers to realize that the behavior of the subsoil of a foundation under load is comparable to that of an elastic solid. In accordance with this conception he introduced the practice of prestressing the soil beneath newly installed temporary or permanent supports for a structure in order to prevent settlement during the transfer of load to the new supports. He vigorously advocated adequate drainage of granular materials during the excavation of cuts below water table and during construction operations within cofferdams. The records of his observations of construction difficulties are an invaluable source of information to the engineering profession.

PLATE 12.

CHAPTER 12

Shoring and Underpinning

12.1. Shoring

Excavation below the foundation level of an adjacent structure usually leads to the necessity for supporting the structure temporarily. The provision of temporary supports is known as *shoring.*

In one method of shoring, notches are cut in the walls of the neighboring building and inclined posts inserted in these notches to carry the weight of that part of the wall above the supports, Fig. 12.1*a*. The lower ends of all the posts must be established on foundation pads similar to those provided for rakes, Art. 6.3. This method is adequate only for relatively unimportant structures.

Better support is provided by cutting the walls or columns at their bases and inserting jacks between the foundations and the walls or columns that they support. As excavation progresses and causes subsidence of the surrounding soil, the jacks may be adjusted to keep the walls or columns at their original level.

12.2. Underpinning

In some instances it is necessary to replace or strengthen the foundation of an existing structure. The operation of providing new permanent foundations is known as *underpinning.* Underpinning is an art practiced by specialists, but an acquaintance with underpinning methods and procedures is an essential part of the knowledge of every foundation engineer. One of the simplest forms of underpinning is the replacement of a wall footing by another footing at greater depth. Masonry or concrete walls are capable of temporarily bridging gaps in their foundation. Hence, it is possible to excavate a pit beside the existing wall to a depth corresponding to the base of a new foundation, to excavate beneath the existing footings for a length of 4 to 8 ft, and to construct a short section of the new footing in this space. After this section is complete, another adjacent section can be done.

If the wall is incapable of supporting itself, holes can be broken through the wall and horizontal beams inserted as shown in Fig. 12.1*b*. These supports, known as *needle beams,* rest at both ends on temporary footings and jacks. Thus, the needle beam is a form of shore. While the wall is supported on the needle beam, its new foundation

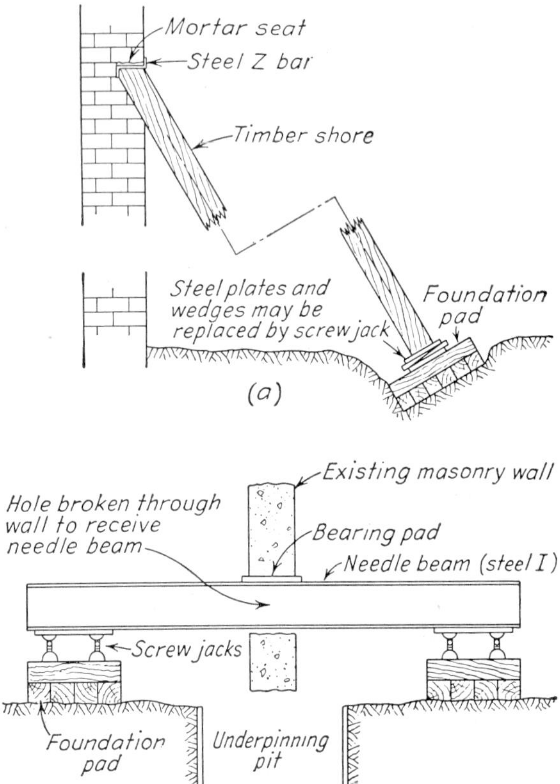

FIG. 12.1. Methods of shoring. (*a*) Notched wall; (*b*) needle beam.

can be constructed. Similarly, needle beams can be used to support columns that are being underpinned.

In many instances the foundations of existing structures have proved inadequate, and it has been necessary to provide new support at a much greater depth than the original footings. The new support usually consists of hand-excavated piers or of piles. Steel H beams or pipes may be jacked into the ground in short sections against the reaction of the weight of the footing.

The transfer of load from shores or from existing supports to new foundations inevitably causes a deformation or settlement of the new foundations. This settlement may crack the structure. The cracking may be reduced or eliminated by prestressing the new foundations, usually by jacking them down against the reaction of the existing structure. While the load is still on the jacks, wedges are inserted to maintain the units in their stressed condition. The jacks are then removed, and the space that they occupied is filled with concrete. This operation is commonly known as *pretesting*.

Methods of underpinning are often elaborate and extremely expensive. The possibility that a structure may have to be underpinned as a result of adjacent construction often leads initially to the use of deep foundations for a structure that could be supported adequately at a higher level if it would not be disturbed.

Part C

SELECTION OF FOUNDATION TYPE AND BASIS FOR DESIGN

Part C contains the information necessary for choosing the most appropriate type of foundation for given soil conditions, and for selecting suitable values of the allowable soil pressure or allowable load per pile.

Gen. Wm. SooySmith (1830–1916)

Eminent American foundation engineer during the era of the westward expansion of the railways and the development of the skyscraper. He employed a wide variety of foundations, depending upon the requirements of his projects. He was first in this country to use pneumatic caissons (Wagoshance Lighthouse, Straits of Mackinac, 1867); he was a leader of the transition in Chicago from shallow footings to long piles; and he was the inventor of the "Chicago method" of pier construction. He also conducted one of the earliest tests on a group of piles (Chicago Public Library, 4-pile group, 1893).

Plate 13.

CHAPTER 13

Factors Determining Type of Foundation

13.1. Steps in Choosing Type of Foundation

The type of foundation most appropriate for a given structure depends upon several factors: the function of the structure and the loads it must carry, the subsurface conditions, and the cost of the foundation in comparison with the cost of the superstructure. Other considerations may enter into the selection, but these three are commonly the principal ones.

On account of the interplay of these several factors, there are usually several acceptable solutions to every foundation problem. Faced with a given situation, different engineers of long experience may arrive at somewhat different conclusions. This fact is an indication that judgment plays a primary part in foundation engineering. It is doubtful that a strictly scientific procedure can ever be established for foundation design, although scientific developments have contributed greatly to the advancement of the art.

When an experienced engineer begins to study a new project, he almost instinctively discards the most unsuitable types of foundation and concentrates on the few most promising ones. When he has narrowed his choice to a few alternatives well fitted to the subsurface conditions and to the function of the structure, he studies the relative economy of these selections before making his final decision.

Engineers of less experience can follow a similar procedure without danger of serious error if they avail themselves of the results of scientific studies and the collective experience of others. To be useful, however, this information must be organized in a logical fashion. Part C of this book represents a digest of experience with various types of foundations and subsurface conditions, arranged in such a manner that the student can become familiar with the mental processes used by men who have achieved success in practice. By doing so, the student himself can use similar techniques and can expect to arrive at reasonable solutions for all but the most complex foundation problems.

In choosing the type of foundation, the engineer successively performs five essential steps: (1) He obtains at least approximate information concerning the nature of the superstructure and the loads to be transmitted to the foundations. (2) He determines the subsurface conditions in a general way. (3) He then briefly considers each of the customary types of foundation to judge whether or not they could be constructed under the existing conditions, whether they would probably be capable of carrying the required loads, and whether they might experience detrimental settlements. In this preliminary way, he is able to eliminate the types obviously unsuitable. (4) The next step is to make more detailed studies and even tentative designs of the most promising types. These studies may require additional information concerning the loads and subsurface conditions and generally must be carried far enough to determine the approximate size of footings or piers or the approximate length and number of piles required. It may also be necessary to make more refined estimates of settlement in order to predict the behavior of the structure. (5) Finally, the engineer prepares an estimate of the cost of each promising type of foundation and chooses the type that represents the most acceptable compromise between performance and cost.

Steps 3 and 4 described in the preceding paragraph require a knowledge of the probable behavior of each type of foundation for each type of subsurface condition. The presentation of such information is the principal object of Part C.

13.2. Bearing Capacity and Settlement

On the assumption that it is practicable to construct a given type of foundation under the conditions prevailing at a site, the probable performance of the foundation must be judged with respect to two types of unsatisfactory behavior. On the one hand, the entire foundation or any of the elements of which it is composed may break into the ground because the soil or rock is incapable of supporting the load without failure. On the other hand, the supporting soil or rock may not fail, but the settlement of the structure may be so great or so uneven that the superstructure may become cracked and damaged. These two types of unsatisfactory behavior have almost independent causes and can usually be investigated separately. The first is a function of the strength of the supporting soil or rock, and is known as a *bearing-capacity failure.* The second depends upon the stress-deformation characteristics of the soil or rock, and is known as *detrimental settlement.*

In each of the following chapters, one of the principal kinds of natural deposits will be considered. For each kind, the various types of foundations will be listed, and methods will be described for determining the load that can safely be transferred to the soil by the foundation. Methods will also be discussed for evaluating the settlement. Finally, for each kind of natural deposit, there will be given a summary of construction difficulties that may have a bearing on the suitability of each type of foundation.

To satisfy the requirements of safety, it is recommended that the factor of safety against a bearing capacity failure under the probable maximum load be at least 3.0. Smaller factors are sometimes justified if the structure is of minor importance or if the subsurface conditions are so uniform or so well known that the variations in the soil constants can be predicted with unusual confidence. A factor of 3.0 is sufficiently conservative to take into account the ordinary but unpredictable variations in the subsurface conditions, and it is comparable with the factor of safety commonly used for the design of the superstructure. On the other hand, for the extreme conditions of loading postulated for the design of foundations in most building codes, a factor of safety of 2.0 is adequate.

The permissible settlement depends upon the type of structure and on its function. The load transmitted to the soil at the base of a bridge pier may be due largely to the dead weight of the pier itself, and the corresponding settlement may be several inches. If the settlement occurs during construction of the pier, it is of no practical significance. If it takes place over a long period of time, it may be of no consequence provided the superstructure consists of simple trusses or cantilevers, but it may be of serious consequence if the superstructure is a continuous girder or truss. Few concrete building frames can stand a differential settlement between adjacent columns of more than about ¾ in. without showing some signs of distress. A steel structure can stand somewhat more, and a brick masonry structure can stand three to four times this amount without serious damage. Irregular or erratic settlement is more harmful to a structure of any type than uniformly distributed settlement.

Inasmuch as the cost of a foundation is appreciably influenced by the amount of differential settlement considered tolerable, the engineer should not underestimate the settlement that his structure can stand.

Baton Rouge Bridge (1940)

One of several outstanding structures spanning the lower Mississippi River, founded on lenses of dense alluvial sand embedded in other alluvial deposits of clay and silt. The piers were established by means of caissons started on artificial sand islands and sunk by open dredging. The final depth of the deepest piers was 180 ft. Sound rock is probably at a depth of several thousand feet.

PLATE 14.

CHAPTER 14

Foundations on Sand

14.1. Significant Characteristics of Sand Deposits

If the site is underlain by sand, the foundation may consist of footings, rafts, piles, or piers. The choice depends primarily on the relative density of the sand and on the position of the water table. The relative density determines the bearing capacity of footings or rafts, the resistance of piles, and the supporting capacity of piers. It also is the principal factor in determining the settlement. The position of the water table is important chiefly because excavation below water level requires drainage and increases the cost of the foundation. However, it also has an appreciable influence on the bearing capacity and settlement.

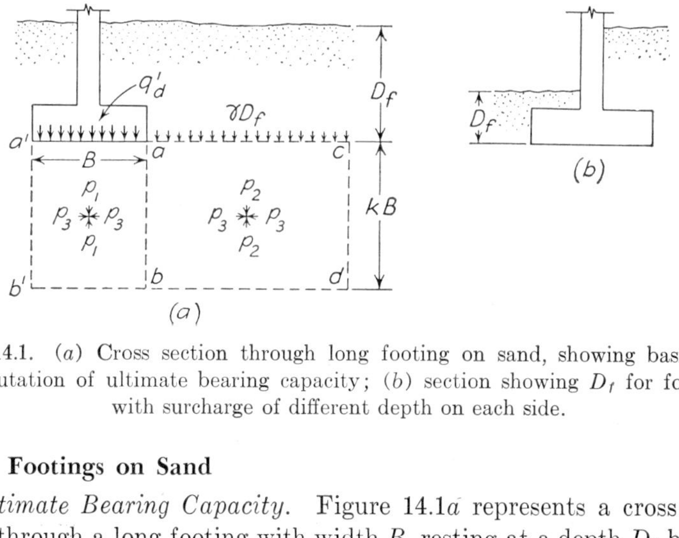

FIG. 14.1. (*a*) Cross section through long footing on sand, showing basis for computation of ultimate bearing capacity; (*b*) section showing D_f for footing with surcharge of different depth on each side.

14.2. Footings on Sand

Ultimate Bearing Capacity. Figure 14.1*a* represents a cross section through a long footing with width B, resting at a depth D_f below the ground surface. The soil below the base of the footing consists of sand to a very great depth. If the footing fails by breaking into the ground, the block $aa'b'b$ must fail like a specimen in a testing machine, and section *ab* must be displaced toward the right. However, *ab* cannot move to the right unless the block *abdc* also fails, whereupon *ac* rises and lifts the surcharge γD_f. Figure 14.1*b* shows the meaning of D_f for a footing with different depths of surcharge on the two sides.

A rough computation of the ultimate bearing capacity of the footing can be made by considering the stresses at mid-height of each of the two blocks that fail. If we assume that the failure occurs within an arbitrary depth kB below the base of the footing, we may write according to the principles of statics

$$p_2 = \gamma D_f + \tfrac{1}{2}kB\gamma \qquad 14.1$$

By analogy to eq. 2.36, we may write for the failure conditions of the right- and left-hand blocks, respectively,

$$p_3 = p_2 N_\phi \qquad 14.2$$

$$p_1 = p_3 N_\phi \qquad 14.3$$

where $N_\phi = \tan^2 (45° + \phi/2)$, Art. 2.9.
The intensity of loading that can be supported by the soil at the base of the footing can also be expressed in accordance with the laws of statics. It is

$$q_d' = p_1 - \tfrac{1}{2}kB\gamma \qquad 14.4$$

By combining these equations, we obtain

$$q_d' = \tfrac{1}{2}kB\gamma(N_\phi^2 - 1) + \gamma D_f N_\phi^2 \qquad 14.5$$

in which q_d' is the ultimate bearing capacity, or the maximum pressure that can be sustained by the soil at the base of the footing.

The factor k is only a rough measure of the depth of sand involved in the failure. Advanced theoretical studies show that this depth is a function of the relative density of the sand and, consequently, of ϕ. If we let $k(N_\phi^2 - 1)$, which is a function only of ϕ, be denoted by N_γ, we obtain

$$q_d' = \tfrac{1}{2}B\gamma N_\gamma + \gamma D_f N_\phi^2 \qquad 14.6$$

In more rigorous solutions, the true shape of the mass that experiences failure is taken into consideration and N_ϕ^2 is replaced by a nearly equal but more correct factor N_q which also is a function only of ϕ. Hence

$$q_d' = \tfrac{1}{2}B\gamma N_\gamma + \gamma D_f N_q \qquad 14.7$$

This equation demonstrates that the bearing capacity of a footing on sand is derived from two sources, the friction due to the weight of the sand below the level of the footing, and the friction due to the weight of the surrounding surcharge or backfill.

The unit weights of most sands, whether dry, moist, or saturated, lie within a fairly narrow range of values. Therefore, the unit weight of the sand is in itself not an important variable in the determination of the bearing capacity of a footing. However, if the sand is located below the free water surface, only its submerged weight is effective in producing friction. The submerged weight is about one half of the moist, dry, or saturated weight. The value of ϕ is not appreciably changed by submergence. Hence, it may be concluded that a rise of the water table from a depth greater than about B below the base of the footing up to the top of the surcharge would have the effect of reducing the bearing capacity to about one half of its value for moist, dry, or saturated sand. Thus, the position of the water table is of outstanding practical importance in establishing the bearing capacity of a footing on sand.

The values of N_γ and N_q increase rapidly as ϕ increases (see Fig. 14.2). Since ϕ depends to a considerable extent upon the relative density of the sand but is practically independent of the grain size, it may be concluded that the bearing capacity is greatly influenced by the relative density but hardly at all by the grain size.

According to eq. 14.7, the portion of the bearing capacity due to the weight of the soil beneath the footing increases directly with the width of the footing. The portion due to the surcharge, however, is independent of the width of the footing.

In summary, eq. 14.7 shows that the ultimate bearing capacity of a footing on sand depends mainly upon four variables: the position of the water table, the relative density of the sand, the width of the footing, and the depth of the surcharge surrounding the footing.

Bearing-capacity Chart. In determining the area required for a footing, it is most convenient to base the computation on the *net soil pressure:* that is, the pressure at the base of the footing in excess of that at the same level due to the surrounding surcharge. Thus, the net ultimate soil pressure q_d is

$$q_d = q_d' - \gamma D_f = \tfrac{1}{2}B\gamma N_\gamma + \gamma D_f(N_q - 1) \qquad 14.8$$

The factors N_γ and N_q are functions only of ϕ which, in turn, is primarily a function of the relative density of the sand. According to Art. 3.4, the most expedient procedure for investigating the relative density is the standard penetration test. The N values as determined by means of this test can be fairly reliably correlated with the values of ϕ, N_γ, and N_q. The results of such correlations are shown in

Fig. 14.2. Hence, if the N values are known, eq. 14.8 can be evaluated with the aid of Fig. 14.2.

It should be understood that the relationship between N and ϕ, Fig. 14.2, is only approximate; the limitations of the standard penetration test have been pointed out in Art. 3.4. On the other hand, the

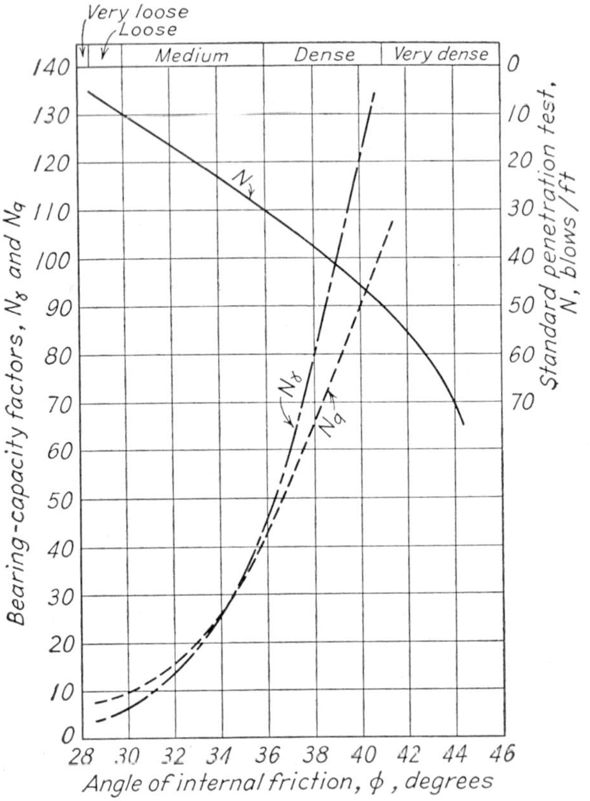

FIG. 14.2. Curves showing the relationship between ϕ, bearing capacity factors, and values of N from the standard penetration test.

relationship between ϕ and N_γ or N_q is based primarily on theory and is much more reliable. Hence, if ϕ is determined by a procedure more refined than use of the standard penetration test, Fig. 14.2 may still be utilized to evaluate N_γ and N_q.

The net soil pressure at the base of a footing should not exceed about one third of the net ultimate soil pressure. The soil pressure at which the factor of safety against breaking into the ground is equal to 3.0 is referred to as the *safe soil pressure.* It may be determined most readily for a footing of given width by means of the diagrams, Fig. 14.3, based on eq. 14.8 and Fig. 14.2, and on the assumption that the sand weighs 100 lb/cu ft. The first diagram, Fig. 14.3*a*, furnishes that part of the safe soil pressure corresponding to the term $\frac{1}{2}B\gamma N_\gamma$ in eq. 14.8; the second, Fig. 14.3*b*, corresponds to the term $\gamma D_f(N_q - 1)$. By entering Fig. 14.3*a* with the width of footing B and the value of N, one obtains the safe soil pressure for a footing surrounded by no sur-

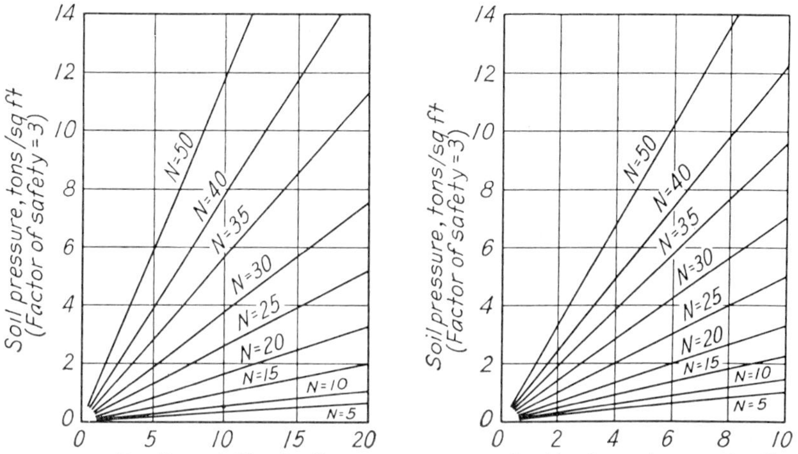

(a) Safe soil pressure without surcharge, $D_f = 0$

(b) Additional safe soil pressure due to surcharge

Charts based on water table not closer than B below base of footing

FIG. 14.3. Safe soil pressures beneath footings on sand, as determined by bearing capacity.

charge. If a surcharge exists, Fig. 14.3*b* indicates the additional safe soil pressure due to the surcharge.

The diagrams are applicable without modification if the groundwater level is at a depth of B or more below the base of the footing. If the groundwater is at the base of the footing, the safe soil pressure obtained from Fig. 14.3*a* should be divided by 2. If the groundwater is at the top of the surcharge surrounding the footings, the increment of safe soil pressure due to the surcharge, as given in Fig. 14.3*b*, should also be divided by 2. If the groundwater is at some intermediate position, sufficiently accurate values can be obtained by interpolation.

The penetration resistance for the sand at a given site should be determined by making standard penetration tests at a number of points, preferably at least one for every 4 to 6 footings. Values of

N should be determined at intervals of $2\frac{1}{2}$ ft in the vertical direction, and the average N value beneath each point should be determined between the level of the base of the footing and a depth B below this level. The smallest average value of N obtained in this manner should be used for computing the safe load on all the footings.

As mentioned in Art. 3.4, the values of N may be too great if the subsoil consists of very fine sand below the water table. These excessive values are referred to as N'. Before the diagrams, Fig. 14.3, are entered, the equivalent value of N, for values of N' greater than 15, should be determined from the expression

$$N = 15 + \tfrac{1}{2}(N' - 15) \tag{14.9}$$

Figures 14.3*a* and 14.3*b* may be used not only for long continuous footings but also for square or rectangular footings, provided the width B is taken as the least dimension of the footing.

Finally, if the value of ϕ has been determined by some means other than the standard penetration test, an equivalent value of N may be obtained from Fig. 14.2 for use in the bearing-capacity charts, Fig. 14.3.

Settlement. The settlement of footings on sand is governed by the stress–deformation characteristics of the material. According to Art. 2.7, the rigidity of sand as measured by the initial tangent modulus increases markedly with increase in relative density and is approximately proportional to the confining pressure.

The confining pressure in a mass of sand is at least roughly proportional to the vertical pressure and is, therefore, also roughly proportional to the unit weight of the sand immediately beneath and beside the footing. The most important factor that has an influence upon the unit weight of the sand is the position of the water table. If the water table is near the ground surface, the effective vertical pressure in the sand is due only to its submerged weight. Hence, if the water table is raised to the ground surface from a great depth, the settlement of a footing is likely to be approximately doubled.

The preceding paragraph leads to the conclusion that, for a given soil pressure, the settlement of a footing on sand depends upon the relative density and on the position of the water table. Theory, as well as laboratory and field investigations, shows that the settlement for a given soil pressure also increases with increasing width of footing. The known facts concerning all these relationships are summarized in Fig. 14.4. From this figure can be determined the soil pressure which, for a given value of N and a given size of footing, will cause the footing to settle 1 in. If the pressure corresponding to some other amount of settlement is desired, it may be computed from the value given by the chart by assuming that the settlement varies directly as the soil pressure.

Figure 14.4 has also been constructed on the assumption that the water level is at least a distance B below the base of the footing.

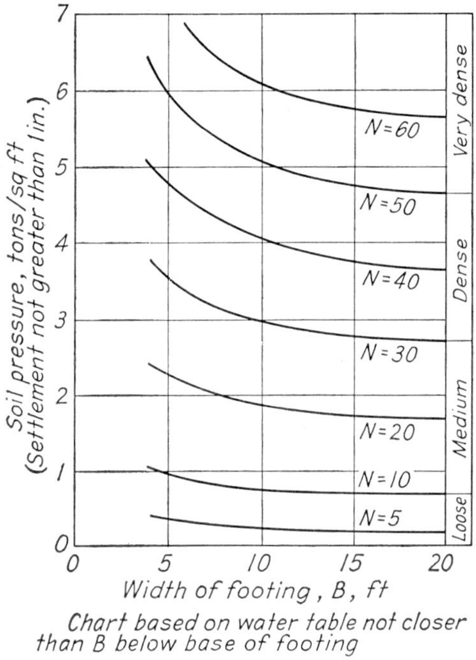

FIG. 14.4. Soil pressure corresponding to 1-in. settlement of footings on sand.

If the water level is near or above the base of the footing, the pressure corresponding to a 1-in. settlement should be taken as half the value given by the chart. For intermediate positions, the proper pressures may be obtained by interpolation.

Values of N with which to enter the chart in Fig. 14.4 should be determined by the procedure advocated in connection with Fig. 14.3. If the footing is not square, the least dimension should be used for B.

Since the soil pressure that produces a settlement of 1 in. varies with the width of the footing, it is apparent that the same soil pressure cannot be used beneath all footings of a given structure, even on sand of perfectly uniform relative density, without differential settlement. However, if the soil pressure beneath all footings is taken to be the same as that beneath the largest one, the settlement of no footing should exceed 1 in. Furthermore, even a fairly uniform sand deposit exhibits variations in relative density, and the footings located above dense sand are likely to settle less than those above loose sand. This tends to cause differential settlements. However, if it is assumed that the largest footing may be located above the loosest part of the deposit and if the soil pressure is chosen so that the settlement of this footing will not exceed 1 in., then the differential settlement between any two of the footings will not exceed 1 in. and, in reality, will rarely exceed ¾ in. This value is generally considered tolerable.

Use of Charts to Select Allowable Soil Pressure. A comparison of Fig. 14.3 and Fig. 14.4 indicates that the pressure that can safely be carried by most footings at a factor of safety of 3.0 is considerably greater than the pressure that will produce a settlement of 1 in. Therefore, in connection with the design of footings for a building, it is rarely possible to make use of the full strength of the sand. Only if the footings are very narrow, the water table high, and the sand loose, is the soil pressure permitted by the settlement criterion likely to be greater than that corresponding to a factor of safety of 3 against a bearing-capacity failure. Both requirements must be checked if a satisfactory foundation is to be achieved.

In practical problems involving building foundations, the following procedure is suggested. The appropriate value of N should be determined according to the rules previously stated. The soil pressure corresponding to a 1-in. maximum settlement should be determined from Fig. 14.4 for the largest footing in the structure. This value should be compared with the safe soil pressure, Fig. 14.3, for this footing and all the narrower footings. Whenever the safe soil pressure is less than that given by Fig. 14.4, it should be used as the basis for design of the corresponding footings. Whichever of the two values governs the design is considered to be the *allowable soil pressure.*

It may be noted that the procedure recommended in the preceding paragraphs for selection of the allowable soil pressure inherently takes into account the normally erratic variations in the relative density of natural sand deposits. If there is no significant trend in the average N values from one part of a site to another, but only erratic variations, the procedure may be used without modification. On the other hand, if the site is underlain by several deposits of significantly different average relative density, the site should first be subdivided into regions each of which may be regarded as having reasonably consistent properties, and the soil pressures for the footings in each region should be determined separately.

Design Loads for Footings on Sand. Considerable judgment is required in choosing the loads that should be considered in proportioning the areas of footings on sand. When settlement governs the allowable soil pressure, the design load should include the full dead load on the footing, plus the pressure due to the weight of the footing reduced by that of the surrounding surcharge, plus the maximum live load including wind and snow load that may ever reasonably be expected to act upon the footing at any one time. This value

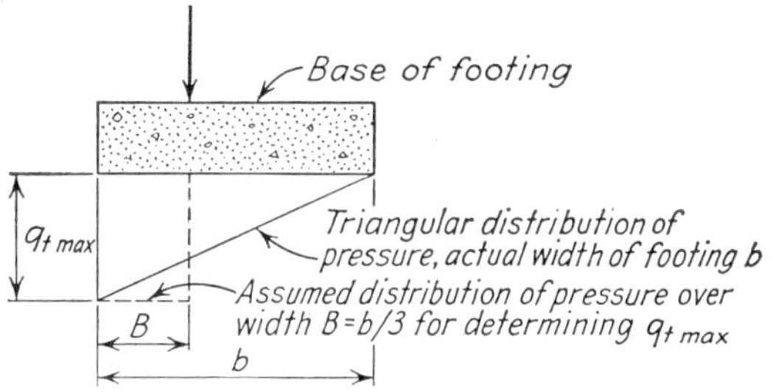

FIG. 14.5. Diagram illustrating method for estimating allowable soil pressure beneath toe of eccentrically loaded footing.

of live load should be used because most of the settlement of a footing on sand takes place as soon as the load is applied. Hence, a single application of the full load will produce the full settlement upon which the design is based. However, the live loads should not be overestimated, because by this action the differential settlement may be increased on account of the disproportionately small settlements that some footings would thereby experience.

If bearing capacity governs the design, the safe soil pressure given by Fig. 14.3 should not be exceeded for the loads listed in the preceding paragraph. Moreover, the most unfavorable conceivable combination of loadings, as prescribed by the building code, should not produce a pressure greater than 1.5 times the values given in the chart. This provides a minimum factor of safety of 2.0 under the severe loadings specified in the code, and a factor of 3.0 under the greatest loads actually likely to be experienced.

Footings with Eccentric Loading. Beneath retaining walls and some other types of structures subjected to lateral loads, the calculated distribution of pressure against the base may be triangular,

Fig. 14.5. The soil pressure q_t at the toe is then twice the average pressure. Therefore, the toe pressure is likely to govern the design. Inasmuch as relatively high pressures act over only part of the base, the allowable soil pressure can be estimated on the assumption that the effective width B of the footing is less than the real width b.

Little theoretical and no experimental information is available upon which a rational analysis can be made. It would appear conservative and reasonable, however, to select the allowable soil pressure by means of the procedures given in this article for concentrically loaded footings, on the assumption that the footing has a width $B = b/3$. The toe pressure q_t should not exceed the allowable pressure determined in this manner.

Illustrative Designs. DP C-1 and C-1a. Footings on Sand

To illustrate the application of the foregoing discussion of footings on sand, two sets of design computations (DP C-1 and DP C-1*a*) are given in the following pages.

The design procedure used in both examples can be divided into two steps:

1. The largest footing is proportioned by trial, because the allowable soil pressure depends on the dimensions of the footing. Considerations of settlement determine the allowable pressure for the largest footing, as shown on Sh. 2 of DP C-1 and Sh. 1 of DP C-1*a*.
2. The smaller footings are proportioned on the basis of the allowable pressure determined for the largest footing. A check is always made to insure adequate safety against bearing capacity failure.

For the two largest footings considered in DP C-1, the soil pressure to insure a factor of safety of 3 against bearing-capacity failure is considerably higher than the pressure determined on the basis that a maximum settlement of 1 in. is tolerable. On the other hand, for the smallest footing (Sh. 3 of DP C-1), the allowable soil pressure is governed by bearing capacity rather than settlement.

DP C-1*a* illustrates the method for considering the effect of the water table. The water table is located about 5 ft below the base of the footings; consequently, the allowable soil pressures given by the curves of Fig. 14.3*a* and 14.4 have been reduced 25 per cent. This reduction represents an interpolation between a 50 per cent reduction if the water table were near the base of the footing and no reduction if the water table were deeper than the width of the footing below the base.

DP C-1
Warehouse Footings
Sh. 1 of 3

Footing Design. General Data

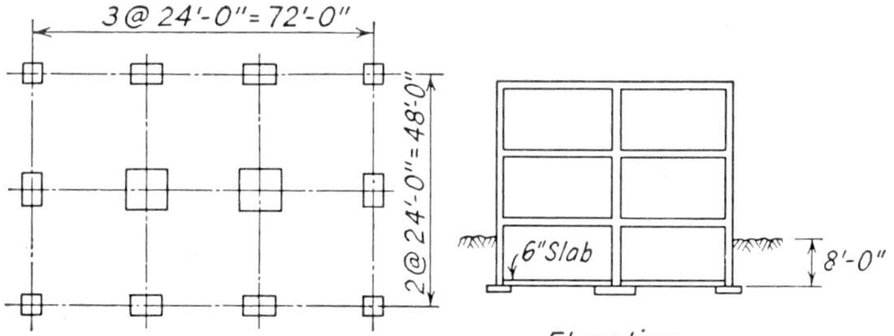

Footing Plan　　*Elevation*

Data:

Preliminary design computations indicate that column loads to footings will be as follows:

	DL	LL	Total
Interior	126^k	372^k	$498^k = 249^T$
Wall	82^k	176^k	$258^k = 129^T$
Corner	43^k	85^k	$128^k = 64^T$

The building will be used as a warehouse; consequently, the full live load is likely to be applied at least once. Since the subsoil is sand, almost the full settlement will occur the first time the load is applied. No reduction in live load will be considered in proportioning the footings.

The tolerable differential settlement between footings = $\frac{3}{4}$".

Borings carried to a depth of 50' within the building area give the following information:

(1) Medium to coarse sand from ground surface to bottom of hole.
(2) Water table at a depth of 34 ft.
(3) Minimum average N value from base of footings to a depth of 20' = 35 blows/ft.

No soft stratum exists below the sand.

DP C-1
Warehouse Footings
Sh. 2 of 3

Footing Design. Required Areas

Interior: 249^T

@ say 3.5 = 71 sq ft

Try 8'-6" x 8'-6". A = 72.3 sq ft Assume depth = 2'-6"

Column load = 249^T

$\div 72.3 = 3.45^T$/sq ft

Additional load:

Floor slab = $0.5 \times \frac{0.15}{2}$
Footing = $2.5 \times \frac{0.15}{2}$ } = 0.22

3.67 ‡

Surcharge load:

Floor slab = $0.5 \times \frac{0.15}{2}$
Soil = $2.5 \times \frac{0.10}{2}$ } = 0.16

3.51^T/sq ft

From Fig. 14.4, allow 3.6^T/sq ft

From Fig 14.3, allow 4.8 + 2.9* = 7.7^T/sq ft

Use 8'-6" x 8'-6" for interior footings

‡ Note that unit pressure due to additional load exceeds unit pressure of surcharge by the product of footing depth and difference between unit weights of concrete and soil; that is, $2.5 \left(\frac{0.15-0.10}{2}\right) = 0.06^T$/sq ft. Hereafter, in this design plate, the computations will be simplified in this manner

* Additional bearing capacity due to surcharge. Assuming slab equivalent to equal thickness of soil, enter Fig. 14 3b with D_f = 3 ft.

DP C-1
Warehouse Footings
Sh. 3 of 3

Footing Design. Required Areas

Wall: 129^T

@ 3.51 = 36.7 sq ft

Try 5'-3" x 7'-0". A = 36.8 sq ft Assume depth = 2'-0"

Column load = 129^T ÷ 36.8 = 3.51^T/sq ft

Additional load minus surcharge $\}$ $2.0\left(\frac{0.15-0.10}{2}\right) = 0.05$

3.56^T/sq ft

Fig. 14.4, allow 4.1^T/sq ft

Fig. 14.3, allow 2.9 + 2.4 = 5.3^T/sq ft

Use 5'-3" x 7'-0" for wall footings

Corner: 64^T

@ 3.51 = 18.2 sq ft

Try 4'-3" x 4'-3". A = 18.1 sq ft Assume depth = 1'-6"

Column load = 64^T ÷ 18.1 = 3.53^T/sq ft

Additional load minus surcharge $\}$ $1.5\left(\frac{0.15-0.10}{2}\right) = 0.04$

3.57^T/sq ft

Fig. 14.4, allow 4.4^T/sq ft

Fig. 14.3, allow 2.4 + 1.9 = 4.3^T/sq ft

Use 4'-3" x 4'-3" for corner footings

DP C-1a
Warehouse Footings
Sh. 1 of 2

Footing Design. Required Areas. Effect of Water Table

The following computations illustrate the effect of the water table on footing areas. The water table is located approximately 5 ft below the base of the footings. Hence, a decrease of 25 per cent must be made in the allowable soil pressures given in Fig. 14.3 and Fig. 14.4. All other data are as given on Sh. 1 of DP C-1.

Interior: 249^T

@ say $3.4 \times 0.75 = 98$ sq ft

Try 10'-0" x 10'-0". A = 100 sq ft Assume depth = 2'-6"

Column load = 249^T ÷ 100 = 2.49^T/sq ft

Additional load minus surcharge: $2.5\left(\frac{0.15-0.10}{2}\right) = 0.06$

2.55^T/sq ft

Fig. 14.4, allow $3.5 \times 0.75 = 2.6^T$/sq ft

Fig. 14.3, allow $(5.6 \times 0.75) + 2.9 = 7.2^T$/sq ft

Use 10'-0" x 10'-0" for interior footings

DP C-1a
Warehouse Footings
Sh. 2 of 2

Footing Design. Required Areas. Effect of Water Table

Wall: 129^T

@ 2.55 = 50.6 sq ft

Try 6'-0" x 8'-6". A = 51 sq ft Assume depth = 2'-0"

Column load = 129^T

÷ 51 = 2.53^T/sq ft

Additional load minus surcharge } $2.0\left(\frac{0.15-0.10}{2}\right)$ = 0.05

2.58^T/sq ft

Fig. 14.4, allow 4.0 x 0.75 = 3.0^T/sq ft

Fig. 14.3, allow (3.3 x 0.75) + 2.4 = 4.9^T/sq ft

Use 6'-0" x 8'-6" for wall footings

Corner: 64^T

@ 2.55 = 25.1 sq ft

Try 5'-0" x 5'-0". A = 25 sq ft Assume depth = 1'-6"

Column load = 64^T

÷ 25 = 2.56^T/sq ft

Additonal load minus surcharge } $1.5\left(\frac{0.15-0.10}{2}\right)$ = 0.04

2.60^T/sq ft

Fig. 14.4, allow 4.1 x 0.75 = 3.1^T/sq ft

Fig. 14.3, allow (2.9 x 0.75) + 1.9 = 4.1^T/sq ft

Use 5'-0" x 5'-0" for corner footings

14.3. Rafts on Sand

Soil Pressure. Because of the large size of rafts compared to that of footings, the factor of safety against a bearing-capacity failure of the underlying sand is always very great. This can be seen from Fig. 14.3. If the width of the raft is only 20 ft, the surcharge only 10 ft, and the number of blows N equal to 10 or more, the safe soil pressure on submerged sand exceeds 1 ton/sq ft. With increasing width of the raft or increasing relative density of the sand, the safe soil pressure increases rapidly. Hence, it may be concluded that the danger of a large raft breaking into a sand foundation is too remote to require consideration.

On account of the large size of rafts, the stresses in the underlying sand are likely to be relatively high to considerable depth. Therefore, the influence of local loose pockets distributed at random throughout the sand is likely to be about the same beneath all parts of the raft, and the differential settlement is likely to be smaller than that of a footing foundation designed for the same soil pressure. Although it is conceivable that a single footing may rest entirely above a pocket of loose sand and experience large settlement, the effects of such a loose pocket beneath part of a raft would be much less important.

Because the differential settlements of a raft foundation are likely to be less than those of a footing foundation designed for the same soil pressure, it is reasonable to permit larger allowable soil pressures on raft foundations. Experience has shown that a pressure approximately twice as great as that allowed for individual footings may be used because it does not lead to detrimental differential settlements.

The shape of the curves in Fig. 14.4 shows that the soil pressure corresponding to a given settlement is practically independent of the width of the footing or raft, provided the width is greater than about 20 ft. This fact provides the basis for the following simple expression for determining the allowable soil pressure:

$$q_a = \frac{N - 3}{5} \qquad 14.10$$

in which q_a is the allowable pressure in tons per square foot, and N is the number of blows in the standard penetration test.

The values of q_a given by eq. 14.10 serve as a rational basis for the design of a raft foundation on sand under most conditions encountered in the field. They may be increased somewhat if bedrock is encountered at a depth less than about one-half the width of the raft. On the other hand, they should be reduced 50 per cent if the water table is at the base of the raft or higher; if the water table is at a depth intermediate between the base of the raft and a depth B below the base, an appropriate reduction between 0 and 50 per cent should be made. If the subsoil consists of very fine sand below the water table, the standard penetration test provides values of N' which must be corrected to the equivalent values of N before use in eq. 14.10. The correction is given by eq. 14.9.

If the value of N is less than 5, the sand is too loose for the successful use of a raft foundation. It must be compacted by some artificial means until the N value is greater than 10 or else the foundation must be established on piles or piers.

The loads that should be considered in computing the soil pressure on the raft for comparison with the allowable pressure are the dead load of the structure including the raft and the maximum live load that is really likely to be active, less the surcharge due to the weight of the soil between the surrounding ground surface and the base of the foundation.

Settlement. If the soil pressure is chosen on the basis of eq. 14.10, and if the value of N is not less than 5, the differential settlements between adjacent columns on a raft foundation on sand will not exceed about ¾ in., provided the base of the raft is located at a depth of at least 8 ft below the surrounding ground surface. Experience has shown that, if the surcharge is less than this amount, the edges of the raft settle appreciably more than the interior because of the lack of confinement of the sand.

Illustrative Design. DP C-2. Raft Foundation on Sand

The application of the principles of Art. 14.3 is illustrated in this design plate. As shown on Sh. 1, five borings have been made at the site of the proposed building. The results of these borings are tabulated on Sh. 2. Below el. 829 the borings indicate the presence of very fine silty sand. For reasons explained in Art. 14.2, the field N values in this material are corrected by means of eq. 14.9 before they are averaged with the N values for the sand above elevation 829.

The allowable soil pressure is determined from the minimum average N value for a depth equal to the width of the raft below the elevation of the base of the foundation. This least favorable condition is found at boring $B3$. The pressure obtained by means of eq. 14.10 may be verified by entering Fig. 14.4 with the maximum width (20 ft) shown in the chart. An appropriate correction is made

DP C-2
Raft Foundation
Sh. 1 of 2

Raft Foundation Design. Determination of Allowable Soil Pressure

Sketches Showing Location of Borings and Typical Cross Section of Building:

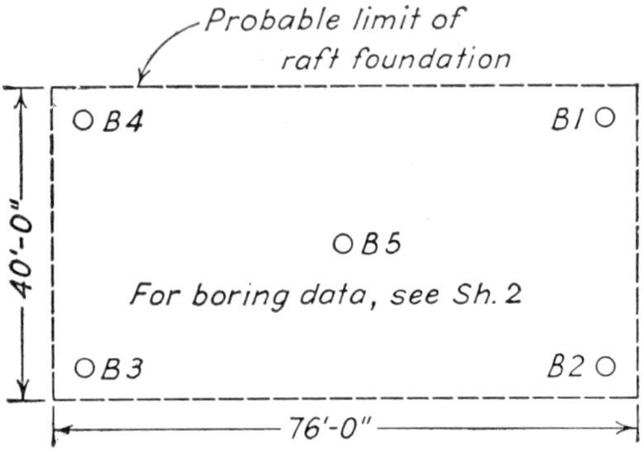

Boring Plan

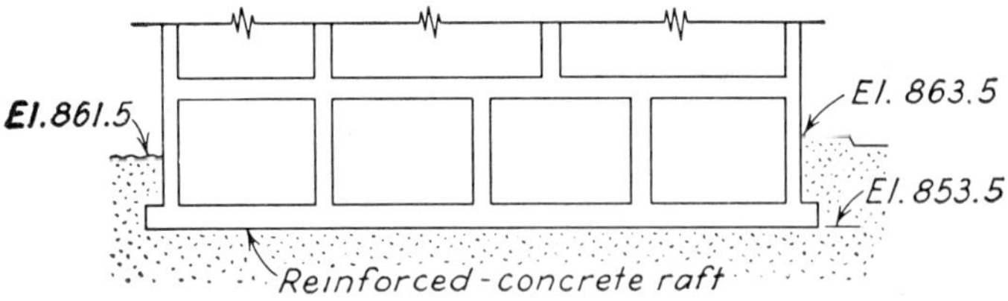

Typical Section

DP C-2
Raft Foundation
Sh. 2 of 2

Raft Foundation Design. Determination of Allowable Soil Pressure

Boring Data:

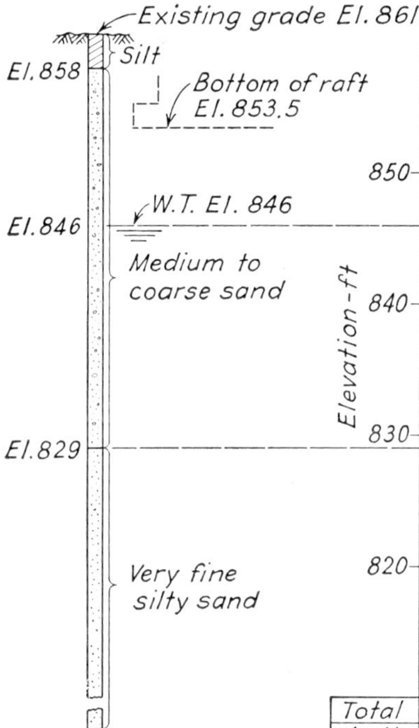

Typical Boring

Blows/ft - Standard Penetration Test				
B1	B2	B3	B4	B5
12	14	8	13	10
10	9	9	19	12
16	10	15	24	16
19	22	19	26	16
14	17	15	14	26
16	12	20	16	28
26	24	21	23	22
22	23	16	31	30
*39(27)	18	16	*29(22)	*30(22)
28(21)	*26(20)	*24(19)	35(25)	32(23)
27(21)	35(25)	24(19)	40(27)	41(28)
36(25)	42(28)	37(26)	38(26)	28(21)
40(27)	27(21)	39(27)	32(23)	26(20)
28(21)	25(20)	28(21)	41(28)	36(25)
	39(27)	27(21)		
Total 277	290	273	317	299
Av. N 19.7	19.3	18.2	22.7	21.3

Allowable Soil Pressure:

By eq. 14.10, $\frac{18.2-3}{5} = 3.0\ ^T/\text{sq ft}$ } without W.T. corr.

By Fig. 14.4, $1.5 \times 2 = 3.0\ ^T/\text{sq ft}$ }

Reduction due to W. T. $\frac{32.5}{40} \times 50\,\% = 41\%$

Allowable soil pressure $= 3.0(1.00 - 0.41) = 1.8\ ^T/\text{sq ft}$

*N′ values corrected by means of eq. 14.9

for the position of the water table. Finally, it is important to note that the allowable soil pressure is 1.8 tons/sq ft in excess of the pressure at el. 853.5 due to the surcharge of soil surrounding the raft.

14.4. Piles in Sand

Use of Piles in Sand. (1) Piles may be driven through various kinds of soft or compressible materials to a layer of dense sand to which they may transfer the weight of the structure. (2) They may be driven into loose sand for the purpose of compacting it and increasing its bearing capacity. (3) Finally, they may be driven into a bed of sand in order to establish the foundation below the greatest depth to which the sand may be removed by scour.

Piles Driven to Bearing in Dense Sand. When the material directly beneath foundation level is too compressible or unstable for the support of the foundation, the weight of the structure should, if possible, be transferred to a better stratum at a lower level. If this stratum consists of sand, piles are often driven through the soft materials and into the sand for a sufficient distance to develop adequate carrying capacity. If the sand is moderately loose, the piles may have to be driven into it for a considerable distance. If it is fairly dense, it may not be possible to obtain more than a few feet of penetration. In either event, the piles derive their support partly from the resistance of the point to further penetration and partly from skin friction between the lower part of the piles and the sand.

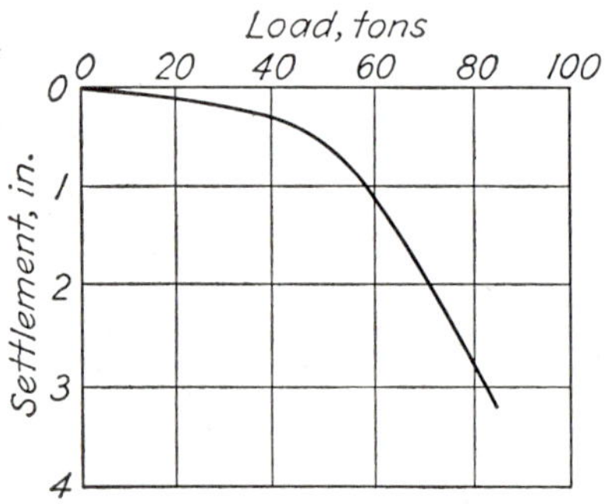

FIG. 14.6 Typical result of load test on pile driven through soft deposits into moderately dense sand.

It is not usually possible on the basis of the results of soil tests to make an accurate estimate of the load that can be carried by a pile driven through soft materials into a deposit of sand. The most reliable information is that determined by means of load tests. In deposits of this type, the shape of the load-settlement curve for a single pile is similar to that shown in Fig. 14.6. Unless the pile fails structurally, which is very unlikely, the load-settlement curve approaches an inclined tangent. Hence, the pile does not fail by plunging into the ground, but merely continues to settle or to penetrate into the sand as the load is increased. The limiting load for design must therefore be based upon a value of settlement that may be considered tolerable. It is commonly required that the pile be subjected to twice the load that it is likely to experience beneath the structure. Under this load, the penetration should not exceed 2 in., and under the design load it should not exceed about 0.75 in.

If the soft materials above the sand contain a few thin stiff layers or a few sand layers, the skin friction of the upper materials may carry a considerable part of the test load. Beneath the completed structure, however, the friction of the upper materials would soon disappear as the soft materials consolidated, and all the load would be transferred to the lower end of the pile. If the subsoil conditions are of this type, it is essential to make sure that the resistance of that part of the pile which is buried in the sand is adequate. This may be done by performing the load tests in pairs. One pile should be driven through the soft material into the sand and tested in the usual manner. The other should be driven into the soft material and stopped when its point is 3 or 4 ft above the top of the sand layer upon which the piles are expected to bear. A test to failure of the latter pile will indicate the amount of load carried by skin friction in the upper layers. The load-settlement curve of the first pile should not be used for determination of the safe pile load until a correction of this amount has been made.

If the number of piles on a given job is relatively small, the cost of load tests may be out of proportion to the cost of the entire foundation. Under these circumstances, it is common practice to determine the safe load per pile by means of a pile-driving formula. Many such formulas have been proposed, some of which are very elaborate. There is no evidence to indicate, however, that the simpler formulas are necessarily less valid than the more complicated ones. In the United States, the most popular pile-driving formula is the Engineering News formula

$$Q_a = \frac{2W_H H}{s + c} \tag{14.11}$$

in which

Q_a = safe design load on the pile in pounds

W_H = weight of the ram of the pile driver in pounds

H = height of fall of the ram in feet

s = penetration of the pile in inches under the last blow of the hammer

c = a constant equal to 1.0 for a drop hammer and 0.1 for a steam hammer

The theoretical factor of safety of this formula, according to its derivation, is 6 but statistical studies have indicated that the actual factor of safety has an average value of about 4. The deviation from the average may be very great so that the real factor of safety on a given job may occasionally be close to or even less than 1, whereas on others it may be occasionally as great as 8 or 10.

If part of the resistance to penetration of a single pile is developed in the overlying soft layers, this temporary resistance may lead to the conclusion that the bearing capacity of the pile is greater than it actually will be under a static load. A very crude estimate of the point resistance can be made by using the formula to compute the safe load on the pile just before it enters the stratum of sand and subtracting this value from that obtained after the point has reached its ultimate position.

The pile foundation may be designed on the assumption that each pile can be assigned a working load equal to that determined by the load test or formula. The differential settlements then do not ordinarily exceed those for properly designed footings on sand of the same relative density. These statements are limited by the condition that the bed of sand on which the piles rest is of great thickness or that it is underlain by no excessively loose or compressible materials.

If a pile cluster is surrounded by a fresh fill after the piles are driven, it is likely that the compressible materials above the sand will settle progressively for a considerable period of time on account of the weight of the fill. Under these conditions, the piles may be acted upon by an additional force due to the skin friction of the subsiding material. This force is known as *negative skin friction* or *drag*. The magnitude of the drag per unit of area cannot exceed the shearing strength of the compressible soil which may usually be considered as one-half the unconfined compressive strength. The area upon which the drag acts is the vertical surface that surrounds the entire group of piles or the entire pile foundation.

Although it is not possible to estimate the additional pile load due to drag with great accuracy, a rough computation can be made to indicate whether or not the added load will be of serious consequence, and appropriate measures can be taken. Several examples of unexpected settlement of large magnitude have been attributed to neglect of negative skin friction.

If the piles are driven through a recently placed fill which has initiated a process of consolidation in the underlying compressible layers, the weight of the fill within the pile cluster is also likely to be transferred to the piles. The load per pile would then be increased still further, by an additional amount approximately equal to $A\gamma H/n$, where A is the area of soil within a horizontal section bounded by the perimeter of the cluster, γ and H are the unit weight and thickness of the fill, and n is the number of piles in the cluster.

Compaction Piles. Driving piles into a bed of loose sand compacts the material partly on account of the decrease in void ratio necessary to compensate for the volume of the piles and partly on account of the compacting effect of the vibrations produced by pile driving. Once the piles are driven, the settlements of the structure will be approximately the same as those estimated according to the procedures described for footings and rafts on relatively dense sand.

Inasmuch as the purpose of compaction piles is merely to increase the density of the sand, the structural strength of the piles themselves is of little consequence. Indeed, it is not uncommon in Europe to make a hole by driving and removing a spud having the shape of a pile, and to fill the hole with compacted sand. By this procedure the density of the foundation material is increased but no piles remain in the ground.

Ordinarily, piles with a heavy taper are most effective and economical for compaction of loose sand. If the structure is to rest on pile footings, the piles should be driven in groups and the innermost piles in each group driven first to achieve the maximum and most uniform compaction beneath the location of each footing. If the structure is to rest on a pile-supported raft, the piles should be spaced uniformly over the entire area.

Compaction piles are usually driven until they encounter a resistance corresponding to a safe load of 20 to 30 tons according to the Engineering News formula, eq. 14.11. The lower value is generally considered appropriate for wood piles, and the higher for concrete. Use of the pile-driving formula does not imply a knowledge of the actual carrying capacity of the piles, but merely provides a convenient criterion to indicate when adequate compaction has been achieved and to guard against overdriving. The length of compaction piles is difficult to predict. Piles having a taper of 1 in. in 2.5 ft rarely penetrate more than 25 ft even in loose sands.

Bulbous uncased piles of the Franki type are widely used for compacting loose sands, especially in Europe.

Piles for Prevention of Failure Due to Scour. In connection with bridge piers located near the channels of rivers, it is necessary to make sure that the base of the foundation is below the level to which the river bottom is removed by scour during floods. In many streams, the depth of the river increases during floods at a rate greater than that at which the crest of the river rises. Moreover, since bridges are often located where the channel is narrow, the depth of scour is likely to be greater than average. Finally, the construction of the piers and abutments usually causes additional constriction of the channel and itself increases the depth of scour.

No universal rule can be given for estimating the maximum depth of scour in a stream with a bottom of sand or silt. Several records indicate depths of as much as 4 ft for each foot of rise of the water, and in some streams the ratio is known to be as great as 7. On the other hand, in the majority of rivers the depth is much less. Experience with a given stream is the best guide in estimating the depth of scour.

In many instances, the sands in the bottom of a river channel are relatively dense, and it is difficult or impossible to drive piles to adequate depth without the use of a water jet.

Uplift Resistance of Piles in Sand. The resistance to uplift of a pile driven into sand depends upon the relative density of the sand, upon the length, diameter, and taper of the pile, upon the use or absence of jetting, and upon several other factors. For this reason reliable values of the uplift resistance can be obtained only by means of load tests.

14.5. Piers on Sand

Conditions for Use of Piers. Piers may be established on the surface of a bed of dense sand at a considerable depth below the ground surface if the upper materials are too soft or compressible for the support of the structure or if the overlying materials may be removed by scour. In these respects, the conditions under which pier foundations are appropriate are similar to those for piles, and the choice depends primarily upon economy and upon certain details that influence the construction procedure. For example, if the compressible material contains organic deposits that include trunks of trees, or if there are likely to be numerous large boulders above the bearing stratum, it may not be possible to drive piles through the unstable soils. Under these conditions, a pier foundation would be indicated. On the other hand, the excavation for a pier is likely to lead to some loosening of the sand deposit upon which the structure is to rest, whereas driving piles into the sand tends to compact the bearing stratum. On account of these conditions, a pile foundation may often be preferable.

It is also sometimes necessary to excavate piers through sand to reach bearing on rock or other firm material. The conditions to be considered under these circumstances will be discussed in the next article.

Bearing Capacity of Piers. The ultimate bearing capacity of piers is somewhat greater than that of footings located upon sand of equal density. The reason for the increase is the fact that the shearing resistance of the surcharge above the level of the base of a pier is not negligible, as was assumed in the derivation of the equations for footings in Art. 14.2. However, if the surrounding material is compressible, the increase in bearing capacity may be very small. Moreover, if there is a possibility that the surrounding material may occasionally be removed by scour, the influence of this surcharge must be neglected. Therefore, in general, it is conservative and justifiable to determine the safe bearing capacity by means of the procedures already described for footings.

The settlement of a pier under a given soil pressure is less than that of a footing on sand of comparable density because of the influence of the weight of the surrounding material. However, the reduction in settlement is not so great as might be anticipated on the basis of the increased confining pressure because the excavation for a pier can rarely be made without loosening the sand at the base. In general, it has been found that the settlement of a pier is about half that of a footing of the same dimensions, provided the soil pressures and relative densities are equal. The allowable soil pressure can, therefore, be determined from Fig. 14.4 by multiplying the values given on the figure by 2 and making suitable correction for the position of the water table. If scour may remove most of the surcharge, however, the soil pressure should not be increased above that permissible for footings.

If several piers on a given stratum have approximately the same dimensions, the differential settlements are not likely to exceed about ½ in. if the soil pressures are selected in accordance with the preceding paragraph. If greater settlements can be tolerated, the allowable pressure may be increased in proportion to the increase in settlement. The maximum value is determined by the requirement that the factor of safety against bearing capacity failure should not be less than 3.

14.6. Excavation in Sand

Sand above Water Table. No construction problems of consequence are encountered in making excavations for footing or raft foundations in sand above the water table. Slopes of 1 vertical to

$1\frac{1}{2}$ horizontal are stable under all circumstances unless the sand deposit is underlain at shallow depth by soft clay, and steeper slopes are commonly used if the sand possesses a small amount of cohesion due to capillary moisture or cementing agents.

Large Excavations below Water Table. If excavations for a basement or for the establishment of a foundation must extend below water table in sand, the water must be removed. This may be done either by pumping the water out of the excavation itself, or by predrainage of the site.

When the water is removed by pumping from an open excavation, ditches must be cut in the bottom to lead the water to a sump at a lower level than the rest of the excavation. The water level must be maintained in the sump at a low enough elevation to keep the free water surface in the surrounding sand below the bottom of the excavation at every stage. If this cannot be accomplished, the bottom of the excavation becomes *quick.* Water appears in the form of springs, the sand begins to boil, the slopes may begin to slough, and the entire base of the excavation may rise. The ditches around the edge of the hole must be kept clear to prevent water from emerging near the toe of the slopes and causing the banks to collapse.

The preceding statements indicate that pumping from sumps may be a hazardous procedure. In loose or fine sands, it sometimes cannot be accomplished at all. In coarse or dense sands, excavations can often be made with success, but, if conditions get out of control and the bottom becomes quick, the sand beneath the entire foundation may be loosened and its bearing capacity permanently decreased. Therefore, on large jobs, it is becoming standard practice to drain the sand before excavation, either by means of well points or deep-well pumps, Chapter 7.

The quantity of water that must be pumped is a matter of importance because sufficient equipment should be provided at the beginning of the job to guarantee efficient removal of the water without the necessity for making additions and alterations during the progress of the job. Inadequate pumping capacity may lead to sand boils and instability of the bottom of the excavation.

The quantity of water that must be pumped depends upon the coefficient of permeability of the soil, the distance the water level must be lowered, and the dimensions of the area to be dewatered. If the excavation is fairly narrow and if the water table does not need to be lowered more than 15 ft, medium to fine sands can usually be drained by a series of well points spaced at about 3 ft with one 6-in.

pump for every 500 ft of cut. Each pump requires a motor of about 20 hp, and drainage is likely to take from 2 to 6 days.

Estimates of the quantity of seepage under more severe conditions may require pumping tests to determine the coefficient of permeability and demand the exercise of considerable judgment based on experience.

The time required to accomplish drainage depends upon the coefficient of permeability of the sand deposit. Only 2 or 3 days are necessary for relatively permeable soils (k greater than 10^{-3} cm/sec) whereas several weeks may be required for soils of moderate permeability ($k = 10^{-3}$ to 10^{-5} cm/sec). Silty sands and other materials having an effective size smaller than about 0.05 mm cannot be drained by ordinary well-point equipment, although somewhat finer soils can be drained if the pumps are capable of maintaining a vacuum in the system.

Establishment of Piers below Water Table. Piers may be carried below water level in open shafts or by means of caissons. The general techniques for making the excavations have been described in Chapter 10. Excavation of an open shaft through sand below water table may often be accomplished by pumping from sumps if the sand is of medium to high density. However, the process is likely to be associated with loosening of the sand. As a consequence, the bearing capacity may be impaired. Use of well points or drainage before excavation has given successful results in a number of instances but is commonly expensive.

When shafts are excavated by hand and kept dry by the use of sumps, the lining is commonly made of pervious material so that the small amount of water that seeps to the excavation from the sides can enter the shaft and be collected in the sump. This prevents the development of a head of water outside the shaft that may lead to the formation of boils or springs at the bottom of the lining.

Open caissons are commonly used to establish bridge piers. The sand is removed from the caissons by dredging. To prevent the rise of sand in the bottom of the caissons, the water level inside should be maintained at a considerable height above the free water surface outside. By this procedure, the flow can be maintained downward through the sand at the bottom of the caisson, and the material will not be loosened by the process of excavation.

If the excavation cannot be made by means of an open caisson, it may be carried out under compressed air. Since the pressure of the air in the working chamber is always greater than that of the water at a given depth, the sand is not loosened during the procedure, and the bearing capacity of the sand is not impaired.

14.7. Effect of Vibrations

Importance. Vibrations constitute the most effective means for compacting deposits of loose sand. By the same token, they constitute one of the most serious causes of excessive settlement of foundations located upon loose sands.

Compaction. If the relative density of a deposit of sand is too low for the establishment of a raft or footing foundation, it may be increased by several means. One of these is the driving of compaction piles. Another is the use of a patented method known as the *vibroflotation* process. According to this method, a heavy steel capsule containing an internal vibrator is lowered into the deposit of sand. At the same time, powerful jets of water are forced into the sand beneath the capsule. Under the combined action of jetting and vibration the device sinks rapidly and creates a crater at the ground surface. As the crater develops, it is filled with sand. Compaction to the depth of penetration for a diameter of 6 to 8 ft is generally quite satisfactory. The procedure may, under some circumstances, be less expensive than driving piles. It is most effective in clean medium to coarse sands, and is not effective in very silty sands or silts.

Settlement. The detrimental effects of vibrations are most apparent when the frequency of the vibrations is of the order of 1200 per minute. This frequency corresponds to the speed at which the rotating parts of steam turbines commonly operate. As a consequence, the settlement of foundations for such machines is a matter of concern and requires special considerations. Moreover, repeated vibrations or impact at lower frequencies may eventually produce large cumulative settlements even in sands of moderately high relative density.

The design of substructures to reduce settlement due to vibrations calls for much judgment and experience and is not yet on a scientific basis. Reduction of the allowable soil pressure is relatively ineffective in decreasing the ultimate settlement. In some instances, machine bases are constructed with very large dimensions so that they are capable of absorbing much of the energy. In other instances, piles or piers may be preferable. No uniformly reliable procedures for reducing the settlement have been found.

Frederick Baumann (1826–1921)

Prominent early Chicago architect, and author of a pamphlet entitled "The Art of Preparing Foundations for All Kinds of Buildings, with Particular Illustration of the 'Method of Isolated Piers' as Followed in Chicago." This pamphlet, which appeared in 1873, contains the first statements published in the United States that the areas of footings should be in proportion to the column loads, and that the centroid of the area of the base of each footing should coincide with the center of gravity of the loads acting on the footing. It also contains values of soil pressure recommended for the Chicago clays.

PLATE 15.

CHAPTER 15

Foundations on Clay

15.1. Significant Characteristics of Clay Deposits

Under various circumstances, footings, rafts, piles, and piers may be used to support structures on deposits of clay. In connection with each type, independent investigations are required to determine the factor of safety against a bearing-capacity failure and the amount of settlement to be expected.

The bearing capacity depends primarily on the shearing resistance of the clay, which may conservatively be taken as one-half the unconfined compressive strength, provided the clay is free of secondary structural defects such as hair cracks or slickensides. Unless the clay is very sensitive, the unconfined compressive strength can be investigated economically and with sufficient accuracy by means of laboratory tests on 2-in. samples taken continuously from drill holes by means of thin-walled tube samplers. On relatively small projects, the compressive strength can be estimated on the basis of the N values determined in the standard penetration test, with the aid of the correlations given in Tables 3.1 and 1.9. However, the standard penetration test is not so reliable an indicator of the compressive strength of clays as it is of the relative density of sands, and the tests on tube samples are preferred.

The settlement depends primarily on the compressibility of the clay which is intimately related to its history of loading. The compressibility can be evaluated most reliably by means of consolidation tests on representative undisturbed samples 4 to 6 inches in diameter, but the cost of obtaining such samples commonly precludes their use. In most instances, the compressibility of soft clays can be estimated with sufficient accuracy from the results of determinations of the Atterberg limits and natural water content, or tests on completely remolded samples. Large-diameter samples may be required from extrasensitive clay deposits and, occasionally, from preloaded deposits.

The techniques for estimating the settlement of a structure on a clay deposit are essentially the same, regardless of the type of foundation. Therefore, they are discussed under a single heading. Considerations of bearing capacity, on the other hand, differ somewhat for the various types of foundations. Consequently, the safe load for each of the various types is discussed separately in the articles that follow. The procedure for making settlement forecasts is then described. Finally, consideration is given to the forces that cause lateral movements of structures founded on clay.

15.2. Footings on Clay

Ultimate Bearing Capacity. Figure 15.1 represents a cross section through a long footing with width B, resting at depth D_f below the ground surface. The soil beneath the base of the footing consists of

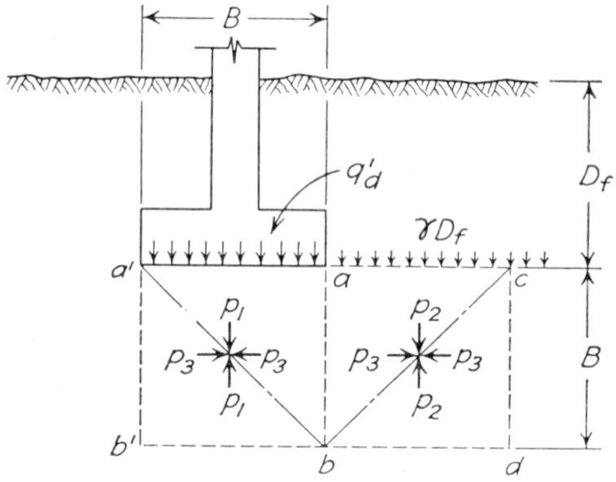

FIG. 15.1. Cross section through long footing on clay, showing basis for computation of ultimate bearing capacity.

intact clay without structural defects such as slickensides or cracks, and the degree of saturation is practically unity. Under these circumstances, the value of ϕ, eq. 2.32, may be considered as zero and the angle α, eq. 2.34, between the principal planes and the plane of failure as 45°. Equation 2.32 then reduces to

$$p_1 = p_3 + 2c = p_3 + q_u \qquad 15.1$$

This is the relation between the principal stresses at failure.

If the footing, Fig. 15.1, fails by breaking into the ground, the block $aa'b'b$ must fail and section ab must be displaced toward the right. However, this cannot occur unless the block $abdc$ also fails, whereupon ac rises and lifts the surcharge γD_f. The failure planes $a'b$ and bc are inclined at 45° to the horizontal. Hence, the depth of the zone of failure is equal to the width B of the footing.

On these assumptions, we may consider the stresses at mid-height of each of the two blocks that fail. Statics leads to the expression

$$p_2 = \gamma D_f + \tfrac{1}{2}B\gamma \qquad 15.2$$

and eq. 15.1 to the expressions for the failure conditions of the right- and left-hand blocks, respectively,

$$p_3 = p_2 + q_u \qquad 15.3$$

$$p_1 = p_3 + q_u \qquad 15.4$$

The intensity of loading that can be supported by the soil at the base of the footing is, therefore,

$$q_d' = p_1 - \tfrac{1}{2}B\gamma = \gamma D_f + \tfrac{1}{2}B\gamma + 2q_u - \tfrac{1}{2}B\gamma = \gamma D_f + 2q_u \qquad 15.5$$

If again we define the net bearing capacity q_d as the pressure that can be supported at the base of the footing in excess of that at the same level due to the surrounding surcharge, then

$$q_d = q_d' - \gamma D_f$$

and

$$q_d = 2q_u \qquad 15.6$$

The derivation given in the preceding paragraph is based on the simplifying assumption that the surface of failure consists of the planes $a'b$ and bc. In reality, the surface is curved. Moreover, the base of a real footing is rough, and shearing stresses are developed that alter the characteristics of the failure surface. The implications of these facts have been taken into account in advanced theories and in experimental studies, and the results condensed into the equation

$$q_d = 2.85q_u\left(1 + 0.3\frac{B}{L}\right) \qquad 15.7$$

which applies to a rectangular footing with a rough base having a width B and a length L.

Safe Soil Pressure. Under the maximum sustained loads that can reasonably be expected, the factor of safety against a bearing-capacity failure should be at least 3. Hence, the safe soil pressure q_a is equal to

$$q_a = 0.95q_u\left(1 + 0.3\frac{B}{L}\right) \qquad 15.8$$

Depending on the shape of the footing, the value of q_a may vary from $0.95q_u$ to $1.24q_u$. Hence, as a rough guide, the safe soil pressure may be considered equal to the unconfined compressive strength.

Because of the variations that normally occur in clay deposits, the value of q_u in eq. 15.8 should represent the average for a depth B below the base of the footing. The unconfined compressive strength should be determined at 6-in. intervals in the vertical direction. The strength of some of the samples may be estimated if the technician is experienced, but at least one out of every five samples should be tested.

According to eq. 15.8, the safe soil pressure on clay is practically independent of the width of the footing. This is in contrast to the safe soil pressure on sands, which increases with the width of the footing. Therefore, at a given soil pressure large footings on clay are no safer than small ones, and, if the strength of the soil decreases with depth, large footings are less safe. This fact leads to the warning that eq. 15.8 can be used for selecting the safe load on a single footing only if that footing is located sufficiently far from its neighbors so that their influence on the stresses in the subsoil beneath the footing is negligible. If one or more soft layers is located in the subsoil even at a depth greater than B below the base of the footing, a computation should be made to ascertain whether the pressure at the top of any of the soft layers exceeds the safe value for that layer. If it does, the design must be revised.

The pressure at the top of a deep layer of clay can be estimated by means of the influence chart described in Art. 15.6. However, it is often more expedient and sufficiently accurate to assume that the pressure at the base of the footing spreads out uniformly within the confines of a truncated prism with sides sloping outward from the edges of the footing at an angle of 60° with the horizontal.

If the clay contains a network of cracks or slickensides, the procedure based upon the unconfined compressive strength cannot be used because the strength of the clay depends upon the spacing and nature of the defects rather than upon the shearing resistance of the intact fragments themselves. Under these circumstances it may be necessary to resort to load tests.

The details of the load-test arrangement depend on the spacing of the cracks, the size of the footing, and the degree of uniformity of the clay. To make sure that a representative number of cracks is included, the test plate, Fig. 15.2, should be at least 2 ft square. The effect of surcharge around the plate should be eliminated. Hence, the test should be conducted in a pit with a width at least three times that of the loading plate. The load should be applied in increments, and after each increment the deformation should be observed until its rate becomes very small. The load-settlement relation should be determined, either to the point of failure or to a value of load three times that to be used for design.

One load test should be made at the level of the base of the footings, and enough additional tests at greater depth to establish the average value of the bearing capacity within the depth B. If the clay weakens consistently with depth, the safe load should be based on the lowest values of bearing capacity. On the other hand, if the variations are erratic, the average value may be used.

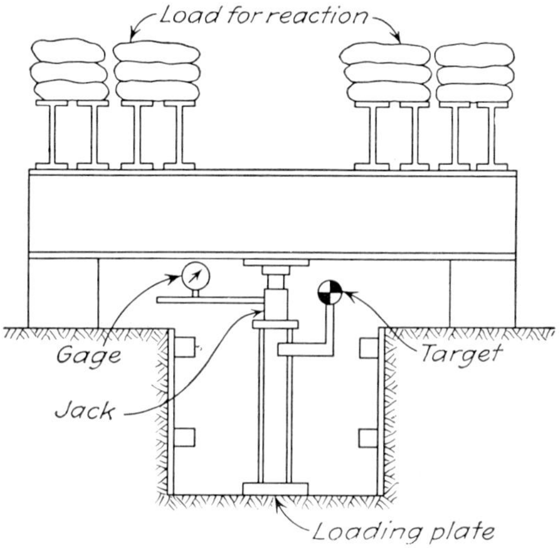

FIG. 15.2. Arrangement for load test on fissured clay.

The safe soil pressure q_t at the toe of a footing beneath which the soil pressure has a triangular distribution may be considered equal to q_a in eq. 15.8, provided one-third the width of the base is substituted for B, and provided q_u is the average unconfined compressive strength within a depth below the footing equal to one-third the width of the base. However, the influence of any soft layers at greater depth must not be overlooked.

Design Loads. Footings resting on clay should be proportioned to provide a factor of safety of 3 against a bearing capacity failure. At this value of the factor of safety the stresses in the clay immediately beneath each footing are not likely to lead to differential settlements among adjacent footings in excess of 0.75 in., provided the footings are sufficiently far apart that the action of each footing is independent of the others, and provided the soil directly beneath the footings is not a soft or very soft normally loaded clay. The loads that should be considered in selecting the sizes of the footings are those that normally act for a large proportion of the time, because the settlement of a footing on clay does not increase significantly during the short periods of loading. Exceptional values of live load due to improbable combinations of loading should not be included, nor should wind load, snow load in temperate climates, or earthquake forces. On the other hand, the factor of safety should not be less than 2 under the most severe conditions and combinations of loading specified in the building code. Hence, the maximum soil pressure under the most unfavorable loading combination should not exceed 1.5 q_a, eq. 15.8.

In many instances, except where stiff or hard clays are involved, the action of any one footing in a group is significantly influenced by the presence of the neighboring footings or other loaded areas. In this event the differential settlement of a footing foundation may be much greater than 0.75 in. Consequently, an investigation of the settlement of the entire foundation must always supplement the determination of the allowable soil pressure for the individual footings. These matters are considered in Art. 15.6.

Illustrative Design. DP C-3. Footings on Clay

The purpose of this design plate is to illustrate the use of various principles discussed in Art. 15.2. It is important to recognize that the dimensions of the footings are determined exclusively on the basis of the strength of the clay. The subject of settlement of foundations above clay is discussed in Art. 15.6.

The data obtained from a boring near the location of the footings are given on Sh. 1. The loads on the footings are also listed. It may be noted that for both footings the magnitude of the loads is such that, when a factor of safety of 3 is provided for dead load plus normal (reduced) live load, the factor of safety for the most unfavorable conditions is greater than 2. Had this condition of loading not existed, the footings would have been proportioned for the maximum loads at a factor of safety of 2.

The safe soil pressure is determined from eq. 15.8, in which the value of q_u is taken as the average unconfined compressive strength of the clay for a depth below the base equal to the width of the footing. Since eq. 15.7 furnishes the value of the bearing capacity in excess of the weight of the surcharge immediately surrounding the footing, the weights of the floor slab and soil above the elevation of the base are subtracted from the load on the footing. However, it is important to note that this subtraction very nearly offsets the weights of the floor slab and footing which were previously added to the

DP C-3
Building Footings
Sh. 1 of 2

Footing Design. General Data

Boring Data:

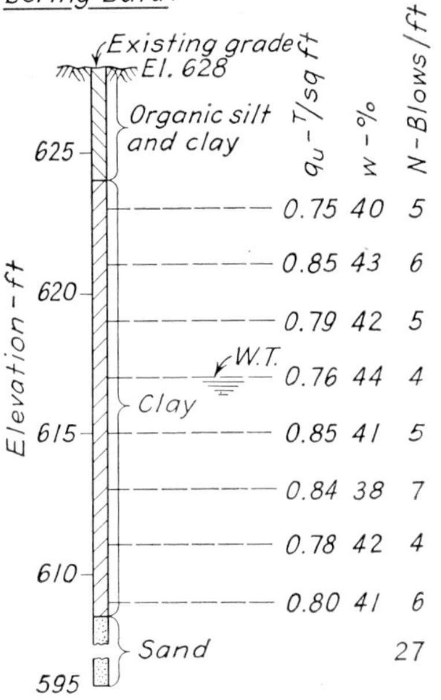

Typical Boring

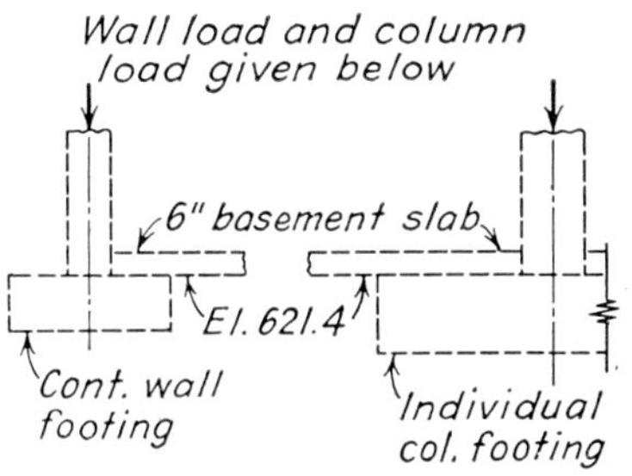

Wt of clay = 113 lb/cu ft

Loads on Continuous Wall Footing

DL	= 3.8 k/ft of wall	Factor of safety ≥ 3
Normal LL	= 2.4 " " "	" " " ≥ 2
Maximum LL	= 4.2 " " "	

Loads on Individual Column Footing

DL	= 92.0 k	Factor of safety ≥ 3
Normal LL	= 84.2 k	" " " ≥ 2
Maximum LL	= 140.1 k	

DP C-3
Building Footings
Sh. 2 of 2

Footing Design. Required Areas

Continuous Wall Footing:

DL = $3.8^k/ft$ DL = $3.8^k/ft$ $\frac{8.0}{6.2} = 1.29 < 1.5$, so, if
Normal LL = 2.4   Max. LL = 4.2   F.S. = 3 for $6.2^k/ft$
$6.2^k/ft$ $8.0^k/ft$ F.S. > 2 " $8.0^k/ft$

$3.1^T/ft$
@ say 0.75 = 4.1 ft

Try width = 4'-3" Assume depth = 1'-6"

Wall load = $3.1^T/ft$
÷ 4.25   = $0.73^T/sq\ ft$

Additional Load:
Floor slab = $0.5 \times \frac{0.15}{2}$
Footing = $1.5 \times \frac{0.15}{2}$
} = 0.15
$0.88^T/sq\ ft$

Surcharge Load:
Floor slab = $0.5 \times \frac{0.15}{2}$
Soil = $1.5 \times \frac{0.11}{2}$
} = 0.12
$0.76^T/sq\ ft$

Av. $q_u = \frac{0.79 + 0.76}{2} = 0.775$, so allow 0.95(0.775) = $0.74^T/sq\ ft$
say ok

Use 4'-3" width of wall footing

Individual Column Footing.

DL = 92.0^k DL = 92.0^k $\frac{232.1}{176.2} = 1.32 < 1.5$
Normal LL = 84.2   Max. LL = 140.1
176.2^k 232.1^k

88.1^T
@ say 1.0 = 88.1 sq ft

Try 9'-6" x 9'-6". A = 90.2 sq ft Assume depth = 2'-6"

Column load = 88.1^T
÷ 90.2   = $0.98^T/sq\ ft$

Additional load minus surcharge } $2.5\left(\frac{0.15 - 0.11}{2}\right) = 0.05$
$1.03^T/sq\ ft$

Av. q_u: $0.79^T/sq\ ft$
0.76
0.85
0.84
0.78
5 | 4.02
$0.80^T/sq\ ft$, so

allow 1.24(0.80) = $0.99^T/sq\ ft$
say ok

Use 9'-6" x 9'-6" col. footing

pressure produced by the load of the wall. Therefore, it is sufficiently accurate to divide the load from the wall or column by the area of the footing and to compare this pressure with the safe value determined by eq. 15.8. If, however, the computations for additional load minus the surcharge are not omitted, they may be simplified as explained on Sh. 2 of DP C-1. This procedure is followed for the column footing on Sh. 2 of DP C-3.

15.3. Rafts on Clay

Ultimate Bearing Capacity. The maximum pressure that can be sustained by the soil at the base of a raft on a deep deposit of clay is given by eq. 15.7. According to this equation, the load that produces failure is practically independent of the width of the raft. The unconfined compressive strength required to support with no margin of safety a soil pressure of 1 ton/sq ft is on the order of 0.3 ton/sq ft. To provide a factor of safety of 3, the required unconfined compressive strength is on the order of 1 ton/sq ft. This suggests that the factor of safety of most large rafts on soft clay deposits is relatively small.

The quantity q_d in eq. 15.7 is the pressure at the elevation of the base of the raft in excess of that exerted by the surrounding surcharge. By increasing the depth of excavation, the pressure that can safely be exerted by the building is correspondingly increased. This can be accomplished by increasing the number or depth of basements. On the other hand, the area of a raft cannot usually be enlarged appreciably in an attempt to reduce the soil pressure because it is not feasible to extend a raft more than a few feet beyond the building proper. Therefore, if a raft foundation is to be constructed at a site underlain by clay too soft to provide support at the normal basement level, the only practicable method for safely constructing the raft is to lower the elevation of its base.

Safe Soil Pressure on Rafts. The factor of safety against failure of the subsoil beneath a raft on clay should not be less than 3 under normal loads, or less than 2 under the extreme loads specified by the building code. Even with this requirement, the danger of breaking into the ground is many times greater than that for a similar structure on sand. In several instances complete failures have occurred because of overloading the clay subsoil of large rigid structures.

15.4. Piles in Clay

Conditions for Use. If the subsoil consists of clay too soft or too compressible for the establishment of footings or of a raft, the weight of the structure may be transferred to piles.

The manner in which a pile driven into a clay deposit can be expected to carry its load may be inferred from the characteristics of the curve showing the resistance of the pile to penetration at different depths. Typical penetration diagrams of this type are shown in Fig. 15.3. Curve *a* is characteristic for friction piles. The number of blows per foot is practically constant with depth and remains relatively small. On the other hand, diagram *b*, for point-bearing piles,

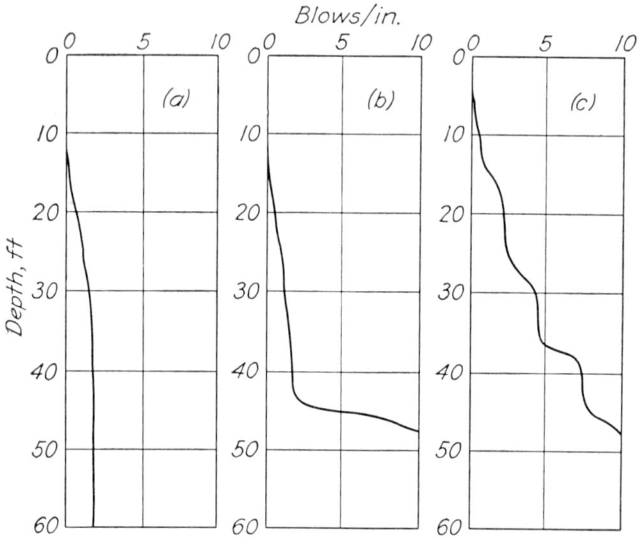

FIG. 15.3. Relation between blows per inch of penetration and total depth of penetration for wood piles driven into (*a*) soft clay of great depth; (*b*) soft strata to firm bearing in very stiff or hard clay; (*c*) clay strata increasing in stiffness with depth.

indicates relatively small resistance until the point encounters the firm layer whereupon the resistance increases very greatly. A typical diagram for a pile driven into a deposit of clay in which the strength increases with depth is shown in *c*. The resistance also increases with depth. If the pile has at least a small taper, the resistance is likely to become almost as great as that encountered by point-bearing piles driven into a layer of hard clay.

Safe Load on Friction Piles. The load that can be carried by a single friction pile in a deposit of soft to medium clay can best be determined by means of a static load test. The shape of the load-settlement curve is characterized by Fig. 15.4. The curve approaches a vertical asymptote corresponding to a sudden plunge of the pile into the ground or to a steady penetration of the pile under the applied load.

The load that causes failure of a friction pile is represented by the abscissa of the vertical tangent to the load-settlement curve. It has a well-defined meaning. The safe load for the single pile may be taken as one third of the failure load for the maximum probable loading, or as one half of the failure load for the extreme maximum loading specified by the building code, whichever is the smaller.

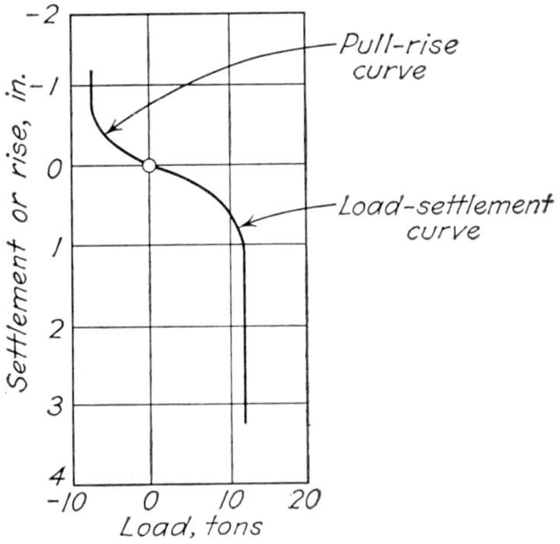

FIG. 15.4. Typical results of load test on friction pile in clay.

The load test may be supplemented by a pulling test. The corresponding pull-rise curve is also shown in Fig. 15.4. It has the same characteristics as the load-settlement curve. If the pile is cylindrical, the difference between the failure loads determined from the load-settlement and pull-rise branches is a rough measure of the point resistance.

Since load tests on piles are relatively expensive, numerous efforts have been made to compute the safe load on friction piles on the basis of pile-driving records or laboratory tests. Experience has demonstrated that pile-driving formulas give entirely erroneous results when the soil consists of clay, and the use of such formulas in this connection should be eliminated. According to the resistance diagram, Fig. 15.3*a*, the penetration per blow remains practically constant for the length of the pile. This implies, on the basis of pile-driving formulas, that the ultimate bearing capacity is independent of the length. In contrast, experience shows that the bearing capacity of a friction pile is approximately proportional to its length.

For cylindrical piles, it has been found that the ultimate bearing capacity is approximately equal to the product of the shearing resistance of the soil and the area of the embedded portion of the pile. The shearing resistance is approximately half the unconfined compressive strength. This relationship has been found more satisfactory than the pile-driving formulas, but it cannot be considered sufficiently reliable to eliminate the necessity for load tests on any but the smallest jobs.

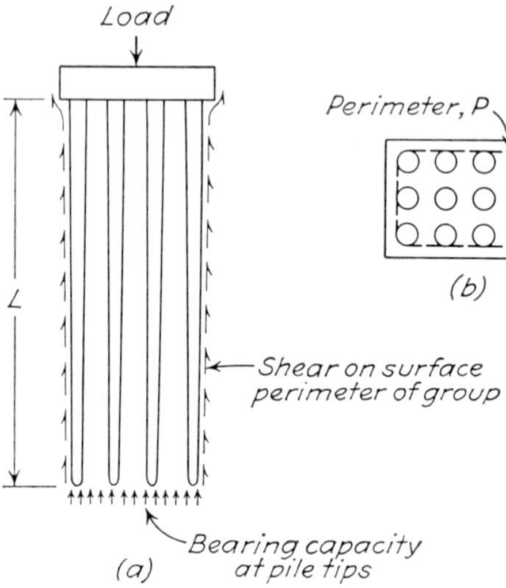

FIG. 15.5. (*a*) Section and (*b*) plan of group of friction piles in clay.

Friction piles are commonly driven in groups or clusters beneath individual footings or as single large groups beneath mats or rafts. The bearing capacity of a pile cluster may be equal to the number of piles multiplied by the bearing capacity per pile, but it may also be much smaller. Therefore, there is a possibility that a group of piles may fail when the load per pile is less than the safe load determined on the basis of load tests on a single pile. This condition must be investigated.

Figure 15.5 shows a cluster of friction piles in a deep clay subsoil. The piles are connected at the top by a rigid cap. If they fail, they must fail as a group, and the failure is likely to occur along a surface such as that indicated by the dash line. The area of the surface corresponds approximately to the product of the perimeter of the pile group and the length of the piles. The ultimate load that can be supported by the group is derived from two sources. The shearing resistance of the soil around the boundary of the pile group is equal to $\frac{1}{2}\ PLq_u$, where P is the perimeter of the group, L is the length of the piles, and q_u is the unconfined compressive strength. To this must be added the bearing capacity of the soil on the plan area of the group at the level of the tips of the piles. If the subsoil is uniform, this may be evaluated approximately by means of eq. 15.7. For long piles in small groups, this quantity is small compared to the shearing strength of the soil surrounding the pile group. On the other hand, as the number of piles increases, the bearing area under the group increases much faster than the shearing surface around the group. The pile spacing which utilizes the full capacity of each pile can be found most easily by trial. A spacing of three times the diameter of the piles is commonly selected as the trial spacing and checked against the criterion that the sum of the shearing and bearing capacities of the group of piles must be at least equal to the capacity of a single pile multiplied by the number of piles in the group.

According to the preceding paragraph, it is apparent that no benefits may be derived from driving additional piles within the boundary of a given group of friction piles if the spacing is equal to or less than that determined by the criterion stated above. On the other hand, it is obvious that the greatest benefits from a friction pile foundation will be obtained if the length of the piles is as great as possible within the limits of economy. In connection with the settlement of such foundations, it will also be found that the effectiveness of a friction pile foundation increases with increasing length of piles.

The ultimate capacity of a group of friction piles or of a single pile varies with time. Usually, the strength is a minimum immediately after pile driving. This is primarily the result of the disturbance of the structure of the clay. On account of thixotropic processes the strength is likely to be partly regained with time. Moreover, under the relatively high pressures produced in the clay by the insertion of the piles, the clay immediately surrounding the piles consolidates rapidly, and its shearing strength increases. This phenomenon is commonly recognized, and it is usually specified that load tests should not be made within 3 days after the piles have been driven. However, the increase in strength may continue for a much longer period of time, depending upon the nature of the clay. The results of a series of load tests on one pile are shown in Fig. 15.6. They represent a rather abnormal increase, but they accentuate the importance of the time effects in connection with clays.

Safe Load on Point-bearing Piles. If a deposit of weak or compressible material is located above a stratum of very stiff or hard clay, the loads may be transferred to the clay by means of piles. The general considerations regarding this type of foundation are the same as those that apply to piles driven into dense sands beneath weak or compressible material. The safe load per pile can be determined most accurately by means of a pair of load tests in one of which a pile is driven to bearing in the firm stratum, whereas in the other the point does not come within 2 or 3 ft of the bearing stratum. The use of a pile-driving formula in place of load tests cannot be recommended because the resistance to rapid penetration into clay

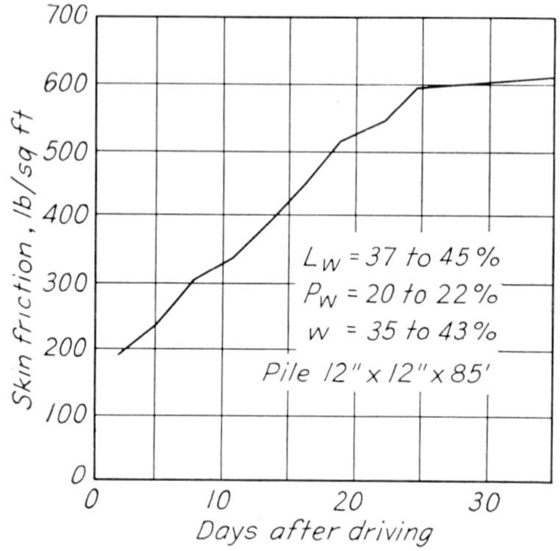

FIG. 15.6. Increase of ultimate bearing capacity of friction pile with time.

under the action of a pile driver is not equal to that under the application of a static load.

The safe load on a group of piles driven through compressible layers to firm material is equal to the number of piles in the group multiplied by the safe load per pile. No reduction need be made on account of close spacing of the piles. In fact, it may be preferable to keep the spacing as small as 2.5 times the diameter of the pile if there is a likelihood that negative skin friction may develop on account of consolidation of the soft deposit. The additional pile load due to the negative skin friction should be taken into consideration.

Safe Load on Piles in Clay of Increasing Stiffness with Depth. In many localities stratified deposits are found in which each member has approximately uniform characteristics but in which the stiffness of the members increases in a general way with depth. Piles are commonly used to transfer the load to a level below the most compressible upper layers. The depth to which the piles are driven depends primarily on the settlement that may be contributed by the materials below the pile tips. This aspect of the problem will be considered later. In general, it is economical to drive the longest piles compatible with the equipment readily available, in order to take advantage of both the increased strength and stiffness of the lower layers.

The load carried by a pile driven into such a deposit is supported partly by the material directly beneath the point of the pile and partly by skin friction and adhesion in the lower few feet where the pile is in contact with relatively stiff materials. The depth to which piles of a given type can be driven with equipment of a particular kind is extremely difficult to predict on the basis of soil investigations and tests. It can be established with reasonable certainty only by driving test piles. Moreover, the load that can be carried by the piles can be determined only by means of load tests on the test piles. During the driving of a test pile, the penetration record should be obtained because it serves during construction as a guide for indicating when the foundation piles encounter material having a resistance comparable to that offered to the piles on which the load tests were conducted.

In some localities where soil conditions are of this character, pile-driving formulas are used. This may be justifiable if the formulas have been adjusted by experience in the locality to indicate safe loads corresponding to various penetrations under the blows of standard hammers. However, use of the formulas should be confined strictly to the regions in which they were developed.

The safe load on a group of piles driven into progressively stiffer materials is practically equal to the number of piles times the safe load per pile.

Uplift Resistance of Piles in Clay. The resistance to uplift of a single pile or of a group of piles in plastic clay is governed by the same considerations that apply to the safe load on friction piles; the bearing capacity of the soil at the points of the piles is, of course, no longer existent.

Illustrative Design. DP C-4. Pile Foundation in Clay

The computations presented in this design plate demonstrate the application of the criterion for the design of groups of friction piles in clay. In the first step, the number of piles has been determined by assuming that the individual capacity of each pile can be developed in the group. Second, the capacity of the group has been investigated and found to exceed the total load on the group with the

DP C-4
Pile Footing
Sh. 1 of 1

Pile Footing Design. Required Number and Spacing

General Data:

Total load on pile group = 220^T including wt. of pile cap

Borings indicate fairly uniform clay to a depth of 100 ft. The clay has an average $q_u = 0.9^T$/sq ft. The sensitivity of the clay is low, and its water content is considerably below the liquid limit

A factor of safety of 3 is desired, using 40 ft piles, having an average diameter of 12 in.

Individual Action:

Surface perimeter = 40 × 3.14 × 1 = 125.6 sq ft/pile

Shearing strength of clay = $\frac{0.9}{2}$ = 0.45^T/sq ft

÷ 3 = 0.15^T/sq ft

Safe pile load = 125.6 × 0.15 = 18.8^T

Piles required = $\frac{220}{18.8}$ = 11.7 Use 12 piles

Group Action:

Try 3 rows of 4 per row @ 3'-0" ctrs. both ways

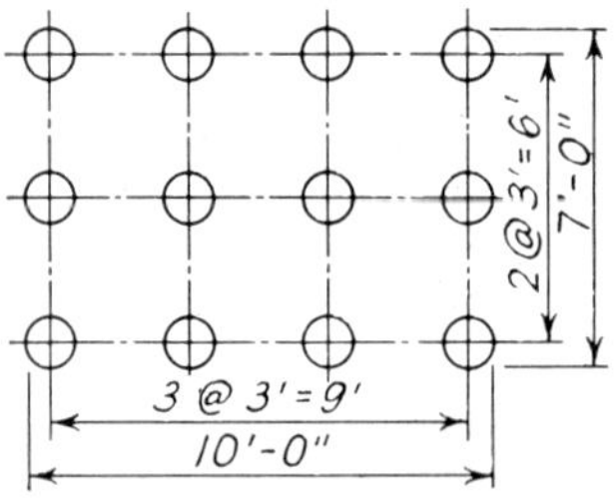

Surface perimeter =
40 × 34 = 1360 sq ft
Total shear = × 0.15 = 204^T

Plan area of group =
7 × 10 = 70 sq ft
$\times 0.95(0.9)(1+0.3\times\frac{7}{10})$ =*

Safe base load = 72^T

Total safe group load = 204 + 72 = 276^T > 220^T ok

Use 12 piles spaced as shown

* Eq. 15.8.

appropriate factor of safety. Hence, a satisfactory design is indicated. In fact, the spacing of the piles could be decreased to $2\frac{1}{2}$ ft before the capacity of this group would become less than 220 tons. However, whether this reduced spacing should be used is a decision that can be made only if one is able to anticipate the behavior of the clay during the driving of the piles.

It should be noted that the strength of the clay provides the basis for the computations in this design plate. The problem of settlement of friction piles in clay is discussed in Art. 15.6 and illustrated in DP C-6.

15.5. Piers on Clay

The safe load on the base of a pier resting on a deposit of clay may be determined by the same rules that apply to footings. The factor of safety against failure should be equal to 3 under the maximum conditions of loading that may normally be expected, and should not be less than 2 under the most severe combinations of loading specified by the building code.

Since the net pressure is a function of the total weight of the pier, it is often advantageous to construct the pier with a hollow space in its interior.

When piers are constructed by sinking caissons through soft clay deposits, it is important to have a knowledge of the skin friction in order that the caisson may not become stuck at some stage of its sinking. Theoretical methods for estimating the skin friction against such caissons have not been found sufficiently reliable for practical use. Most of the available information has been obtained from a knowledge of the loads required to start caissons that had become stuck. The unit skin friction ranges from 150 to 600 lb/sq ft in silt and soft clay to values between 1000 and 4000 lb/sq ft in very stiff clay. Even when data are available from experience on nearby jobs, care should be taken in applying the information to a new project because the total skin friction depends upon a number of factors including the shape of the lower part of the caisson, on its diameter, and on the method of excavation.

15.6. Settlement of Foundations above Clay

Introduction. In connection with all types of foundations on clay soils, it is essential to satisfy the requirement that the factor of safety be adequate against a bearing capacity failure. However, it is equally important to make a reliable estimate of the amount of differential settlement that may be experienced by the structure even if the soil pressure does not exceed the safe load. If the differential settlement is found to be excessive, it may be necessary to change the type of foundation. The purpose of a settlement forecast is to obtain a reliable conception of the differential settlements in order to determine whether or not the foundation layout under consideration is satisfactory.

Settlement forecasts based on the results of soil tests and theory have been found sufficiently reliable if the subsoil contains one or more layers of soft clay located between stiffer layers, provided the clay is normally loaded. If the clay is preloaded, the estimates are somewhat less reliable but are conservative because the estimated settlements are greater than those that may actually develop. Even if the subsoil consists of a fairly homogeneous deposit of clay to considerable depth, fairly reliable results can be obtained by assuming the deposit to consist of a number of separate layers. On the other hand, if the deposit contains various elements of compressible soil arranged in an erratic manner, a settlement forecast can do no more than indicate the maximum differential settlement that is probable without furnishing a conception of the distribution of the settlement.

In most instances a high degree of accuracy is not warranted. The proper choice of foundation can usually be made if it is known whether the differential settlement will be of the order of ½ in., 2 in., or 20 in.

The general method for computing the settlement of a point on the base of a structure due to the consolidation of a thin layer of clay involves four steps. The first of these is the computation of the original intergranular pressure p_0 at mid-height of the layer. The second is the computation of the increase in pressure Δp, directly below the given point and at mid-height of the clay layer, caused by the construction of the building. The third is the determination of the compressibility of the clay as described by the compression index C_c. Finally, these quantities are utilized in a computation of the decrease in thickness of the stratum at the point in question. If there is only one thin layer, the surface subsidence is assumed equal to this decrease in thickness. If there are several layers, similar computations are performed for each layer, and the decreases in thickness of the various layers below the point are added.

The computation of the original intergranular pressure p_0 and the determination of the compression index C_c have been described in Arts. 2.4 and 2.8. In the following paragraphs, methods will be developed for determining the increase in stress Δp and for computing the decrease in thickness of the clay layer.

Computation of Pressure. Since the stresses in the subsoil of a structure having an adequate factor of safety against a bearing capacity failure are relatively small in comparison with the ultimate strength of the material, a reasonable approximation to the manner in which the vertical load is distributed through the subsoil can be obtained on the assumption that the soil is elastic in its behavior. If the nonuniformity of the subsoil is also neglected, the vertical stress produced at any point in the soil mass by the application of a load on the surface of the ground can be computed by one of the equations developed by Boussinesq on the basis of the theory of elasticity. It is very likely that the real pressures in the soil mass differ appreciably from those computed by the means of the theory, but it does not seem likely that direct experimental evaluation of the order of accuracy of the computations can ever be obtained. However, because of the simplicity of the equations and the comparative reasonableness of the assumptions, Boussinesq's method is almost universally used, and the results are justified by the general agreement between observed settlements and those computed with the aid of the theory.

Since the load is commonly applied to the subsoil through footings, piers, or rafts that are assumed to exert a uniform pressure on the soil, the engineer commonly wishes to know the vertical pressure produced at some point on a horizontal plane in the interior of the soil mass by a load uniformly distributed over a portion of the ground surface. To solve this problem, a graphical representation of Boussinesq's equation was developed by N. M. Newmark. It consists of an influence chart, Fig. 15.7. The chart represents a map of the ground surface drawn to such a scale that the distance AB on the map is equal to the depth below the surface at which the value of the pressure is to be determined. When this condition is satisfied, a uniform load on any one of the "square" areas bounded by two adjacent circular arcs and two adjacent radii produces a pressure at a point directly below the center of the chart equal to 0.005 times the intensity of the uniform load. In determining the depth corresponding to the distance AB, the ground surface is always considered to be the plane through the base level of the foundation.

If, for example, the chart is to be used for computing the increased pressure at a depth of 40 ft below footing level due to construction of a building on footing foundations, the footing plan is drawn on a sheet of tracing paper to such a scale that the distance AB equals 40 ft. The point on the building beneath which the pressure is to be determined is placed over the center of the chart. The number

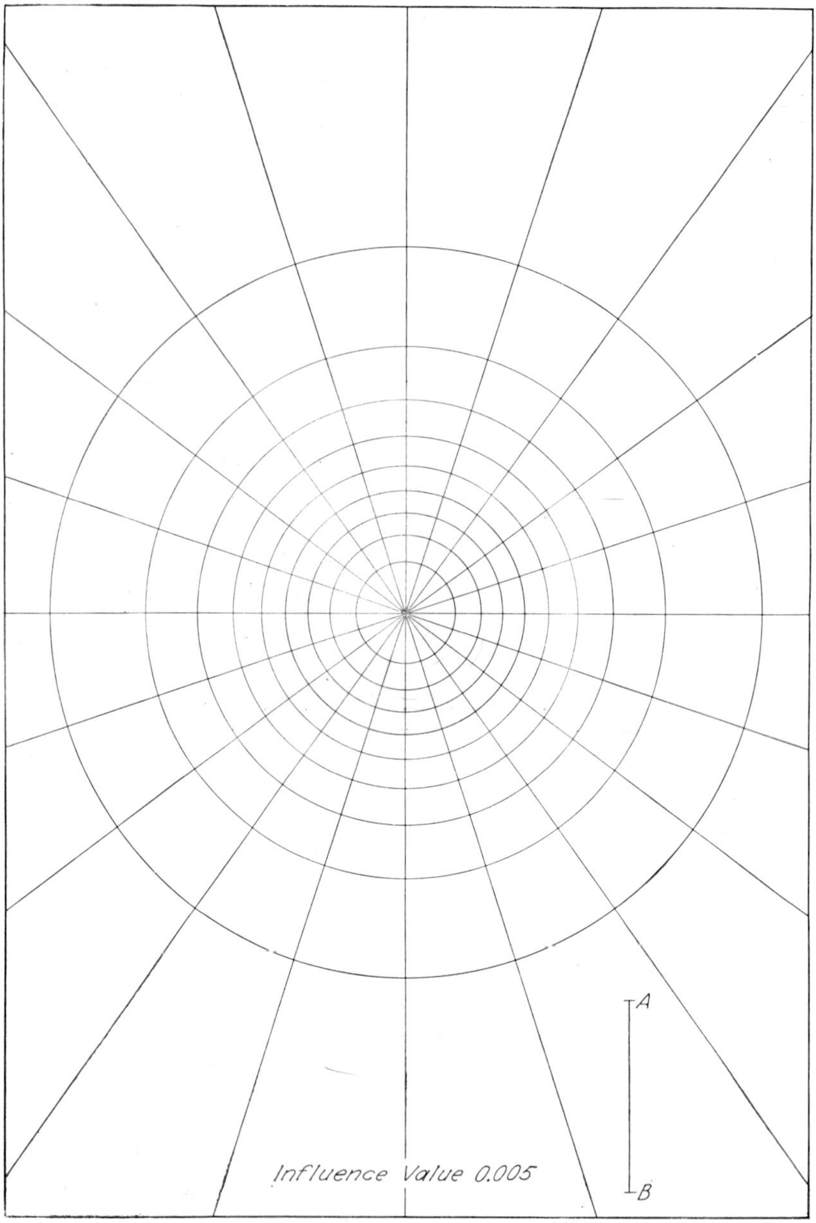

FIG. 15.7. Influence chart for computation of vertical pressure (after N. M. Newmark).

of squares covered by loaded footings is counted. This number multiplied by $0.005q$ equals the increase in pressure Δp due to the uniform load q exerted by all the footings.

To obtain the pressure Δp at various other points in the subsoil, the tracing paper is moved so that each of these points is successively below the center of the chart, and the counting operation is repeated. In this manner, the distribution of the pressure Δp at a given elevation below the footings can be determined very readily. To obtain the pressures at a different depth, a new tracing must be prepared of the foundation plan drawn to such a scale that the new depth is equal to the distance AB.

The reduction in pressure at any point caused by the excavation of the basement may be computed in the same manner on the assump-

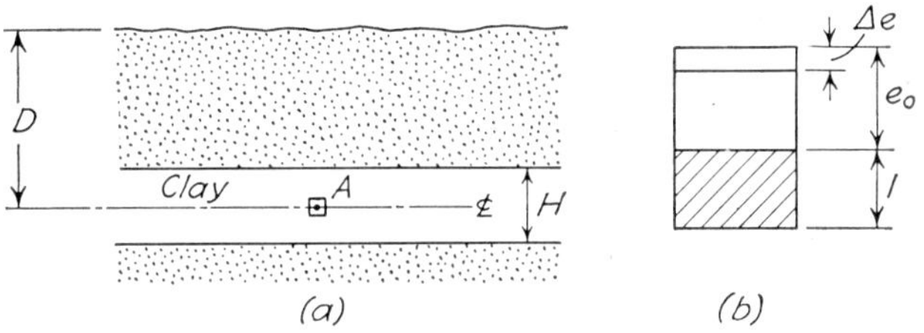

FIG. 15.8. (*a*) Section through compressible clay layer; (*b*) compression of element of layer.

tion that the weight of the excavated soil is a uniformly distributed upward load acting at the level of the bottom of the excavation.

The increase or decrease of pressure Δp at mid-height of a thin clay layer causes consolidation or swelling of the clay. Considerable time is usually required for these processes. Hence, temporarily applied loads have little influence upon the settlement of a structure above clay strata. The loads that should be considered in a settlement computation commonly include the upward pressure due to the weight of the removed soil, the dead weight of the structure and any backfill within it, and the average live load that can be expected over fairly long periods during the life of the structure. If construction of the building permanently changes the water table, the change in intergranular pressure due to this cause must also be incorporated into the quantity Δp.

Computation of Settlement. A cross section through a layer of clay with a thickness H and with its mid-height located at a depth D below the original ground surface is shown in Fig. 15.8*a*. The original intergranular pressure at a point A is equal to p_0, and the increase in pressure is Δp. The void ratio of the clay before construction is equal to e_0.

Figure 15.8*b* shows a prismatic element containing point *A*. The element can be assumed to consist of solid matter with a height equal to unity and a void space with an additional height equivalent to e_0. The total height of the element is, therefore, $1 + e_0$. If the void ratio decreases an amount Δe on account of consolidation, the unit strain of the element is $\Delta e/(1 + e_0)$. On the assumption that this strain is constant from top to bottom of the clay layer, the decrease in thickness of the layer, or the settlement *S* above point *A*, is given by the equation

$$S = H \frac{\Delta e}{1 + e_0} \qquad 15.9$$

According to eq. 2.23, the change in void ratio Δe can be expressed as follows:

$$\Delta e = C_c \log_{10} \frac{p_0 + \Delta p}{p_0} \qquad 15.10$$

By substituting this quantity into eq. 15.9, we find the following expression for the settlement of the ground surface above point *A* due to consolidation of the layer with thickness *H*.

$$S = \frac{C_c}{1 + e_0} H \log_{10} \frac{p_0 + \Delta p}{p_0} \qquad 15.11$$

If the pressure Δp is negative, the clay would be expected to swell and the quantity C_c would be replaced by an index of the swelling. However, as a rule, the swelling of a natural soft clay deposit due to relief in pressure caused by excavation is very small. As a consequence, the value of *S* may usually be taken as zero if Δp is negative.

Reliability of Settlement Forecast. The ultimate differential and total settlement can be predicted with a high degree of reliability if the subsoil consists of normally loaded clay of ordinary sensitivity and if the clay is practically homogeneous or consists of homogeneous layers. The compression index may be determined by means of consolidation tests on undisturbed or remolded samples, or by means of eq. 2.24. If the deposit has an erratic structure the pattern of the differential settlement cannot usually be predicted. It may often be possible to estimate only the maximum and minimum probable settlements, irrespective of the parts of the structure beneath which they will occur. This may be done by assigning to the clay the maximum compression index compatible with the results of the soil exploration within that part of the subsoil where Δp is likely to be a maximum, and the minimum compression index where Δp is likely to be a minimum. The real differential settlement is not likely to exceed that computed on this basis.

If the deposit is preloaded but this fact is not recognized, the computed settlements are likely to be substantially more than the real ones (see Art. 2.8). The discrepancy decreases as the ratio $\Delta p/p' - p_0$ approaches unity. However, even if the computation is based on the e-log p curves for good undisturbed samples, the computed settlements are likely to exceed the real ones, and the amount of the discrepancy cannot often be predicted reliably.

Extrasensitive clay deposits call for the best possible samples. Settlement computations for nonerratic deposits of this type, however, can usually be made with a high degree of accuracy.

Since the principal source of error in connection with settlement computations is the misjudgment of the degree of preloading of clay deposits, all available evidence should be utilized. In many instances it is enlightening to estimate the values of Δp at critical points in a clay deposit beneath an older, neighboring structure of known behavior. For example, if such a structure has not experienced detrimental settlement and if it produces stresses in the subsoil not less than those expected beneath the proposed structure, the satisfactory behavior of the new structure may usually be inferred. Such a conclusion should be verified by investigations to make sure that the subsurface conditions at the two sites are actually comparable.

Rate of Settlement. Equation 15.11 furnishes the ultimate settlement due to the primary consolidation of a clay layer. The rate at which the settlement will occur can be predicted with considerable accuracy if the coefficient of consolidation c_v for the clay is known and if the distance between drainage layers is known. Many clay deposits contain a large number of sand or silt layers or lenses that drain the deposit to a greater or less degree. Unless the continuity of these layers can be established and unless their spacing can be learned, it is not possible to estimate the rate of settlement except within very wide limits. The estimate can be made by the procedures described in Art. 2.8.

In connection with building construction, the rate of settlement is not often a matter of primary concern because the damage will eventually occur if the final differential settlements are greater than those the structure can stand. Occasionally, however, an estimate of the rate of consolidation is necessary. For example, if a structure has been damaged by increasing differential settlement, underpinning operations to arrest the movement may be contemplated. Before undertaking such a costly venture, it is desirable to learn the degree of consolidation that has occurred. In several instances, buildings have been underpinned at great expense at a time when practically all the primary consolidation had occurred. Knowledge of the degree of consolidation would have indicated that the underpinning was unnecessary.

Secondary Consolidation. The foregoing procedures for predicting the settlement of structures are based upon the assumption that the ultimate settlement will correspond to the primary consolidation observed in a consolidation test. The secondary compression is ignored. In most instances the primary consolidation is much larger than the subsequent secondary compression, and the settlement forecasts serve their purpose adequately. However, the slow settlement that occurs after the excess porewater pressure has been dissipated continues at a nearly constant rate for many years. Ultimately the secondary compression may become very important.

There are no theoretical means for estimating the rate of secondary compression because the basic nature of the slow deformation is not yet thoroughly understood. The best procedure available for estimating its magnitude is to observe the behavior of similar buildings on the same deposit. The rate appears to vary from about $\frac{1}{16}$ in. per year to over $\frac{1}{2}$ in. per year in unusual instances.

Organic clays and plastic silts are often so highly compressible that their consolidation does not take place in accordance with the assumptions upon which the theory of consolidation is based. The primary consolidation may occur very rapidly and may be relatively small. All the rest of the settlement may have the characteristics of secondary compression. For this reason, the secondary compression is given particular attention in such countries as Holland and Belgium where large organic deposits are found.

Settlement of Footing and Raft Foundations. Footings located above soft, normally loaded clay deposits commonly settle excessively even if the soil pressures are relatively low. The amount of settlement can be estimated by the procedures described in the preceding sections. The computations are made on the assumption that the superstructure has no rigidity. As a consequence, the differential settlements indicated by the forecast are likely to be somewhat greater than those that actually occur. However, the forecast can be expected to be sufficiently reliable to permit the designer to estimate whether the deformations are too great for his structure. If they are, he must either combine his footings into a raft or provide the structure with pile or pier support.

The settlement of individual footings is likely to exhibit a somewhat erratic pattern because of local soft or stiff spots that may exist beneath some of the footings. If a raft foundation exerts the same pressure per unit of area on the clay, the stresses beneath the foundation are increased to a considerably greater depth than beneath the footing foundation, and the soft and stiff spots largely compensate for each other. As a consequence, the settlement pattern of a raft foundation is likely to be much more uniform than that of a footing foundation on the same material and is likely to have a bowl-shaped character with the largest settlement near the middle.

The settlement of a raft foundation above clay can be reduced by lowering the elevation of the raft and thus increasing the amount of excavation. It is often possible to compensate for the entire weight of the structure by the weight of the soil excavated for the basements. In this manner structures have been built successfully on very soft clay deposits. In spite of the complete load compensation, they are likely to experience at least a small amount of settlement because of the heave and subsequent settlement of the clay during excavation and reloading, and because of the slight increase in compressibility caused by disturbing the structure of the clay during excavation.

In some instances, a layer of dense sand or stiff clay overlies a deep deposit of soft clay. A structure may be founded on footings resting on the sand or stiff clay. The safe load on the footings is determined by the characteristics of the material directly beneath them for a distance equal to the width B of the largest footing. However, the entire layer of sand or stiff clay acts as a natural raft, and the settlement of the structure as a whole depends upon the increase of stress in the soft deposit and on the compressibility of the deposit. The forecast of settlement may be made by the same procedures that are used in connection with rafts. If the differential movements will be too great for the structure, the situation can rarely be improved by changing the size of the footings. It is necessary to resort to a foundation on which the loads are compensated by the weight of the excavated soil, or to use piles or piers extending through the upper crust and the underlying compressible material to a firm base.

Settlement of Friction Piles in Clay. Friction piles may be used in clusters to support footings, or they may be placed beneath an entire raft foundation. Their function is to transmit the loads to a considerable depth in the subsoil where the values of the intergranular pressure p_0 are larger than near the surface. By so doing, the piles reduce the settlement to less than that which would be experienced by footings without piles. Moreover, the footing load is gradually transmitted to the soil throughout the length of the piles by means of skin friction. In this manner, large concentrations of stress are avoided.

As a rough approximation, the settlement of a group of friction piles can be computed on the assumption that the clay contained between the top of the piles and their lower third point is incompressible and that the load is applied to the soil at the lower third point of the piles. The presence of the piles below this level is ignored.

It is apparent that friction piles are most beneficial if their length is great compared to the width of the footing or the raft that they support. Long friction piles beneath a relatively small footing may greatly reduce the settlement whereas a very large number of short friction piles beneath an extensive mat foundation may have practically no beneficial effect.

At a constant load per pile the settlement of a group of friction piles increases with increasing number of piles in the group. This occurs for the same reason that the settlement of a footing on clay increases with the size of the footing. Efforts have been made to compensate for the increase in settlement by reducing the allowable load per pile in some arbitrary manner as the number of piles in the group is increased. The expressions for accomplishing the reduction are known as efficiency equations. None of the existing equations, however, takes into account all the variables that are likely to be associated with clay deposits, and their use is not recommended. It seems preferable to make rough estimates of the settlement on the basis of the rule stated above and to proportion the pile groups with the aid of judgment.

Settlement of Point-bearing Piles and Piers. The load on a pile driven through compressible strata into stiff or very stiff clay is ultimately carried by the point of the pile. This produces a large concentration of stress near the point. If a load test of customary duration is made on a single pile, the results may be encouraging because during the test most of the load is likely to be carried by skin friction in the upper layers and very little load may be transmitted to the point. Nevertheless, the settlement of the same pile beneath a building might eventually become quite large.

It is seldom possible to make a satisfactory estimate of the amount of settlement, although as a rough approximation settlement computations based upon consolidation tests on samples of the stiff clay may be useful. The best results could probably be obtained by performing a load test on a plate resting on the stiff clay at the bottom of a hollow pile. The plate should be porous to permit drainage and the load test should be carried out for a period of several weeks to permit consolidation to occur. Obviously, such tests would be very expensive and could be made only on the most important jobs. If the settlement would be excessive, the load per pile would have to be reduced.

If point-bearing piles are driven into a dense or stiff stratum that overlies a softer deposit, the bearing stratum should be considered as a natural raft and the settlement computed by the procedures described above.

While point-bearing piles are being driven on one part of a job, other similar piles driven earlier may be lifted off the bearing stratum on account of the upward movement of the clay displaced by the additional piles. The lifted piles must be redriven to firm bearing in the supporting stratum or else the settlement may be excessive. To determine whether redriving is necessary, it is advisable to establish the elevation of the top of each pile when it is driven and to check it after all the piles have been driven. If the piles are not rigid units, the movement of the top may not be an indication of the movement of the point. This is the condition, for instance, when the shells for cast-in-place concrete piles are temporarily left unfilled. The heave should then be observed with the aid of pipes known as *telltales* extending inside the shells from the points to an elevation near the ground surface where the movements can be measured.

The settlement of piers on normally loaded soft clays is likely to be large even if the net load at the base is small. Therefore, piers are not economical or satisfactory unless they rest upon stiff or hard preloaded clays. Even on fairly stiff clays, the progressive settlement of very large piers may become important with the passage of time. Settlement computations based upon the results of consolidation tests are likely to indicate greater settlement than will actually occur because of the difficulties associated with determining the compressibility of preloaded clays.

Settlement of Structures on Erratic Deposits. In the preceding discussions, it has been assumed that the consolidating clay occurs in the form of well-defined layers or in fairly uniform masses that can arbitrarily be divided into layers. The properties of any layer have been assumed to vary only slightly from point to point in horizontal directions.

In many instances, deposits of soft clay and plastic silt are heterogeneous in character. Attempts to compute the distribution of settlement over the area to be occupied by a structure are very unlikely to be successful because the location of all the soft spots can hardly be learned by sampling and testing. A preferable procedure is to make a comprehensive series of subsurface soundings at the site to define the outlines of the soft pockets and to get a general conception of the variations in consistency. A few borings are then made at the softest and stiffest places and rough estimates of settlement are based on these extreme conditions. These are generally sufficiently reliable to demonstrate whether or not the proposed type of foundation will be satisfactory.

Illustrative Design. DP C-5. Settlement of Raft Foundation above Clay

The purpose of this design plate is to illustrate the application of some of the principles discussed in preceding sections of Art. 15.6. The computations are for the most part self-explanatory. However, it should be noted that, since the water content of the clay is closer to the liquid than to the plastic limit, it is advisable to assume that the clay is normally loaded. The initial void ratio, the initial pressure at the middle of the clay layer, and the compression index are found in accordance with the discussions in Arts. 1.6, 2.4, and 2.8, respectively. It must be recognized that the differential settlement between the center and a corner of the raft may be considerably less than that indicated by the computation on Sh. 2 because of the stiffness of the superstructure.

Illustrative Design. DP C-6. Settlement of Friction Piles in Clay

An estimate of the settlement of a group of friction piles in normally loaded clay is presented in this design plate. The total settlement of 5.3 in., Sh. 2, should be regarded as only a rough approximation. Nevertheless, the magnitude of the settlement predicted in this manner usually gives the designer a satisfactory basis for deciding whether the foundation can be considered adequate or should be rejected in favor of one of some other type, such as longer piles driven to rock or a caisson resting on rock.

It may be noted that the properties of the clay that affect its compressibility change significantly at elevations 308 and 300. For this

DP C-5
Raft Foundation
Sh. 1 of 2

Raft Foundation Design. Computation of Settlement

General Data:

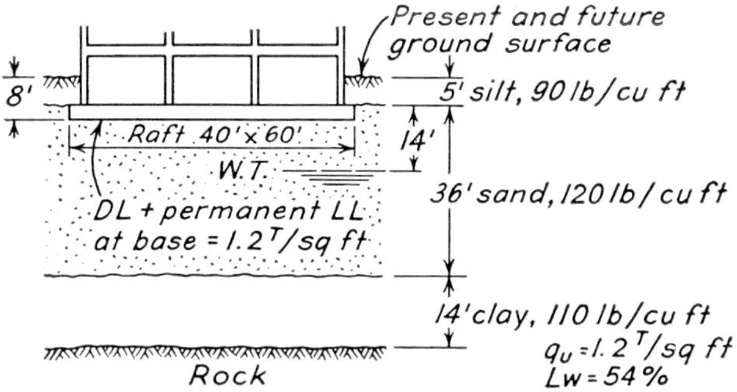

$q_u = 1.2\ ^T/sq\ ft$
$Lw = 54\%$
$Pw = 26\%$
$w = 44\%$
$s_s = 2.7$

Computations for p_o at Mid-depth of Clay:

$5 \times 0.090 =$	$0.45\ ^k/sq\ ft$
$14 \times 0.120 =$	1.68
$22 \times (0.120 - 0.0625) =$	1.26
$7 \times (0.110 - 0.0625) =$	0.33
	$3.72\ ^k/sq\ ft = p_o$

Compression index $= C_c = 0.009\,(54 - 10) = 0.396$

Initial void ratio $= e_o = 1.19$

(Note that for saturated soil, $e = ws_s/100$)

Net Load at Base of Raft:

DL + permanent LL $= 2.40\ ^k/sq\ ft$

Soil removed:

Silt $= 5 \times 0.090$
Sand $= 3 \times 0.120$ } $= 0.81$

$1.59\ ^k/sq\ ft$ = net pressure

DP C-5
Raft Foundation
Sh. 2 of 2

Raft Foundation Design. Computation of Settlement

Computations for Δp at Mid-depth of Clay:

Use sketch of 40′ x 60′ raft to scale 1″ = 40′
Use Fig. 15.7

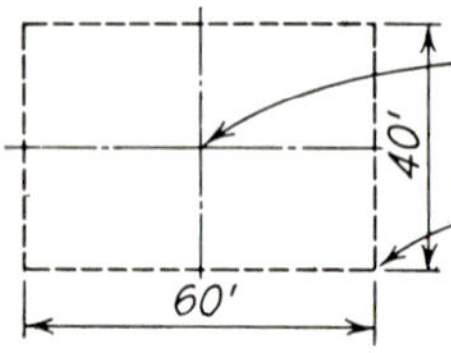

85.5 squares $\times 1.59 \times 0.005 = \underline{0.68\ ^k/sq\ ft}$

38.3 squares $\times 1.59 \times 0.005 = \underline{0.30\ ^k/sq\ ft}$

Settlement Computations:

At center $\dfrac{14' \times 12''}{1 + 1.19} \times 0.396 \times \log \dfrac{3.72 + 0.68}{3.72} = \underline{\underline{2.2''}}$

At corner: $\dfrac{14' \times 12''}{1 + 1.19} \times 0.396 \times \log \dfrac{3.72 + 0.30}{3.72} = \underline{\underline{1.0''}}$

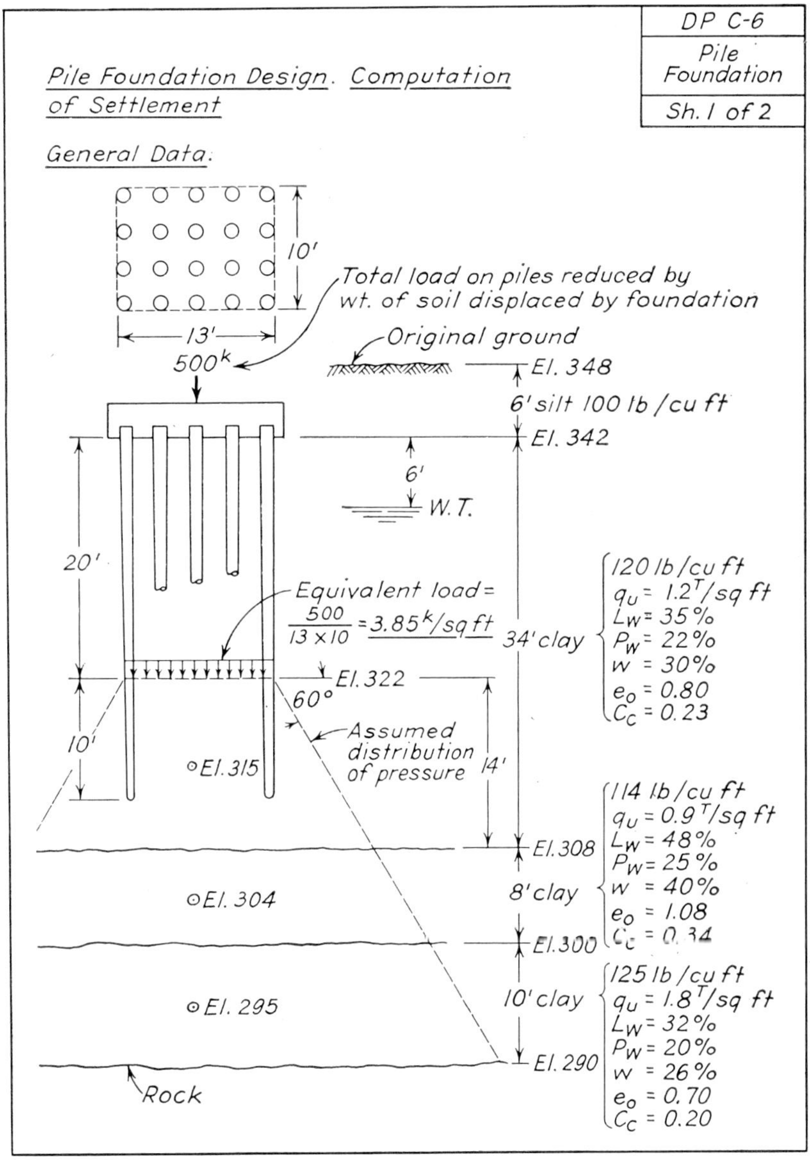

DP C-6
Pile Foundation
Sh. 1 of 2
Pile Foundation Design. Computation of Settlement
General Data:
10'
13'
Total load on piles reduced by wt. of soil displaced by foundation
Original ground
500k
El. 348
6' silt 100 lb/cu ft
El. 342
6'
W.T.
20'
Equivalent load = 500/(13 × 10) = 3.85k/sq ft
34' clay
120 lb/cu ft
qu = 1.2T/sq ft
Lw = 35%
Pw = 22%
w = 30%
eo = 0.80
Cc = 0.23
El. 322
60°
Assumed distribution of pressure
14'
10'
El. 315
114 lb/cu ft
qu = 0.9T/sq ft
Lw = 48%
Pw = 25%
w = 40%
eo = 1.08
Cc = 0.34
El. 308
8' clay
El. 304
El. 300
125 lb/cu ft
qu = 1.8T/sq ft
Lw = 32%
Pw = 20%
w = 26%
eo = 0.70
Cc = 0.20
10' clay
El. 295
El. 290
Rock

DP C-6
Pile Foundation
Sh. 2 of 2

Pile Foundation Design. Computation of Settlement

Computations for Values of p_o:

$$
\begin{aligned}
6(100) &= 600 \text{ lb/sq ft} \\
6(120) &= 720 \\
21(120-62.5) &= 1207 \\
&\overline{2527} \text{ lb/sq ft} = p_o \text{ at El. 315} \\
7(120-62.5) &= 403 \\
4(114-62.5) &= 206 \\
&\overline{3136} \text{ lb/sq ft} = p_o \text{ at El. 304} \\
4(114-62.5) &= 206 \\
5(125-62.5) &= 312 \\
&\overline{3654} \text{ lb/sq} = p_o \text{ at El. 295}
\end{aligned}
$$

Computations for Values of Δp:

$\tan 30° = 0.58$

A at El. 315 $= [13 + 2(7)(0.58)][10 + 2(7)(0.58)] = 382$ sq ft

Δp at El. 315 $= 3.85 \times \dfrac{130}{382} = 1.31^k/\text{sq ft}$

A at El. 304 $= [13 + 2(18)(0.58)][10 + 2(18)(0.58)] = 1041$ sq ft

Δp at El. 304 $= 3.85 \times \dfrac{130}{1041} = 0.48^k/\text{sq ft}$

A at El. 295 $= [13 + 2(27)(0.58)][10 + 2(27)(0.58)] = 1830$ sq ft

Δp at El. 295 $= 3.85 \times \dfrac{130}{1830} = 0.27^k/\text{sq ft}$

Settlement Computations:

El. 322 to 308 $\quad S = \dfrac{14 \times 12}{1 + 0.80} \times 0.23 \times \log \dfrac{2.53 + 1.31}{2.53} = 3.9$ in.

El. 308 to 300 $\quad S = \dfrac{8 \times 12}{1 + 1.08} \times 0.34 \times \log \dfrac{3.14 + 0.48}{3.14} = 1.0$

El. 300 to 290 $\quad S = \dfrac{10 \times 12}{1 + 0.70} \times 0.20 \times \log \dfrac{3.65 + 0.27}{3.65} = 0.4$

Total settlement = 5.3 in.

reason, the clay is divided into three layers below the lower third point of the piles where the load is assumed to be delivered to the soil. The settlements contributed by these separate layers are added to obtain the total settlement. The increase in pressure at mid-thickness of each layer is computed on the assumption that the load is spread at an angle of 60° from the horizontal, starting at the lower third point of the piles. This assumption is justified if the rigidity of the pile cap is such that a dish-shaped pattern of settlement is impossible. The 60° spread of the load results in a reasonable approximation to the average increase in pressure at any level under the entire group of piles rather than to the increase in pressure under a specific point, such as would be obtained by the use of Fig. 15.7.

An investigation of the adequacy of the pile group as determined by the strength of the clay is omitted from this design plate because it would duplicate many of the computations given in DP C-4.

15.7. Lateral Forces and Displacements Due to Vertical Loads on Clay

Retaining walls, abutments, and wing walls are acted upon not only by vertical loads but also by the lateral pressure of the supported soil. The computation of the lateral earth pressure against such structures is a routine procedure and is discussed in Chapter 22. However, additional lateral forces are developed in the subsoil beneath the structures on account of the vertical load produced by the backfill. These lateral forces may be large and may cause outward movement of the structures.

If the subsoil consists of clay or plastic silty soils, the outward movement may be progressive and may become very large. In several instances the movement has been rapid and of sufficient magnitude to be regarded as a failure. Figure 15.9 shows the pertinent data concerning a typical example of progressive forward movement of large magnitude. The piles penetrated the soft material and rested on the medium clay at a depth of 40 ft. The two abutments and two intermediate piers for the bridge were completed, and backfilling was in progress, when movements of all four elements of the substructure were observed. Measurements yielded the results indicated on the figure. The distance L_2 between the intermediate piers decreased almost as much as the distance L_1 between abutments. This means that the movement was deep-seated and involved the lateral squeeze or flow of clay toward the center of the bridge from each end. Such a movement could hardly have been caused otherwise than by overloading the clay stratum by the weight of the backfill. Under the weight of the fill the bearing capacity of the soft reddish clay and overlying material was probably approached, and a slow lateral flow or creep was initiated by the excessive shearing forces. The lateral pressure against the abutments themselves was probably not excessive, because they were of the open type and the fill was allowed to extend through them.

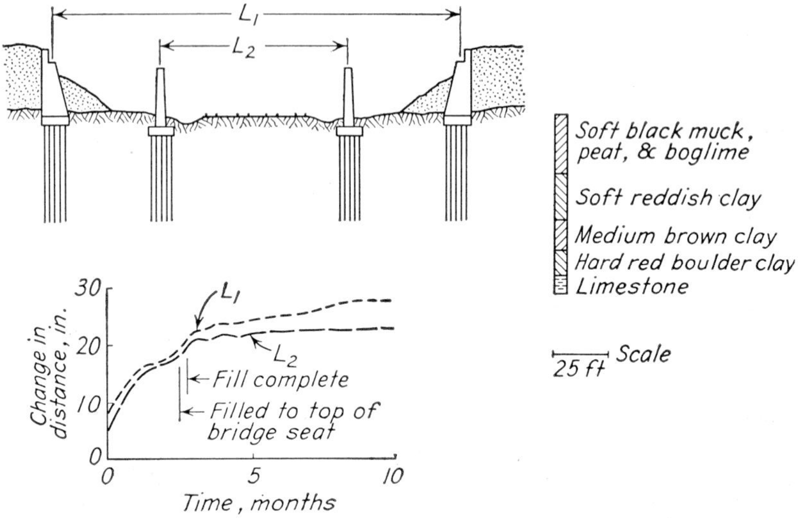

FIG. 15.9. Example of progressive movement of bridge piers due to lateral forces caused by weight of backfill.

The bearing capacity of a clay subsoil is approximately $3q_u$. Creep in many clays is likely to occur at loads greater than about half the ultimate strength. Therefore, as a very rough approximation, progressive movement of a wall or abutment may take place if the pressure in the clay is increased on account of the weight of the backfill by more than $1.5q_u$. If the backfill rests directly on the surface of the clay, the maximum height H_c of fill that is unlikely to be associated with a slow outward movement is roughly

$$H_c = 1.5q_u/\gamma \qquad 15.12$$

where γ is the unit weight of the backfill. In this equation q_u and γ must be expressed in consistent units.

If consolidation of the clay is facilitated by drainage layers or by installing sand drains, the value of q_u may increase appreciably and the height H_c increase correspondingly.

Vertical piles beneath the wall or abutment can provide little restraint against the horizontal movements, because the bending strength of piles is relatively small, and can be developed only if the lower part of the piles is firmly embedded in highly resistant material. Batter piles may be somewhat more efficient, but their effectiveness is likely to be considerably smaller than is generally believed. As the wall begins to move forward, the load on the batter piles increases greatly and the points penetrate further into the ground. Even a small penetration at the pile points permits appreciable lateral movement of the base of the structure. In any event, if the batter piles do not reach firm bearing in an underlying dense stratum, they are of little or no value.

Numerous abutments have experienced progressive movement in spite of the presence of batter piles. If the subsoil is too weak to support the weight of the backfill without progressive movement, it may be necessary to support the approach to the structure on a viaduct or trestle and to eliminate the backfill.

Canal Excavated in Loess

This canal, built in central Nebraska in 1938, was excavated with banks standing at the nearly vertical slope of ¼ horizontal to 1 vertical and rising to heights as great as 55 ft. An experimental cut in the locality demonstrated that steeper slopes would be subject to local slumps, whereas flatter slopes would quickly be attacked by erosion. The ability to stand on very steep slopes is a distinguishing characteristic of true loess.

Plate 16.

CHAPTER 16

Foundations on Silt and Loess

16.1. Significant Characteristics of Silt and Loess

Silt. Deposits of silt vary in character from somewhat plastic soils to materials possessing no plasticity whatsoever. Silt of the latter type, for which the plasticity index is close to zero, is commonly known as *rockflour*. Its characteristics are essentially those of fine sand, and the rules for establishing foundations above such deposits are essentially the same as those governing foundations on sand. Silts of the former type may derive their plasticity either from a high percentage of plate-shaped particles or from an appreciable organic content. These materials have many of the characteristics of soft and medium clays, and the design of foundations is based upon considerations similar to those that govern structures on clay.

If the N value of a silt deposit according to the standard penetration test is less than about 10, the silt may be regarded as loose or soft. Such a deposit is commonly unsuited for the direct support of foundations. For N values greater than 10, a preliminary conception of the probable behavior of the deposit can be obtained on the assumption that the material is either clay or sand, depending upon whether it is of the plastic or nonplastic type, using the appropriate relationships between N and the other physical characteristics of the soils.

Deep beds of silt, often more or less organic, are encountered near the present or former shores of oceans and lakes and in the beds of present or ancient rivers. When these deposits are below water table and have never had the opportunity to become dried, they are likely to be as soft and compressible as normally loaded clays near the liquid limit.

Loess. The grain size of loess is within the range assigned to silts. However, the characteristics of true loess deposits are likely to be extremely different from those of waterlaid silts, and the design of foundations on such deposits requires special consideration. On account of the calcareous or clayey binder present in most loess deposits, the material is likely to have appreciable cohesion in spite of its relatively high void ratio. It may be capable of sustaining loads of several tons per square foot without appreciable settlement. However, the quality of the binder is likely to differ from point to point, partly on account of erratic variations in the leaching action of groundwater. Hence, the strength of the deposit is likely to vary widely within short distances.

Deposits of true loess are unsaturated. If they become saturated, some of the binder is likely to dissolve or soften, and the deposit may lose its cohesion. In this event the structure of the soil collapses, and the void ratio decreases significantly. This is likely to cause settlement of the ground surface. Such settlement may be sufficient to damage a building, irrespective of the soil pressure for which its foundation was designed. Once the characteristics of a true loess have been destroyed, the material behaves like an ordinary silt.

16.2. Footings and Rafts on Silt

Soft or loose silt is not capable of supporting footing foundations. Structures above such deposits must be founded on piles or piers, or else on rafts established at such a level that the weight of the excavated material is approximately equal to that of the building.

A tentative design of footing or raft foundations above deposits of silt having an N value greater than 10 may be based on the assumptions that the plastic varieties of silt behave like clay, whereas rockflour behaves like sand. There are relatively few extensive deposits of silt having N values greater than 10. As a consequence, foundation experience in connection with such deposits is meager, and simple procedures for design cannot be established for use with confidence.

In reality, the assumption that a deposit of silt behaves either like sand or like clay is an oversimplification because most silts possess an appreciable amount of internal friction as well as some cohesion. Attempts to evaluate the cohesion by means of unconfined compression tests are likely to lead to values that are not conservative because the capillary tension of the water in the specimens produces an effect similar to that of a confining pressure. Hence, the test results indicate that the soil possesses appreciable cohesion. This condition may be investigated by testing some specimens completely immersed in water. If their strength is considerably below that of specimens tested in air, it may be taken for granted that most of the strength of the silt is derived from friction.

Reliable values of cohesion and friction for design under these circumstances can be obtained only from triaxial compression tests. The performance of the tests requires much experience, and the results can be interpreted successfully only with the exercise of mature judgment. For these reasons, the design of important foundations resting upon silts cannot be considered a routine matter and should not be undertaken without the assistance of experts.

16.3. Piles in Silt

Piles are commonly used to transfer loads from above a deposit of soft silt to firmer strata below. Friction piles are often driven in soft silt deposits of great depth. The considerations governing the driving of piles in silt are similar to those that pertain to driving in soft clay, with certain exceptions.

When piles are driven into soft clay, it is commonly observed that the skin friction decreases immediately during the process of driving and thereafter gradually increases. On the other hand, after piles are driven into some silts or silty clays, the decrease in bearing capacity of the silt continues for several days. Thereafter, the strength is likely to increase. In many saturated silts, the driving of piles temporarily liquefies the material for a considerable distance, but within a few days or weeks the strength of the silt may be largely regained. On account of the influence of these various factors, the bearing capacity of a single pile driven entirely into silt can be predicted only on the basis of load tests, and the tests should be made at several different times after the pile has been driven. In this manner, the trend of the relationship between time and strength can be established.

The ultimate bearing capacity and the settlement of a group of friction piles in a deep deposit of silt may be investigated by the same procedures used for friction-pile clusters in soft clay. In general, the bearing capacity of such a cluster depends to a considerable extent upon the perimeter of the group and cannot be increased appreciably by driving additional piles within the boundaries of a given cluster. Likewise, at a constant load per pile, the settlement of a pile group increases with increasing number of piles.

The use of pile-driving formulas to estimate the bearing capacity of a pile in silt cannot be recommended.

16.4. Piers in Silt

The excavation of piers through silt to a firm stratum commonly presents a number of construction difficulties if the silt is soft and below the water table. The difficulties of excavation increase with decreasing plasticity of the silt. Some rockflours have become notorious for their troublesome characteristics and are known locally as *bull's liver.* When encountered, they flow into an excavation like a viscous liquid and require tight sheeting. It may even be necessary to resort to compressed air to make any progress.

Open excavation below water table in plastic silts involves much less difficulty but is likely to result in considerable loss of ground and settlement of the adjacent ground surface due to drainage of the silt and the accompanying consolidation.

16.5. Footings and Rafts on Loess

The values of N in the standard penetration test do not appear to be a sufficiently reliable indication of the characteristics of loess to serve as a basis for design. In general, the N values for typical loess deposits seem to vary from about 4 to 20. It is probable that the strength of the material is greater than that indicated by the N values because the uniform consistency and high void ratio of loess facilitate penetration of the sampling spoon.

The standard penetration tests, however, may prove useful in establishing the uniformity or nonuniformity of a deposit in both horizontal and vertical directions. After a site has been explored by borings and penetration tests, it is advisable to perform load tests for determination of the allowable soil pressure. If the material appears to be uniform, three or four load tests may be sufficient for jobs of ordinary size. Otherwise, several tests should be made at locations where the number of blows is a maximum or a minimum, as well as where average conditions prevail.

The load tests should be made in pits not smaller than 4 by 4 ft in plan, and the bearing plate should be 1 ft square. No surcharge should be placed around the loaded area. The load should be increased until failure occurs or until the pressure beneath the test plate is at least three times that contemplated beneath the footings or raft. A load-settlement curve should be plotted on the basis of the results of each test.

If the load-settlement curve exhibits a point at which the settlement begins to increase rapidly for small increments of load, the allowable soil pressure should not exceed one third of the load corresponding to this point. Moreover, in any event, the allowable soil pressure should not exceed half the value at which the settlement of the test plate is equal to ½ in.

In using the results of such load tests, care must be taken to ascertain that the strength of the deposit does not decrease appreciably within the depth in which the stresses due to the building loads are significant. If such a decrease occurs, the pressures at any depth in the foundation should not exceed the allowable soil pressure for that depth as determined by means of the load tests.

Upon completion of a footing or raft foundation on loess, special precautions must be taken to make sure that the water table will not be raised or the loess become saturated by any subsequent activities. In a number of instances, small structures founded on loess in semi-arid climates have settled on account of saturation of the subsoil due to sprinkling or irrigation required to maintain lawns or landscaping. In other instances, wash water or other waste fluids have been allowed to accumulate beneath structures and have led to large subsidences.

16.6. Piles and Piers in Loess

If the bearing capacity of a loess subsoil is inadequate or if the loess is too nonuniform in its characteristics, structures may be established on piles. If the deposit is relatively thin, the piles may pass through the loess and into an underlying stratum where firm support may be obtained. If the deposit is thick, the piles may behave as friction units. Piles may sometimes be driven into loess without difficulty because the large percentage of voids filled with air can easily be reduced. However, the length of piles that can be driven can be determined reliably only by means of test piles. Likewise, the bearing capacity of an individual friction pile in loess can be determined reliably only by test. Tapered piles are usually advantageous.

In some instances when the supporting capacity of loess is inadequate, the loads are transmitted by piers through the loess to underlying firmer materials. Excavation can usually be carried out with ease because the material stands on vertical faces with little lateral support. It is essential to prevent the accumulation of surface water in the bottom of the excavations because such water may cause a softening of the cohesive bond and lead to construction difficulties.

Tower Latino Americana, Mexico City (1952)

This 43-story structure, tallest in Mexico City, has a basement with an excavation 42 ft deep. The substructure rests on cast-in-place button-bottom piles founded at a depth of 110 ft on a thin but very dense sand layer located between deposits of exceptionally compressible lacustrine clay. To prevent the excessive heave usually associated with excavation in this city, the hydrostatic pressure in the underlying clay was reduced by pumping from wells draining thin sand layers in the clay; to prevent settlement of the surrounding areas the water was fed into a gravel-filled trough and into injection wells just outside a sheet-pile wall enclosing the site. The successful completion of this foundation without damage to adjacent structures demonstrates the power of soil mechanics in a city noted for spectacular foundation problems.

PLATE 17.

CHAPTER 17

Foundations on Nonuniform Soils

17.1. Introduction

In the preceding chapters of Part C, it has generally been assumed that the subsoil is relatively uniform either to a very great depth or else to a limited depth where a firm base is encountered. In reality, such conditions are so unusual as to be considered rare exceptions. Hence, the procedures described in the preceding chapters are not often directly applicable to the solution of practical problems. Nevertheless, they are of value because they can be modified to give reliable indications of the probable behavior of foundations on nonuniform materials.

Most subsoils consist either of definite strata or of more or less lenticular elements. Some of the components of the deposit may consist of fairly resistant and incompressible material, whereas others may be relatively weak and compressible. On the basis of preliminary information, such as that from exploratory borings together with standard penetration tests and simple laboratory tests, one can usually conclude at once whether some parts of the subsoil are sufficiently strong and incompressible to be of no concern. Attention can then be concentrated on the weaker or more compressible members.

The principal task of the designer before he can select the appropriate type of foundation is to determine the influence of the presence of the elements believed to be weak. In general, this may be done by estimating or computing the stresses in the subsoil on the assumption that the subsoil is uniform and elastic. After the physical properties of the doubtful materials have been evaluated on the basis of the exploratory data, the capacity of the doubtful materials to resist the stresses without failure or excessive compression can be determined. The result of this investigation is usually sufficient to permit selection of the appropriate type of foundation. Occasionally, more elaborate exploratory procedures and soil tests may be required to provide the basis for a sound decision.

The computation of stresses may be made by means of Newmark's chart or, under many conditions ordinarily encountered, by some simplified procedure. Although the chart is based upon the assumption that the material is homogeneous, the errors in the stresses due to stratification or other irregularities are not likely to be great enough to invalidate the predictions of the probable behavior of the soil.

In the following sections, the more important kinds of nonuniform soil deposits will be discussed.

17.2. Soft or Loose Strata Overlying Firm Strata

When the upper part of the subsoil consists of soft or loose soils, the unsatisfactory character of the materials is likely to be apparent, and the necessity for providing adequate support is rarely overlooked. The principal decision is whether or not a footing foundation can be used. This may be determined by computing the safe load for the upper material on the assumption that it extends to great depth and by estimating the settlement that would arise from the compression of the soft part of the deposit. If the computed safe load is too small or the computed settlement too great, footings must be eliminated from consideration. One alternative is to provide pile or pier support. Another possibility is to reduce the excess load on the subsoil by excavation and to construct a raft foundation.

If piles or piers are adopted, their bearing capacity and behavior may be judged on the basis of the considerations discussed in Chapters 14, 15, and 16.

17.3. Dense or Stiff Layer Overlying Soft Deposit

Choice of Foundation. The implications of the presence of a soft deposit at some depth below firm strata are not so obvious as if the soft materials were at shallow depth. If the firm deposit is relatively thin, footings or rafts may exert sufficient pressure to break into the underlying soft soil. A number of failures of this type have occurred. Even if the overlying firm layer is of sufficient thickness to prevent such a failure, the settlement of the structure due to consolidation of the soft deposit may be excessive.

The factor of safety against breaking through the stiff crust may be conservatively estimated by determining the pressures at the upper surface of the soft deposit. The maximum pressure should not exceed the safe load for the soft material as determined by the procedures discussed in the preceding chapters.

If the footings are widely spaced, the stress at the top of the soft layer can be decreased to some extent by increasing the size of the footings. However, if the footings are spaced rather closely, the distribution of pressure at the top of the soft layer may remain practically independent of the soil pressure beneath the footings themselves. If the pressure cannot be reduced to the safe load for the soft material, pile or pier support is required, or else material must be excavated to compensate for part of the weight of the building.

Even if the safe load on the soft soil beneath the firm layer is not exceeded, the settlement of a footing or raft foundation may be excessive. The computation of settlement may be made by the procedures already described in connection with confined layers of clay. If the settlement is excessive, one of the other types of foundation mentioned in the preceding paragraph must be adopted.

If the computed settlement is not excessive, the footings can be designed as if the soft deposit were not present, by means of the rules given previously for the kind of soil constituting the firm layer

Shortcomings of Load Tests. The use of load tests has been very common as a means of estimating the allowable soil pressure for a given subsoil. Examples of conditions appropriate for the use of the method have been given in several of the preceding chapters. However, the method is likely to be misleading and dangerous if the load tests are made on a firm stratum underlain by softer materials. The reason is illustrated is Fig. 17.1. In this figure, A represents a test plate 1 ft square and B a footing 10 ft square. Both rest on the surface of a stratum of stiff clay 3 ft thick. The stiff clay rests upon a deep deposit of soft clay having an unconfined compressive strength of 0.3 ton/sq ft and a compression index equal to 0.27. The stiff clay itself has an unconfined compressive strength of 2 tons/sq ft and may be considered practically incompressible. The failure load for the test plate is approximately 7.4 tons/sq ft, provided the entire surface of sliding is located within the stiff clay layer. Even under this load on the test plate, the maximum stress at the top of the soft layer is only 0.37 ton/sq ft, which is considerably less than the bearing capacity of the soft material. Therefore, the soft material would not influence the failure of the test plate.

The load that may be carried safely by the footing may be estimated in the following manner. The perimeter of the footing is 40 ft and the thickness of the stiff layer is 3 ft. If the footing were to break into the soft soil, a shearing force of $3 \times 40 \times 1.0 = 120$ tons would have to be overcome. This is equivalent to 1.2 tons/sq ft. In addition, the bearing capacity of the soft clay beneath the footing would have to be exceeded. This is equal to about 1.1 tons/sq ft. Hence, the ultimate bearing capacity of the large footing is not greater than approximately 2.3 tons/sq ft, which is very much less than that of the test plate.

With respect to settlement, the discrepancy is even more striking. At a soil pressure of 1 ton/sq ft, the stress at the top of the soft layer directly beneath the test plate is approximately 0.05 ton/sq ft, and the settlement of the underlying soft clay is practically zero. On the other hand, the stress at the top of the soft clay beneath the center of the large footing is approximately 0.88 ton/sq ft. It decreases with depth as indicated in the figure. The corresponding settlement

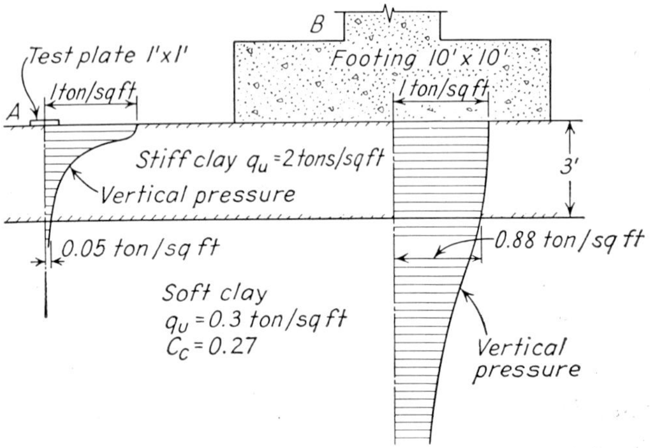

FIG. 17.1. Diagram illustrating reasons for difference in behavior of test plate 1 ft square and footing 10 ft square if both rest on upper surface of stiff clay crust underlain by soft clay.

is about 10 in. Hence, the settlement of a structure may be excessive even if practically no settlement occurs during a load test in which the same soil pressure is used that will exist beneath the structure.

Several serious accidents and many examples of excessive settlement have occurred as a result of selecting the allowable soil pressure on the basis of load tests on a stiff crust. Hence, before the load-test method can be used, it is essential to investigate whether the strength of the soil decreases with depth. If it does, load tests must be performed at such levels that the capacity of the softest layers may be investigated. It is generally preferable to determine the safe load on clay soils by means of unconfined compression tests.

Illustrative Design. DP C-7. Footings on Sand above Clay Layer

This design plate illustrates the computations necessary to proportion the footings in accordance with the properties of the sand

DP C-7
Building Foundation
Sh. 1 of 4

Footing Design. General Data

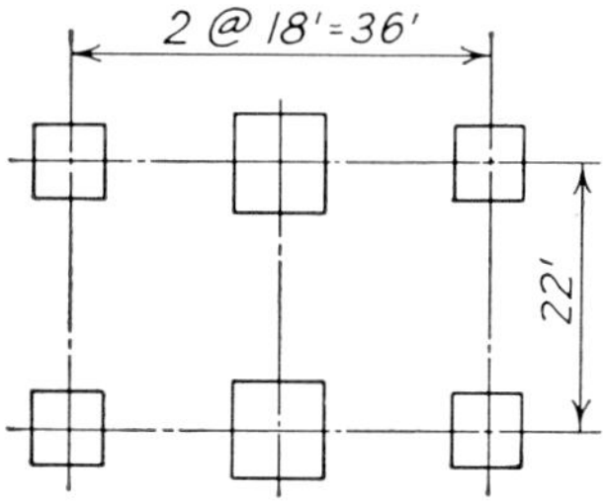

Columns Loads to Footings:

Side: DL + normal LL = 240^k; DL + max LL = 334^k

Corner: DL + normal LL = 162^k; DL + max LL = 222^k

Assume permanent LL on basement floor = 50 lb/sq ft

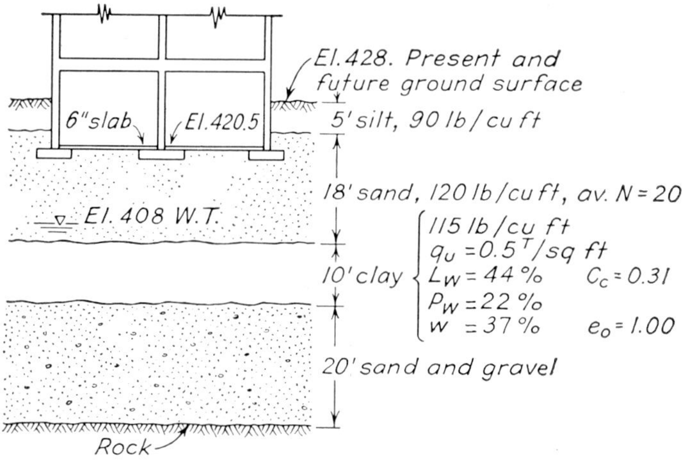

Footing Loads and Factors of Safety:

Side footings, $\frac{334}{240} = 1.39 < 1.5$ so if F. S. = 3 for 240^k, F. S. > 2 for 334^k

Corner footings, $\frac{222}{162} = 1.37 < 1.5$

Use 240^k for determining area of side footings.
Use 162^k for determining area of corner footings.

DP C-7
Building Foundation
Sh. 2 of 4

Footing Design. Required Areas

Side Footings:

$240^k = 120^T$

@ say 2 = 60 sq ft

Try 8'-0" × 8'-0". A = 64 sq ft Assume depth = 2'-0"

Column load = 120^T

÷ 64 = 1.88^T/sq ft

Additional load minus surcharge } $2.0\left(\frac{0.15-0.12}{2}\right) = 0.03$

1.91^T/sq ft

Fig. 14.4, allow 2.0^T/sq ft

Fig. 14.3, allow 1.3 + 0.9 = 2.2^T/sq ft

Use 8'-0" × 8'-0" for side footings

Corner Footings:

$162^k = 81^T$

@ 1.91 = 42.4 sq ft

Try 6'-6" × 6'-6". A = 42.3 sq ft Assume depth = 2'-0"

Column load = 81^T

÷ 42.3 = 1.92^T/sq ft

Additional load minus surcharge } = 0.03

1.95^T/sq ft

Fig. 14.4, allow 2.1^T/sq ft

Fig. 14.3, allow 1.1 + 0.9 = 2.0^T/sq ft

Use 6'-6" × 6'-6" for corner footings

DP C-7
Building Foundation
Sh. 3 of 4

Footing Design. Settlement Computations

Computations for p_o at Mid-depth of Clay Layer:

$5 \times 0.090 = 0.45^{k}/\text{sq ft}$
$15 \times 0.120 = 1.80$
$3 \times (0.120 - 0.0625) = 0.17$
$5 \times (0.115 - 0.0625) = 0.26$
$2.68^{k}/\text{sq ft} = p_o$

Net Load Removed at Bottom of Basement Slab:

Basement excavation { 5′ silt @ 0.090 = $0.45^{k}/\text{sq ft}$; 3′ sand @ 0.120 = 0.36 } = 0.81

Basement slab = 0.08
Permanent LL on basement floor = 0.05 } 0.13

Net load removed = $0.68^{k}/\text{sq ft}$

Net Load Added at Bottom of Each Footing

Say $\dfrac{1.91 + 1.95}{2} \times 2 = 3.86^{k}/\text{sq ft}$ = net load added

Computations for Δp at Mid-depth of Clay

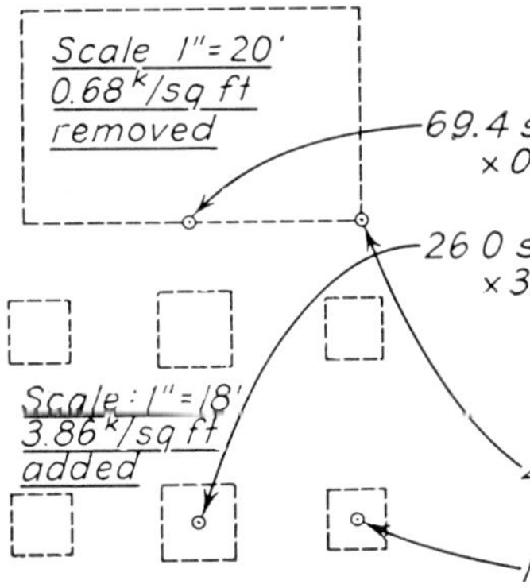

69.4 squares $\times 0.68 \times 0.005 = 0.24^{k}/\text{sq ft}$
26.0 squares $\times 3.86 \times 0.005 = 0.50$
Net $\Delta p = 0.26^{k}/\text{sq ft}$ under side footings

41.0 squares $\times 0.68 \times 0.005 = 0.14^{k}/\text{sq ft}$
18.3 squares $\times 3.86 \times 0.005 = 0.35$
Net $\Delta p = 0.21^{k}/\text{sq ft}$ under corner footings

DP C-7
Building Foundation
Sh. 4 of 4

Footing Design. Settlement Computations

Side footing: $\frac{10 \times 12}{1 + 1.00} \times 0.31 \times \log \frac{2.68 + 0.26}{2.68} = 0.74''$

Corner footing: $\frac{10 \times 12}{1 + 1.00} \times 0.31 \times \log \frac{2.68 + 0.21}{2.68} = 0.60''$

immediately beneath the foundation, and to predict the settlements resulting from the consolidation of the clay that underlies the deposit of sand.

The computations on Sh. 2 are for the most part similar to those in DP C-1. Likewise, those on Shs. 3 and 4 of this design plate duplicate many of the computations given in DP C-5. Therefore, a detailed explanation of the computations is unnecessary. However, it should be noted that two sketches of the foundation plan, Sh. 3, are required for computing the increase in pressure at the middle of the clay layer, because the excavated soil will be removed from the entire area of the building whereas the footings will deliver their loads to the soil beneath six separate areas.

17.4. Alternating Soft and Stiff Layers

If the deposit contains a number of weak layers, bearing-capacity and settlement computations may be made for each. If the structure cannot be supported on footings near the surface of the ground, piles or piers may be used to transmit the loads to one of the firm strata at sufficient depth to provide a satisfactory foundation. This depth can be determined on the basis of the results of the computations. The choice between piles and piers, or of the type of pile to be used, is likely to depend upon the difficulty that may be experienced in driving through the firm layers above the bearing stratum. The depth to which piles can be driven in such a deposit can seldom be predicted with accuracy, and any conclusions must be considered tentative until test piles have been driven.

Excavation to compensate for part or all of the weight of the structure may permit construction of a raft. This alternative should be considered along with the use of piles or piers.

17.5. Irregular Deposits

If the subsoil consists of lenticular or wedge-shaped masses, it is rarely possible to make an accurate estimate of bearing capacity or settlement. In such instances, it is advisable to determine the general character of the deposit by means of numerous subsurface soundings supplemented by a few borings and soil tests. The purpose is to form an opinion regarding the size and distribution of the softer elements of the deposit and to judge the most unfavorable combination of elements that can reasonably be expected. The estimate of settlement should be based on the assumption that the most unfavorable conditions may occur in the most highly stressed portion of the subsoil.

Damage Due to Adjacent Construction

Two-story structure distorted by differential settlement. The settlement had its seat in deposits of highly compressible clay which consolidated partly because of the weight of the building on the right and partly because of a general lowering of the groundwater level. Serious differential movements did not occur, however, until the steel-frame structure on the left was built. The pile foundation of this structure arrested the normal settlement of the edge of the two-story building and caused ruinous distortions. Movements of this magnitude are rare except in Mexico City, where the photograph was taken, but smaller movements of the same character have often occurred elsewhere.

PLATE 18.

CHAPTER 18

Damage Due to Construction Operations

18.1. Settlement Due to Excavation

Introduction. Every process of excavation is associated with a change in the state of stress in the soil. The change in stress is inevitably accompanied by deformations. These commonly take the form of a subsidence of the area surrounding the excavation, an inward movement of the soil at the sides, and an upward movement of the soil beneath the bottom. Structures supported by foundations resting upon the material that deforms experience corresponding movements. They commonly settle and move toward the excavation.

Inasmuch as no excavation can be made without altering the state of stress to some extent, some movement of surrounding structures must be regarded as inevitable. However, to the inevitable movements associated with a given type of construction there may be added others due to poor workmanship. If the bracing of the sides of an excavation is done haphazardly, for example, large displacements may occur. The portion of the movement associated with poor workmanship must be regarded as unnecessary and should be avoided.

Different methods of construction involve different amounts of inevitable movements. In some instances, a given procedure cannot be used because damage to adjacent structures would unquestionably be too great, even if the workmanship were excellent. Therefore, the engineer should be aware of the consequences of using different construction procedures and should be careful not to specify methods of excavating and bracing that cannot be carried out without causing damage to adjacent property.

In order to avoid this danger, the engineer should be familiar with the various procedures for excavating and bracing the sides of excavations. On a given job, he should make use of such observations as are necessary for determining whether the movements are greater than those that must be considered inevitable and for determining how best to reduce any preventable movements.

Excavation in Sand. Sand above the water table is commonly moist and possesses enough cohesion to facilitate excavation. Even in properly braced large cuts, the settlement of the adjacent ground surface does not usually exceed about 0.5 per cent of the depth of the cut, and the influence of the settlement does not extend farther from the sides of the cut than a distance equal to the depth.

When large cuts are made in sand below the water table, it is advisable to attempt to lower the water table before construction. After the water table has been lowered, the excavation can be carried out with no more subsidence than would be associated with a similar cut in moist sand. However, the process of lowering the water table may itself lead to subsidence under some conditions. These will be pointed out subsequently.

Piers are often established in sand below the water table by sinking a caisson and dredging from the inside. If the water level inside the caisson falls lower than that corresponding to the pressure at the cutting edge, sand is likely to flow into the caisson. The volume of sand removed may be several times that of the caisson. Such a procedure results in subsidence in the neighborhood of the excavation. In many instances it may be prevented by keeping the water level inside the caisson higher than that outside. If the removal of the sand by dredging is not practicable, it may be necessary to balance the water pressure by means of compressed air to keep the sand from flowing in.

Excavations in Clay. When large cuts are excavated in soft clay, the weight of the soil surrounding the edge of the excavation acts like a surcharge on the soil at the level of the bottom of the cut. Lateral forces are developed in the subsoil for the reasons that have been discussed in connection with the movements of abutments and retaining walls due to the weight of the backfill. If the depth of the cut becomes so great that the bearing capacity of the soil beneath the sides is approached, large movements are inevitable, irrespective of the care with which the sides of the cut may be braced.

If the bottom of the excavation is underlain at a shallow depth by a firm stratum, the tendency for such a bearing-capacity failure is greatly reduced. The movements can also be decreased by driving piles around the boundary of the cut until they are firmly embedded in the underlying stiff stratum. These piles are braced by struts as excavation progresses.

Field measurements have demonstrated that the volume of soft clay that moves inward at the sides of a cut plus that which moves upward beneath its base is represented by an equivalent volume of subsidence surrounding the excavation. Therefore, any measures that can be taken to prevent lateral movements or heave will be reflected in a decrease in settlement. Hence, it is advisable to keep such cuts braced tightly at all times and to insert braces as soon as possible after portions of the excavations have been completed.

The excavation of open shafts or the sinking of open caissons in clay is likewise associated with loss of ground. The material beneath the bottom of the excavation at any stage rises toward the excavation. If the sides remain unsupported even for a short time, lateral movements also take place. Such movements may not produce appreciable settlement around a single shaft, but the cumulative settlement due to the excavation of many shafts beneath a small area may amount to several inches or feet.

The inevitable settlements may be reduced by choosing construction procedures that permit smaller loss of ground. These include the use of sheet piles or cylindrical shells that eliminate exposed vertical faces. If the movements are still too great, the hole may be filled with a heavy fluid and excavation carried out by a mechanical auger or rotating bit. Still further reductions can be accomplished, but at great expense, by the use of compressed air. In recent years, it has become economical to drive heavy steel pipes of large diameter and to remove the soil by dredging, by jetting, by a process of washing similar to that employed in making test borings, or by means of a blast of compressed air.

Excavation in Stratified Deposits. In some localities, deposits of clay or other impervious materials are underlain by pervious silts, sands, and gravels. Excavation in an open shaft through the impervious soils may be accomplished without difficulty. However, if the voids of the underlying granular materials are occupied by water or gas under pressure, the materials will tend to break into the shaft and flood at least its lower part. In this process, the clay strata may be undermined because of the removal of the underlying sand. This may cause subsidence even at a considerable distance from the point at which the excavation is being made.

18.2. Settlement Due to Vibrations

Vibrations have relatively little effect upon cohesive soils, but they may greatly increase the relative density of cohesionless materials. The moisture contained in many sands above the water table provides sufficient cohesion to prevent a rearrangement of the grain structure. However, relatively dry sands and sands beneath the water table are readily compacted. Any operations that produce vibrations in such materials may result in appreciable subsidence of the ground surface.

The principal sources of vibration due to construction operations are pile driving and blasting. Since both operations have been used for the purpose of compacting loose sands, it is obvious that both are likely to produce subsidence in connection with construction activities.

In one instance about 100 piles were driven to a depth of 50 ft in a loose deposit of sand and gravel. Settlement of the ground surface within the area occupied by the piles was about 6 in. It decreased to about 1/8 in. at a distance of 50 ft. If the driving of piles in such sands is likely to damage adjacent structures, it may be necessary to resort to some other type of construction.

18.3. Settlement Due to Lowering the Water Table

Cause of Settlement. Whenever the water table is lowered, the effective weight of the material between the original and final position of the water table is increased from that of submerged to that of moist or saturated soil. This causes corresponding increase in pressure between the grains of all the soil beneath the original position of the water table, and it is accompanied by strains in accordance with the stress-strain relationships for the material in question. The resulting displacements produce a settlement of the ground surface that is roughly proportional to the descent of the water table.

Effect of Lowering Water Table in Sand. The increase of intergranular pressure in a mass of sand does not ordinarily produce significant settlement because even a loose sand is relatively incompressible. Only if a sand is extremely loose, so that its structure may actually collapse, is there likelihood of important subsidence. On the other hand, fluctuations of the water table may ultimately produce large settlements because the deformation of sand increases perceptibly with each application of load as indicated in Fig. 2.11. During normal construction operations, the water table is usually lowered only once or twice and the cumulative effect is not important.

In some instances, large subsidences have occurred near drained excavations in sand because springs have been permitted to form at or near the bottom of the excavations. Water flowing into springs from the surrounding sand may carry the material into the excavation grain by grain, and a tunnel may be formed beneath some slightly cohesive stratum. When the tunnel becomes sufficiently large, its roof may collapse and the ground surface above it subside. The subsidence may take the form of a sink hole at considerable distance from the edge of the excavation.

Effect of Lowering Water Table in Clay Strata. Lowering the water table within or above a stratum of clay ultimately increases the intergranular pressure precisely as in sand. Moreover, on account of the high compressibility of clay and various organic soils, the resulting settlements may become very large. However, the process of consolidation initiated by lowering the water table may require considerable time for the development of the ultimate amount of settlement. The time depends primarily upon the permeability of the material. The settlement of peat and organic silt may occur very rapidly and may amount to several feet. The settlement of more impervious clays may not be excessive during the relatively short construction period. The rate of settlement and its magnitude can be estimated in accordance with the principles discussed in Arts. 2.4 and 2.8.

If the compressible layers are underlain by a layer of sand, the water table may be lowered for a very large distance from the excavation. The corresponding settlements of the overlying materials may extend far from the site of the construction and may lead to serious structural defects. During the construction of a set of locks in Holland, lowering of the water table through a distance of 21 ft in soil consisting of about 20 ft of clay and peat underlain by sand produced settlements noticeable as far as 2500 ft. Even at a distance of 130 ft, the settlement amounted to 24 in.

18.4. Displacement Due to Pile Driving

When piles are driven into loose cohesionless materials, the ground surface is likely to subside, in spite of the fact that the volume of the piles may represent an appreciable fraction of the volume of the subsoil. In most other materials, the driving of a pile is likely to be associated with the displacement of a volume of material equal to or somewhat less than that of the pile. When several piles are driven in a foundation, especially if the spacing is close, the ground surface may heave as much as several feet. The heave may extend to a considerable distance from the boundaries of the pile group. Structures located within this distance are displaced upward.

In some instances, especially if the soils are silty, the high stresses set up in the mass of soil due to the introduction of the piles causes an intense and rapid process of consolidation, whereupon the heave is followed within a few days by a settlement. In less pervious materials such as clays, a considerable amount of the heave is likely to remain.

The displacement associated with the driving of piles is also accompanied by lateral movements. Structures located near the piles are likely to be displaced away from the operations. In some instances, the displacement may be very important. For example, a large number of foundation piles may be driven behind a dock next to a river with a relatively deep channel. The resistance of a dock to outward movement is usually small. Most of the displacement due to pile driving is then likely to be represented by lateral movements of the dock. Encroachment of dock structures into river channels for this reason have in many instances amounted to a foot or more.

18.5. Importance of Field Observations for Control of Construction Operations

Legal Importance of Observations. Until recently, the inevitability of a certain amount of settlement or heave adjacent to construction operations was ignored. Contractors and engineers alike considered it unwise to admit the influence of their operations upon neighboring structures. Numerous specifications have been written containing the statement that excavation shall be carried out in such a manner that no settlement of the surrounding soil shall occur. If no complaints were received from adjacent property owners, it was generally conceded that these provisions had been complied with. On the other hand, if complaints arose, there were always wide grounds for differences of opinion concerning the magnitude of the settlement and the amount of damage resulting therefrom.

More recently it has become the practice for the engineer responsible for a construction project or for the contractor, or both, to establish reference points not only within the limits of their own property but also on structures belonging to adjacent property owners. Observations are made on these reference points often enough to indicate the trend of any movements. Likewise, the adjacent property owners themselves commonly employ independent engineers to observe the movements of their structures. If damage occurs, the facts of the case can be established readily, and a fair settlement can generally be agreed upon without extensive litigation.

Observations for Improving Technique. Observations of settlement, heave, and lateral movement serve another vital purpose, especially when combined with a careful record of the details of the construction procedure. They serve as an indication of unsatisfactory techniques of excavating, bracing, or pile driving. If settlements are consistently larger than have been observed in connection with similar projects under similar soil conditions, it may be concluded that the method of construction is not being carried out in the most workmanlike manner. Variations in technique may be tried and their merit judged on the basis of any changes in the trend of the settlements. By adopting those techniques that lead to reductions in settlement and discarding those that lead to increases, the settlement can be reduced to the smallest amount compatible with the general method of construction.

In many instances, measurement of the deflections of the members of a system of bracing, or measurement of loads in these members, has led to a substantial reduction in the cost of construction, together with a decrease in the accompanying settlements. Hence, systematic observations of this type are almost always economical. They have the further advantage that unforseen conditions that might lead to disaster are certain to be preceded by apparently unmotivated increases in deflection or pressure. With adequate warning, the cause of the increases may be investigated and remedial measures taken before difficulties arise.

18.6. Influence of Construction Procedures on Design

In the preceding chapters, methods have been described for determining the adequacy of various kinds of foundations. In general, the foundation has been considered satisfactory if it does not transmit pressures to the subsoil that exceed the safe load or that will cause excessive settlement. However, certain types of foundations that may be entirely acceptable from these two points of view may also be extremely difficult or impossible to install. Furthermore, some installations may lead to excessive subsidence of surrounding property. Hence, the practicability of constructing each type of foundation that may be considered is an important matter. In many instances, it is a decisive factor in the final choice.

The choice of the type of foundation for a given structure may also be influenced by the possibility of damage due to adjacent construction operations at a subsequent date by other parties. For example, a satisfactory foundation for a given building may consist of piers excavated through soft clay to a stratum of very hard clay resting upon waterbearing dense sand which in turn rests upon bedrock. Nevertheless, if a new building is constructed on adjacent property and is founded upon piers extending through the hard clay and waterbearing sand to bedrock, there is great likelihood that the sand may flow into the excavations for the new piers, and that the support for the hard clay stratum beneath the older building may be lost. In view of this possibility, it may be preferable to establish the first building on piers extending to rock, even at extra expense. Hence, the possibility of damage due to future construction in the neighborhood may also be a significant factor in the final choice of the type of foundation.

Part D

STRUCTURAL DESIGN OF FOUNDATION ELEMENTS

The final step in the design of a given foundation is the structural analysis and design of its various parts.

Most modern substructures consist of reinforced-concrete elements designed in accordance with the same fundamental principles that apply to reinforced-concrete members in general. Hence, in Part D it is presumed that the reader already possesses a thorough knowledge of statics, strength of materials, and principles of reinforced-concrete design. Only those aspects of design peculiar to foundation engineering will be discussed.

In many instances, at least preliminary structural designs of a foundation must be prepared to serve as a basis for estimates of cost before the most suitable type of foundation can be selected. Therefore, the information in Part D may often be required before all the steps outlined in Part C can be carried out.

The designer is likely to have the impression that those aspects of foundation engineering dealing with soils are based on highly empirical procedures, whereas he may regard the design of reinforced-concrete members as having a more satisfactory theoretical basis. Yet, more mature reflection on the origin of current procedures for analyzing and designing concrete members leads to the conclusion that modern design codes have comparatively little basis in theory and have been derived to a considerable degree from the results of tests in the laboratory and in the field. Without proper appreciation of this fact, the designer is likely to lack a sense of proportion and to devote relatively too little consideration to the characteristics of the soil deposits at a site.

Arthur N. Talbot (1857–1942)

Teacher and research worker associated for over 50 years with the University of Illinois and influential in the formation and development of its Engineering Experiment Station. Beginning in 1903 he directed a comprehensive investigation of reinforced concrete in all its structural forms. His investigation of footings, from 1908 to 1912, furnished the basis for procedures of design used with but slight modification today, and represented the only extensive experimental work on this subject prior to 1944.

PLATE 19.

CHAPTER 19

Individual Column and Wall Footings

19.1. Basis for Design Procedure

The design of ordinary concentrically loaded footings is based on the assumption that the soil pressure against the bottom of the footing is uniformly distributed. There is much indirect evidence that this is a satisfactory and generally conservative assumption, although there have been few if any reliable field observations of the actual distribution of contact pressures.

The rules for design of footings, like those for all types of concrete members, have a primarily empirical basis. Current methods originated with an extensive series of tests made by A. N. Talbot and reported in 1913. Many of Talbot's findings are still reflected in the present code of the American Concrete Institute. The principal changes since 1913 take account of the increased strength of concrete and reinforcing steel.

The first comprehensive re-investigation of footing behavior since Talbot's work was carried out by F. E. Richart at the University of Illinois in 1944. The information gained from these tests has led to some revisions in specifications for design and is likely to lead to others. For example, the tests demonstrated that the allowable bond stresses could properly be increased for recently developed reinforcing bars with more effective types of deformations.

19.2. Critical Sections

The procedure for designing footings, like that for other reinforced-concrete members, is related to the manner in which the member may fail. Experiments have indicated that several different types of structural behavior may be exhibited before the ultimate strength of a footing is reached. At some strength less than the ultimate, the deformations and cracks in the footing are likely to be so great that the footing is no longer useful. This is referred to as the stage of primary failure. The nature of the primary failure is sometimes difficult to determine because of the interdependence of various factors.

If inadequate tensile reinforcement has been provided at the bottom of the footing, the primary failure is evidenced by excessive yielding of the bars. On the other hand, if the bond stresses between the steel and concrete are excessive, the primary failure takes the form of slipping of the bars. In both these instances, the ultimate failure is usually one of diagonal tension.

If the primary failure of a square column footing occurs as the result of diagonal tension, a section of concrete, flaring out at about 45° downward from the column, is punched from the rest of the footing. The initial cracks that eventually develop into the surface of failure form on the under side of the footing while the stresses in the steel are still low. As the shear increases, these cracks take an inclination of about 45° toward the edges of the column, and the footing fails in diagonal tension before the average stress in the tensile steel reaches the yield point. If the shape of the footing is rectangular, failure in diagonal tension becomes more and more likely to take the form of an inclined crack across the full width of the footing as the ratio of length to width increases.

The preceding discussion of the factors that may influence the structural strength of a footing indicates that the design or stress analysis may appropriately be based upon the moment and shear at certain critical sections where failure due to excessive strains in flexure or diagonal tension may be initiated. The critical section for shear (diagonal tension) is commonly located at a distance d from the face of the column, pedestal, or wall, as shown in Fig. 19.1. Hence, the length of the section is twice the depth of the footing plus the width of the column. The total shear used in computing the diagonal tensile stress is the sum of the forces on area *cdef* in Fig. 19.1*a* or on *cdefgh* in Fig. 19.1*b*.

If the part of the structure resting on the footing proper consists of reinforced concrete, the critical section for flexure is ordinarily assumed to extend across the footing at the face of the column, pedestal, or wall, as shown by *ab* and *jk* in Fig. 19.1. The stresses in the concrete and steel, as well as the bond stress, are computed from the forces and moments acting on this section. For footings under masonry walls, it is common practice to investigate flexural and bond stresses at a section under the wall; usually the quarter point is chosen. If the column load is transferred to the footing through a base plate, it seems reasonable, because of the flexibility of the plate, to use a section midway between the edge of the plate and the face of the column.

The foregoing comments concerning critical sections have made it apparent that the stresses computed in an analysis of a spread footing are only average values found from total moments or shears acting on sections many feet in extent. It is impossible to predict, under field conditions, the actual variation of these stresses along any section of the footing. It is fortunate, therefore, that footings,

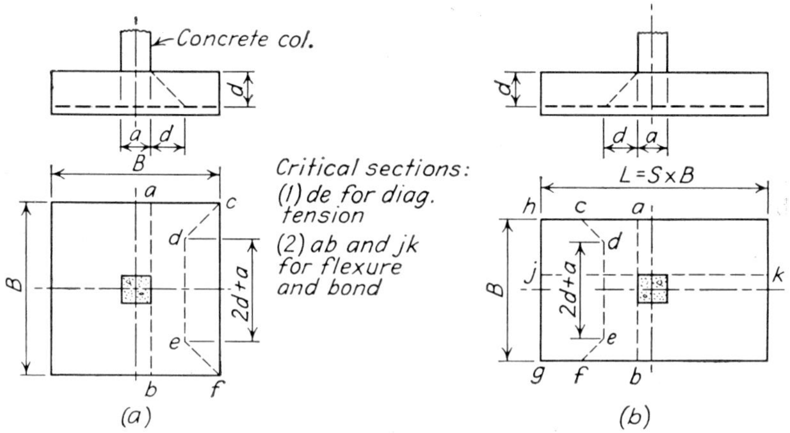

FIG. 19.1. Critical sections for diagonal tension, flexure, and bond in square and rectangular column footings.

like other indeterminate structures, have considerable ability to redistribute the moments and shears before failure occurs at the sections that may be overstressed.

19.3. Placing of Reinforcement

In a wall footing the main reinforcement is placed at right angles to the wall and should be spaced uniformly. Furthermore, some longitudinal reinforcement is desirable to assist in bridging soft spots due to the almost certain variation in soil conditions along the length of the footing. A steel percentage of 0.3 to 0.4 is adequate for this purpose in all except unusual cases.

The reinforcement in square footings is usually placed in directions parallel to the edges. In each direction the bars are commonly spaced uniformly. Likewise, the reinforcement in the long direction of rectangular footings is usually distributed uniformly across the full width. On the other hand, tests on rectangular footings have demonstrated that the bars in the short direction should be spaced more closely near the center than at the outside. The following equation is often used for determining the distribution of the steel:

$$n_b = \left(\frac{2}{S+1}\right) n_t \qquad 19.1$$

in which n_b = number of bars placed in a central width equal to the short dimension of the footing

n_t = total number of bars required by moment at the critical section

S = ratio of long side to short side of the footing

Special care is usually required to insure good concrete at the bottom of footings, particularly if the bottom of the excavation is below the water table. The excavation should be dry when the concrete is placed. Moreover, the reinforcement should be protected by at least 3 in. of concrete.

19.4. Depth of Spread Footings

It is evident that the structural design of footings is largely a matter of trial. Revisions may be necessary as the design progresses. Usually the depth determined by diagonal tension is adequate to satisfy the requirements of flexure and bond. However, if bars capable of developing only low bond stresses are used, the depth of the footing is likely to be governed either by that required to develop the column dowels or that which leads to a reasonable amount of reinforcement as determined by bond. The use of bars with improved deformations, which justify high allowable bond stresses, has increased the relative number of footings in which the depth is controlled by diagonal tension.

If diagonal tension on the critical section *de*, Fig. 19.1, is assumed to control the depth of a square footing, the following equation is useful in selecting the depth d.

$$\frac{d}{B} = \frac{\sqrt{2C + 4C^2 + \frac{k^2}{4}} - \frac{k}{2}(1 + 4C)}{2 + 4C} \qquad 19.2$$

The constant C is evaluated from the net soil pressure q_n (lb/sq ft) and the allowable stress in diagonal tension v (lb/sq in.) by the expression $C = q_n/500v$. The net soil pressure q_n, effective in producing shear and moment in the footing, is the column load divided by the footing area. The value of k is the ratio of the least column or pedestal width a to the footing width B; thus $k = a/B$.

19.5. Procedure for Design and Use of Minimum-depth Curves

The curves of Fig. 19.2 have been developed from eq. 19.2 to give the designer a rapid indication of the minimum depth of footing. If a depth equal to or greater than that indicated by Fig. 19.2 is

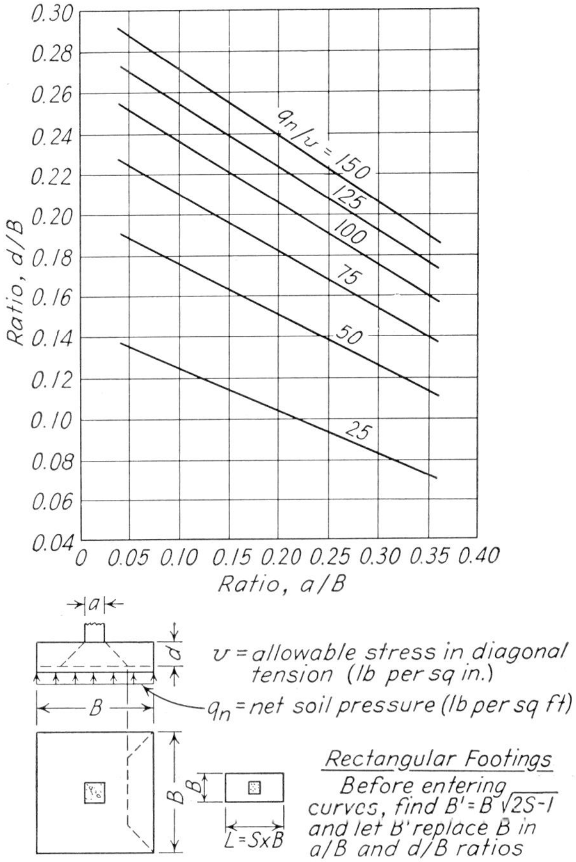

FIG. 19.2. Curves for selecting depth of footing as determined by diagonal tension.

used, the diagonal tensile stress will not exceed the allowable value. Although the curves have been constructed for square footings, they can be utilized equally well for rectangular footings by using an adjusted value of B. The equation for making this adjustment is

$$B' = B\sqrt{2S - 1} \tag{19.3}$$

in which B' is the adjusted width of the footing, which replaces B

in both the a/B and d/B ratios. The value of S is the ratio of the long side L to the short side B.

The complete design procedure can be summarized as follows:

1. Proportion the area of the footing in accordance with the methods given in Part C.
2. Select the trial depth from Fig. 19.2.
3. Check the compressive stress in the concrete due to moment. This value is almost always less than the allowable stress.
4. Select the reinforcement required for moment.
5. Check the bond stress.

Some designers prefer to modify steps 4 and 5 by first selecting the size of the bars and then computing the number of bars required for moment and the number for bond. If the number required for bond greatly exceeds that required by moment, the use of smaller bars, providing more perimeter per area, is indicated.

19.6. Illustrative Designs. DP D-1. Column Footings

The two footings designed in DP D-1 illustrate the application of previous discussions of critical sections, placing of the reinforcement, and design procedure. Both footings were proportioned in DP C-1. The computations in DP D-1 indicate that the previously assumed depths of the footings were 4 to 6 in. too large. However, any revisions of the depths in DP C-1 would not be justified in view of the negligible effect of depth on the required areas.

The design computations and allowable stresses used in these examples are for the most part in agreement with the ACI Building Code adopted in 1951. One deviation from the ACI Code is found in the magnitude of moment and shear for flexure and bond. The ACI Code recommends the use of 85 per cent of the static moment and shear for the determination of the area of the steel and for the computation of the bond stress. This procedure cannot be justified either by statics or by experiment. With the present growing practice of using bars that develop high bond stresses without hooks, it becomes increasingly important that the full static moment and shear be used for investigating stresses at the face of the column.

It should be noted that a bond stress of $0.08f'_c$ has been allowed without hooks. The allowable diagonal tensile stress is $0.03f'_c$. These allowable stresses are in accordance with the ACI Code, except that the Code specifices a maximum of 75 lb/sq in. for diagonal tension. However, other specifications either have no such limit or else place the limit at a different magnitude of stress. In any event,

DP D-1
Warehouse Footings
Sh. 1 of 2

Footing Design. Structural

Interior: (see DP C-1, Sh. 2)

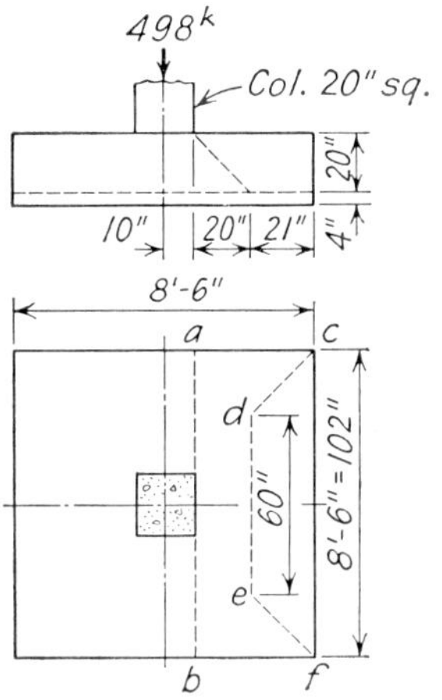

Concrete and Steel Data:

$f'_c = 3000$ psi

$\times 0.45 = 1350$ psi $= f_c$

$f_s = 20$ ksi

Assume $j = \frac{7}{8}$ $\quad k = \frac{3}{8}$

$v = 0.03 \times 3000 = 90$ psi

$u = 0.08 \times 3000 = 240$ psi

Net soil pressure $= 6.90^k/\square' = q_n$

Use Fig. 19.2 $\frac{a}{B} = \frac{20}{102}$ = say 0.20

$\frac{q_n}{v} = \frac{6900}{90} = 77$

Use $d = 0.19 \times 102$ = say 20"

$jd = 17.5"$ $\quad kd = 7.5"$

Check diagonal tension at de.

$V = 6.9\left(\frac{8.5 + 5.0}{2} \times 1.75\right) = 81.5^k$

$v = \frac{81.5}{60 \times 17.5} = 78$ psi ok allow 90 psi

Check concrete at ab.

$V = 6.9(8.5 \times 3.42) = 201^k$

$\times \frac{3.42}{2} = 344'^k = 4130''^k = M$

Total compression $= \frac{4130}{17.5} = 236^k$

$\div \left(\frac{102 \times 7.5}{2}\right) = 617$ psi ok allow 1350 psi

Steel Required at ab.

$A_s = \frac{236}{20} = 11.8\,\square''$ $\quad$ 20-#7 $\quad A_s = 12.0\,\square''$ $\quad \Sigma_o = 55\,\square''/_{''}$

$u = \frac{201}{55 \times 17.5} = 209$ psi ok allow 240 psi

Use 20-#7 @ 5" ctrs. both ways.

DP D-1
Warehouse Footings
Sh. 2 of 2

Footing Design. Structural Wall: (see DP C-1, Sh. 3)

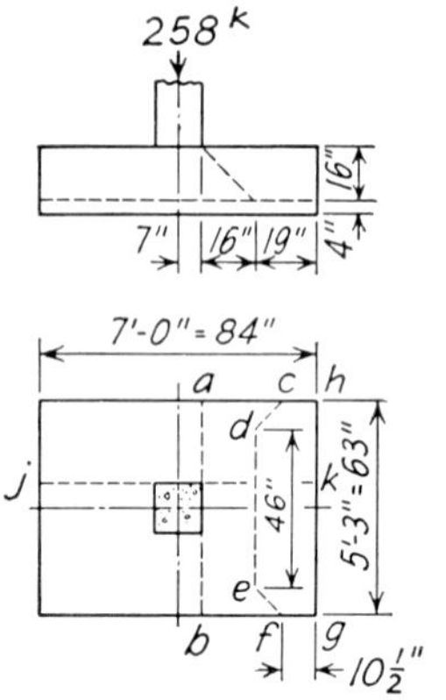

Concrete and Steel Data:
Same as Sh. 1 (interior footing)

Net soil pressure = $7.02^{k}/\square'$ = q_n

Use Fig. 19.2: $S = \frac{7}{5.25} = 1.33$

$B' = 5.25\sqrt{2 \times 1.33 - 1} = 6.78'$

$\frac{a}{B'} = \frac{14}{6.78 \times 12} = 0.17$

$\frac{q_n}{v} = \frac{7020}{90} = 78$

Use $d = 0.195 \times 6.78 \times 12$ = say 16"

$jd = 14.0"$ $kd = 6.0"$

Check diagonal tension at de.

$\frac{3.83 + 5.25}{2} \times 0.71 = 3.22^{\square'}$

$5.25 \times 0.88 = 4.62$

$7.84^{\square'} \times 7.02 = 55.1^{k} = V$

$v = \frac{55.1}{46 \times 14.0} = 86$ psi
ok allow 90 psi

Check concrete stress at ab.

$V = 7.02(5.25 \times 2.92) = 108^{k}$

$\times \frac{2.92}{2} = 157'^{k} = 1885''^{k} = M$

Total compression $= \frac{1885}{14} = 135^{k}$

$\div \left(\frac{63 \times 6.0}{2}\right) = 713$ psi
ok allow 1350 psi

Steel Required at ab.

$A_s = \frac{135}{20} = 6.75^{\square''}$ 16-#6 $A_s = 7.1^{\square''}$ $\Sigma_0 = 37.7^{\square''}/_{''}$

$u = \frac{108}{37.7 \times 14} = 205$ psi ok allow 240 psi

Steel Required at jk.

$V = 7.02(7.0 \times 2.04) = 100^{k}$

$\times \frac{2.04}{2} = 102'^{k} = 1224''^{k} = M$

$A_s = \frac{1224}{20 \times 14} = 4.4^{\square''}$ 13-#6 $A_s = 5.7$ $\Sigma_0 = 30.7^{\square''}/_{''}$

$u = \frac{100}{30.7 \times 14} = 233$ psi ok allow 240 psi

Distribution of Steel at jk (see eq. 19.1).

$n_b = \frac{2}{1.33 + 1} \times 13$ = 11 bars in center 5'-3"

Use 16-#6 long way (equal spacing).

Use 13-#6 short way (10 spa. @ 6" at center).

the reinforcing bars must conform to ASTM Specification A305-49 which deals with the minimum requirements for deformed bars.

19.7. Isolated Column Footings on Piles

Various types of piles have been described in Part B. The problem of selecting the proper type of pile for given soil conditions has been given considerable attention in Parts B and C.

Reinforced-concrete footings are used with all types of piles and serve as pile caps as well as supports for the columns. The piles commonly project 3 or 4 in. into the footing as shown in Fig. 19.3, and about 3 in. of concrete should separate the steel and the tops of the piles.

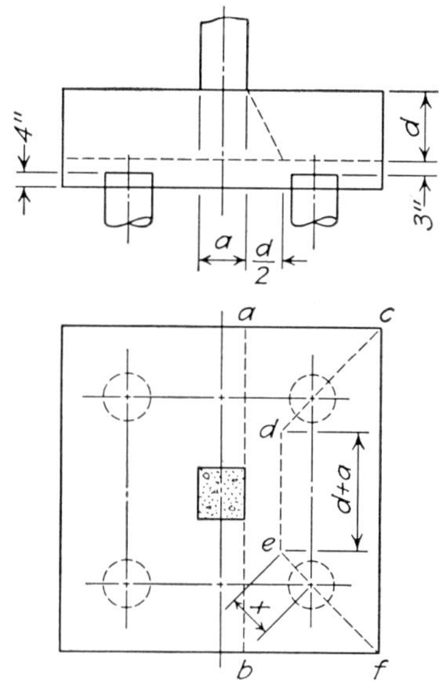

FIG. 19.3. Concrete footing on piles, showing critical sections for design.

In general, the procedure for designing footings supported by piles closely parallels that used for footings on soil. Any differences are due to the fact that pile footings are subjected to concentrated reactions instead of relatively uniform soil pressure. Moreover, the locations of the piles in the field are likely to be at least several inches from their theoretical positions. For these reasons, it is common and conservative practice to take the critical section for diagonal tension closer to the face of the column for pile footings than for footings on soil. As shown in Fig. 19.3, a section *de,* located at a distance equal to one-half the depth of the footing from the face of the column, is often used for investigating diagonal tension. The critical section *ab* for flexure and bond may be assumed at the face of the column as in the case of footings on soil.

If the center of a pile is 6 in. or more outside the critical section, the entire reaction of the pile should be assumed effective in producing moment or shear on the section. The reaction from any pile located 6 in. or more inside the section probably contributes very little to the moment or shear. Hence, the reaction may be considered as zero. For intermediate positions, a straight-line interpolation is

DP D-2
Pile Footing
Sh. 1 of 1

Pile Footing Design. Structural

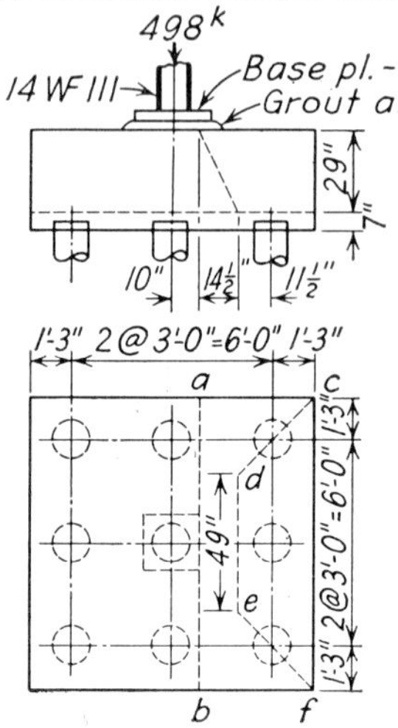

Concrete and Steel Data.

See DP D-1, Sh. 1

Col. load = 498^k

$@\ 60^k/\text{pile} = 8.3$ **Use 9 piles**

Net load per pile $= \frac{498}{9} = 55.3^k$

Assume $d = 29''$ $jd = 25.4''$ $kd = 10.9''$

Check diagonal tension at de

$V = 2 \times 55.3 = 110.6^k$

$v = \frac{110.6}{49 \times 25.4} = 89$ psi

ok allow 90 psi

Check concrete stress at ab

$V = 3 \times 55.3 = 165.9^k$

$\times\ 26'' = 4320''^k = M$

Total compression $= \frac{4320}{25.4} = 170^k$

$\div \left(\frac{102 \times 10.9}{2}\right) = 305$ psi

ok allow 1350 psi

Steel Required at ab

$A_s = \frac{170}{20} = 8.5\ \square''$ $15 - \#7$ $A_s = 9.0\ \square''$ $\Sigma_0 = 41.2\ \square''/''$

$u = \frac{165.9}{41.2 \times 25.4} = 158$ psi ok allow 240 psi

Use 15 - #7 @ 7" ctrs. both ways.

commonly used to estimate the appropriate portion of pile reaction for analysis and design. By reference to Fig. 19.3, it becomes apparent that the shear and moment on *ab* will be produced by two full pile reactions, whereas the shear on *de* will be one pile reaction (two halves), provided the centers of the piles are 6 in. or more from the points *d* and *e*. For example, if the distance *x* is only 3 in., three fourths of a pile reaction are assumed to contribute to the shear on section *de*.

19.8. Illustrative Design. DP D-2. Pile-supported Column Footing

The problem of selecting the allowable pile load of 30 tons, arbitrarily used in DP D-2, has been discussed in Part C. A section halfway between the face of the steel column and the edge of the base plate has been selected as critical for tension and bond in the steel. The section determined in this manner is approximately 10 in. from the center line of the column. Moreover, it appears reasonable to locate the critical section for diagonal tension at a distance of 10 in. plus half the footing depth from the center line.

It is of interest to note that the over-all footing depth required in DP D-2 is 12 in. greater than that required for the interior footing in DP D-1, although the column loads and plan dimensions are equal. Furthermore, the depth to the steel in the footing on piles is 9 in. more than that required when the footing is placed on soil. The factor contributing most to this increase is the different method of locating the critical section for diagonal tension.

Footings Subjected to Moment

Model of concrete arch-rib structure. Buildings consisting of several units of this type have been built with spans of more than 300 ft to serve as hangars and amphitheaters. The footings are subjected to large moments on account of the thrust from the arches and are sometimes made wider at the toes than at the heels to equalize the soil pressures as much as possible. They are very heavily reinforced.

Plate 20.

CHAPTER 20

Footings Subjected to Moment

20.1. Introduction

Chapter 19 dealt with footings supporting concentrically located columns that produced either uniform soil pressures or uniform pile reactions. However, many foundations must resist not only vertical loads but also moment about one or both axes. The moment M may exist at the bottom of a concentrically placed column, Fig. 20.1a,

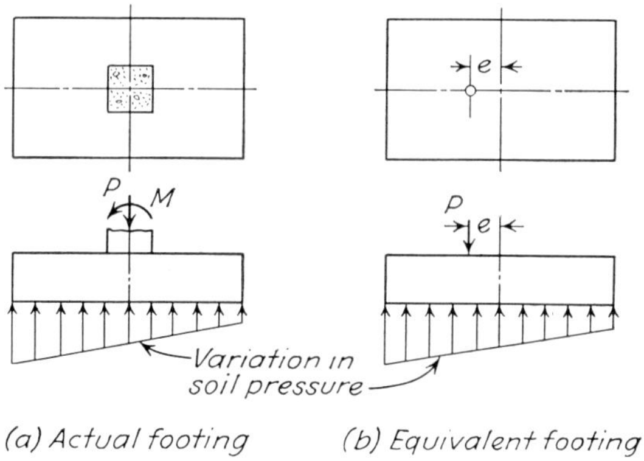

FIG. 20.1. Spread footing subjected to moment.

whence it is transferred to the footing, or it may be produced by a vertical load P located eccentrically at a distance e from the centroid of the base of the footing, Fig. 20.1b. If the footings in a and b have the same dimensions in plan, the soil reactions are identical provided $M = P \times e$. Footing b is then said to be equivalent to a. The substitution of an eccentric load for the real moment and column load usually simplifies the computations.

Other common foundations that must resist moment are those for retaining walls, abutments, and bridge piers. The moment on a retaining wall is due primarily to the active earth pressure, whereas the moments on the foundations for bridge piers are produced primarily by wind and traction on the superstructure. Foundations for these structures are discussed in more detail in Art. 20.9 and in Chapter 22.

In the discussion of concentrically loaded footings, Chapter 19, particular attention was given to the choice of critical sections for investigation of moments and shears; the computations of the moments and shears themselves involved no difficulty because the soil reaction or the load per pile was considered uniform. In contrast, the principal problem in the design of eccentrically loaded footings is the determination of the soil pressure or the load per pile, inasmuch as these quantities are no longer uniform. Once they have been determined, the selection of critical sections and the computation of stresses due to moment and shear are carried out as for concentrically loaded footings. Hence, Chapter 20 is concerned primarily with the computation of soil pressure or load per pile.

Fundamental to all computations in this chapter are the laws of statics. Regardless of the methods of analysis, the distribution of vertical soil pressure on the base of a footing must satisfy the requirements of statics that (1) the total upward soil reaction must be equal to the sum of the downward loads that act on the base, and (2) the moment of the resultant vertical load about any point must be equal to the moment of the total soil reaction about the same point. In addition, an adequate horizontal soil reaction, usually provided by shear along the base, must be available to oppose the resultant horizontal load.

Ordinary footings are commonly assumed to act as rigid structures. This premise leads to the conclusion that the vertical settlement of the soil beneath the base must have a planar distribution because a rigid foundation remains plane when it settles. A planar distribution of soil pressure follows from a second assumption that the ratio of pressure to settlement is constant. Neither of these assumptions is strictly valid, but each is generally considered sufficiently accurate for ordinary problems of design.

20.2. Resultant within Middle Third

A footing subjected to moment is shown in Fig. 20.2. The forces acting on the footing, including the weight of the footing itself, have been resolved into ΣV and ΣH. It is apparent that the moment ΣM acting about the center of gravity of the base may be expressed as $\Sigma M = \Sigma H \times h = \Sigma V \times e$. Solving for e, we obtain

$$e = \Sigma M/\Sigma V \qquad 20.1$$

Equation 20.1 is of fundamental importance because it determines the position of the resultant of all forces acting above (or below) the base of a footing, regardless of how complicated the conditions of loading may be. In some problems, such as the design of retaining walls, it may be more convenient to choose a center of moments elsewhere than at the center of the base; nevertheless, eq. 20.1 remains

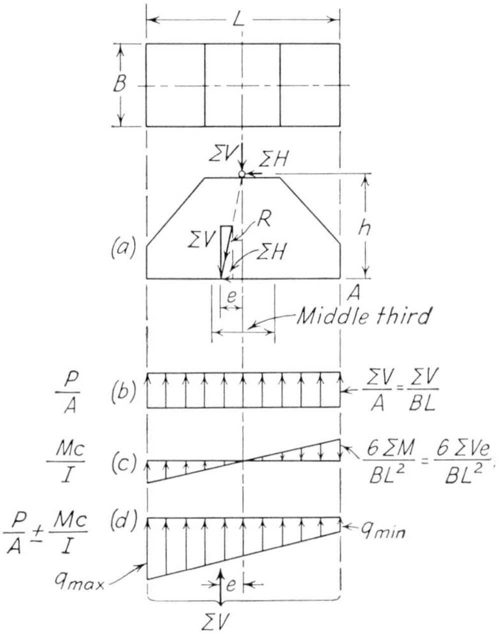

FIG. 20.2. Forces at base of footing. Resultant inside middle third.

applicable. For example, if the point A in Fig. 20.2 is taken as the center of moments, the value of ΣM must include $\Sigma V \times L/2$. When this total moment is divided by ΣV, the position of the resultant is found unchanged at a distance of $L/2 + e$ from A.

If the vertical load ΣV acts alone, the soil pressure is uniform as shown in Fig. 20.2*b*. The horizontal force ΣH, if acting alone, produces a shear that must be resisted by the soil at the base and a moment that corresponds to the distribution of pressure shown in *c*. The maximum pressure caused by the moment is obtained by dividing the moment by the section modulus of the area of the base. The combined effect of ΣV and ΣM corresponds to the distribution of soil pressure shown in *d*. Adding, for each edge of the pedestal, the pressures in b and c, we have

$$q_{max} = \frac{\Sigma V}{BL} + \frac{6\Sigma Ve}{BL^2} = \frac{\Sigma V}{BL}\left(1 + \frac{6e}{L}\right) \tag{20.2}$$

and

$$q_{min} = \frac{\Sigma V}{BL} - \frac{6\Sigma Ve}{BL^2} = \frac{\Sigma V}{BL}\left(1 - \frac{6e}{L}\right) \tag{20.3}$$

Equations 20.2 and 20.3 are often expressed as

$$q = \frac{\Sigma V}{BL}\left(1 \pm \frac{6e}{L}\right) \tag{20.4}$$

Equation 20.4 is no more than a special form of the basic formula for stress on a section subjected to a direct load P and a moment M. The formula is commonly expressed in strength of materials in the form,

$$f = \frac{P}{A} \pm \frac{Mc}{I}$$

20.3. Resultant outside Middle Third

If ΣH, Fig. 20.2, is increased, the moment at the base of the footing becomes larger and increases the soil pressures indicated in c. When the extreme pressures in c become equal numerically to those in b, q_{min} becomes zero. For this condition

$$\frac{\Sigma V}{BL} = \frac{6\Sigma Ve}{BL^2}$$

Solving for e, we find $e = L/6$, which indicates that the resultant must fall within the middle third of the base in order for compression to exist beneath the entire base. This condition is generally considered desirable.

Figure 20.3 shows the distribution of the soil pressure beneath the footing when the resultant is outside the middle third of the base. According to the laws of statics, the total upward force must be equal to and colinear with ΣV. These two conditions may be expressed by

$$\Sigma V = \frac{q_{max}Bx}{2} \tag{20.5}$$

and

$$x/3 = (L/2) - e \tag{20.6}$$

From these expressions the maximum soil pressure q_{max} can be computed.

Inspection of eq. 20.5 reveals that the maximum soil pressure is merely twice the average pressure produced by ΣV acting on the area Bx. The same result may be obtained from eq. 20.2 if the dimension L is replaced by the distance x and, since e must be measured from the center of gravity of the area under compression, if e is replaced by $x/6$. With these substitutions, eq. 20.2 becomes identical with eq. 20.5.

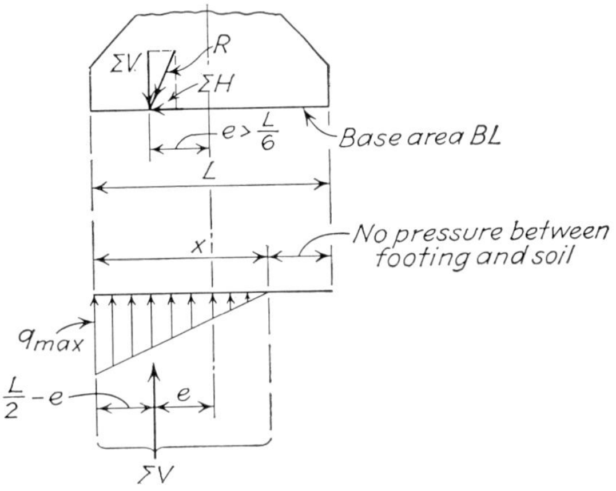

FIG. 20.3. Forces at base of footing. Resultant outside middle third.

20.4. Illustrative Design. DP D-3. Footing Subjected to Moment

The computations in DP D-3 represent common procedure for the design of rectangular footings subjected to moment about one axis. The critical sections are chosen in accordance with the rules in Chapter 19. The shear and moment at the critical sections are computed in accordance with the nonuniform distribution of pressure against the base. This is the principal difference between the computations illustrated in DP D-3 and those in DP D-1.

If the designer wishes to use Fig. 19.2 for determining the depth of the footing, q_n should be taken as the average pressure on the area contributing to the shear at section *de*. The computations in DP D-3, Sh. 1, indicate that the average pressure on area *cdefgh* is about 5.35 kips/sq ft, whence $q_n/v = 59.5$. Since the footing is rectangular, $S = 10/6 = 1.67$, $B' = 6\sqrt{2(1.67) - 1} = 9.14$ ft, and $a/B' = 0.219$. By means of these values and Fig. 19.2, the appropriate value of d/B' is found to be 0.16. Hence, the depth of the footing should not be less than about 18 in.

The need for a concentration of steel in the central region of the footing, as reflected in eq. 19.1 for concentrically loaded footings,

DP D-3
Footing
Sh. 1 of 2

Design of Footing Subjected to Moment

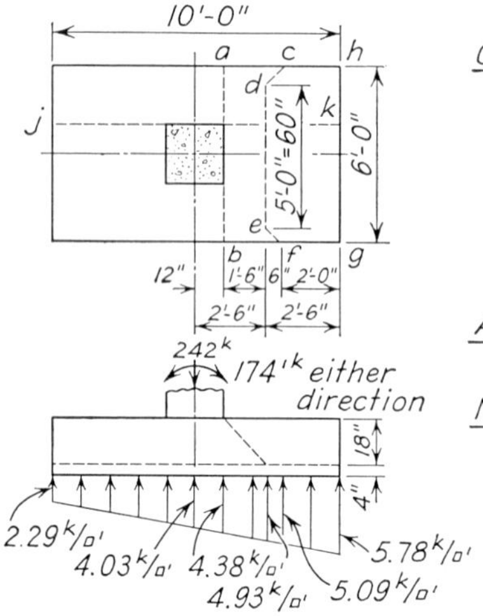

Concrete and Steel Data

$f_c' = 3000$ psi

$\times 0.45 = 1350$ psi $= f_c$

$f_s = 20$ ksi

$v = 0.03 \times 3000 = 90$ psi

$u = 0.08 \times 3000 = 240$ psi

Assume $j = \frac{7}{8}$ $k = \frac{3}{8}$

Assume $d = 18''$

$jd = 15.75''$ $kd = 6.75''$

Net Soil Pressures

$e = \frac{174}{242} = 0.72' < \frac{10}{6}$ so resultant in mid $\frac{1}{3}$

$q = \frac{242}{60}\left(1 \pm \frac{6 \times 0.72}{10}\right) = 5.78\ ^k/_{□'} = q_{max}$

$= 2.29\ ^k/_{□'} = q_{min}$

Check diagonal tension at de.

Average pressure on cdef $= 5.01\ ^k/_{□'}$

$\times 2.75\ ^{□'} = 13.8^k$

Average pressure on cfgh $= 5.44\ ^k/_{□'}$

$\times 12.00\ ^{□'} = 65.3^k$

$79.1^k = V$

$v = \frac{79.1}{60 \times 15.75} = 84$ psi ok allow 90 psi

Check concrete at ab.

Use two triangular pressure variations for convenience.

$\frac{5.78}{2}(4 \times 6) = 69.4^k$

$\times \frac{2}{3}(4) = 185.2'^k$

$\frac{4.38}{2}(4 \times 6) = 52.6^k$

$\times \frac{1}{3}(4) = 70.2'^k$

$V = 122.0^k$ $255.4'^k = 3065''^k = M$

Total compression $= \frac{3065}{15.75} = 194.5^k$

$\div \left(\frac{72 \times 6.75}{2}\right) = 830$ psi

ok allow 1350 psi

DP D-3
Footing
Sh. 2 of 2

Design of Footing Subjected to Moment

Steel Required at ab

$$A_s = \frac{194.5}{20} = 9.73\,\square'' \quad 22\text{-}\#6 \quad A_s = 9.7\,\square'' \quad \Sigma_0 = 51.9\,\square''/'' $$

$$u = \frac{122}{51.9 \times 15.75} = 149 \text{ psi ok allow } 240 \text{ psi}$$

Steel Required at jk

Since $174'^k$ may act in either direction, use average pressure from ℄ to k over entire area

$$\text{Average pressure} = \frac{4.03 + 5.78}{2} = 4.91\,^k/\square'$$

$$V = 4.91(2 \times 10) = 98.2^k$$

$$\times 1.0 = 98.2'^k = 1178''^k = M$$

$$A_s = \frac{1178}{20 \times 15.75} = 3.74\,\square'' \quad 11\text{-}\#6 \quad A_s = 4.8\,\square'' \quad \Sigma_0 = 26.0\,\square''/''$$

$$u = \frac{98.2}{26.0 \times 15.75} = 240 \text{ psi ok allow } 240 \text{ psi}$$

Use 22-#6 long way }
Use 11-#6 short way } uniform spacing.

is partly offset by the increase of the soil pressure near the edges. Consequently, the reinforcement in the short direction is distributed uniformly.

20.5. Moment about Both Axes

Article 20.2 has shown that the soil pressures beneath footings may be estimated by means of the common formula for stress

$$\frac{\Sigma V}{A} \pm \frac{\Sigma Mc}{I}$$

provided the moment acts about only one axis and the entire area of the base is in compression. When moments act simultaneously about both axes, the pressures may be computed by means of eq. 20.7, provided the entire area of the base is in compression.

$$q = \frac{\Sigma V}{A} \pm \frac{\Sigma M_1 c_1}{I_1} \pm \frac{\Sigma M_2 c_2}{I_2} \qquad 20.7$$

The location of the maximum and minimum soil pressures may be determined readily by observing the directions of the moments. Likewise, the proper signs in eq. 20.7 may be determined by inspection for any other point on the base of the footing.

When moments act simultaneously about both axes, the possibility must not be overlooked that only part of the base may be in compression. This condition exists when the minimum soil pressure determined by eq. 20.7 has a negative value indicating tension that cannot be developed between the soil and concrete. Under this circumstance, the computation of the maximum soil pressure must be based, as in Art. 20.3, on the area actually in compression. Equation 20.7 is no longer applicable.

Although several analytical and graphical methods are available for the solution of this problem, the following procedure of trial and error has the advantage of simplicity. It is illustrated in Fig. 20.4.

1. Find e_1 and e_2 by means of eq. 20.1, and let ΣV act through point A. The moments M_1 and M_2 may then be disregarded because the same soil pressure will act against the base of the equivalent eccentrically loaded footing.

2. Use eq. 20.7 to compute fictitious pressures at the corners as if tension could exist at D.

3. Locate the line of zero pressure on the basis of the pressures computed in step 2.

4. Select a trial axis of zero pressure, ZZ, approximately parallel to the line determined in step 3 but somewhat closer to point A.

5. Compute the moment of inertia about ZZ of the area assumed to be in compression.

6. Find $q_{max} = \Sigma Vzb/I_{zz}$. The pressures at the other corners may be computed in like manner by substituting the appropriate dimensions for b.

7. Compare the magnitude of the total soil reaction (volume of the pressure diagram) with the vertical load ΣV, and compare the point

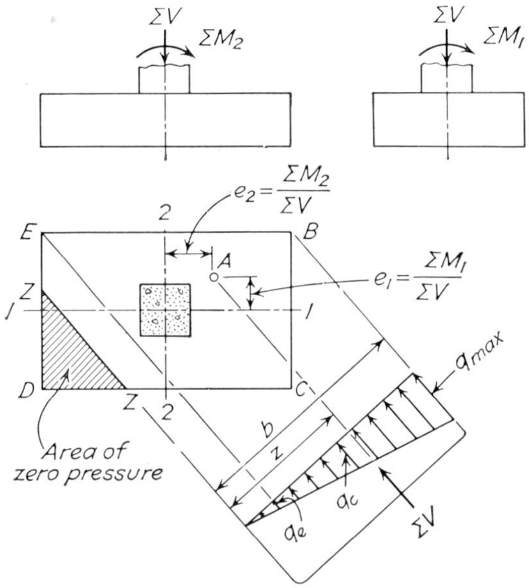

FIG. 20.4. Footing subjected to moments about both axes with zero pressure under part of base.

of application of the total soil reaction (center of gravity of the volume of the pressure diagram) with the position of point A. These comparisons constitute a check on the requirements of statics.

8. If the discrepancies found in step 7 are too large, another axis ZZ may be assumed, and the computations in steps 5 to 7 made again. However, a great degree of refinement cannot be justified in view of the uncertainties associated with the various assumptions that must be made in solving any problem of this type.

20.6. Footings Having Unsymmetrical Shapes

The principles discussed in this chapter, except for the remarks pertaining to Fig. 20.4, apply only to footings having in plan at least one axis of symmetry. Occasionally an unsymmetrical plan may be necessary. If the resultant load does not coincide with the center of gravity of the area of the footing, the computation of the soil pressures becomes a problem involving bending on an unsymmetrical section. Theoretically, eq. 20.7 is not applicable even though the entire base may be in compression. However, unless the footing is greatly unsymmetrical, the errors involved in using eq. 20.7 are tolerable for design. The subject of bending on unsymmetrical sections has received adequate treatment elsewhere.

20.7. Moment on Pile Footings

The reactions exerted by piles beneath a footing subject to moment are calculated in a manner similar to that described in the preceding articles concerning the pressure under soil-supported footings. Pile caps, such as those shown in Fig. 20.5 and Fig. 20.6, are commonly

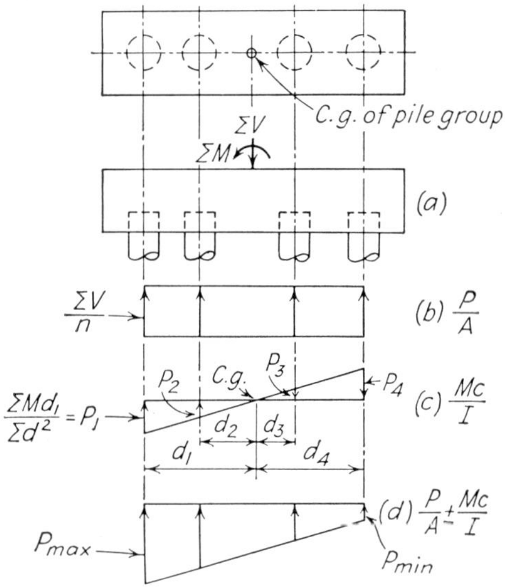

FIG. 20.5. Computation of pile reaction.

assumed to act as rigid structures. A planar distribution of the vertical settlement of the piles follows as a result of this assumption. Finally, if the ratio of reaction to settlement is assumed to be constant, the loads in the piles vary in the same planar fashion. Neither the assumption of the rigid cap nor the supposition that reaction is directly proportional to settlement is strictly valid, but each is generally considered sufficiently accurate for the purposes of design.

The analysis of a moment-resistant group of piles is illustrated in

Fig. 20.5. If there were no moment and ΣV acted through the center of gravity of the four piles, the loads in the piles would be as shown in *b*. On the other hand, if there were no resultant vertical force and only ΣM acted on the group of piles, the loads would be as shown in *c*.

The loads shown in *d* represent the total reaction and are the sums of those in *b* and *c*. The same loads would have been produced by ΣV acting eccentrically at a distance *e* to the left of the center of gravity.

From statics, it is evident that the resisting moment of the reactions furnished by the piles, Fig. 20.5*c*, must equal the applied moment, ΣM. The following equation expresses this relationship if resisting moments at the junctions of the piles and the cap either do not exist or are disregarded.

$$\Sigma M = P_1 d_1 + P_2 d_2 + P_3 d_3 + P_4 d_4 \qquad 20.8a$$

If the variation in pile reactions shown in *c* is assumed to be linear, then

$$P_1/d_1 = P_2/d_2 = P_3/d_3 = P_4/d_4$$

or

$$P_2 = P_1 d_2/d_1; \qquad P_3 = P_1 d_3/d_1; \qquad P_4 = P_1 d_4/d_1$$

Substituting these values of P_2, P_3, and P_4 in eq. 20.8*a*, we have

$$\Sigma M = P_1 d_1{}^2/d_1 + P_1 d_2{}^2/d_1 + P_1 d_3{}^2/d_1 + P_1 d_4{}^2/d_1 \qquad 20.8b$$

Solving for P_1,

$$P_1 = \frac{\Sigma M d_1}{d_1^2 + d_2^2 + d_3^2 + d_4^2} = \frac{\Sigma M d_1}{\Sigma d^2} \qquad 20.9$$

Similarly, the part of the load on any other pile due to moment may be computed by means of eq. 20.9 if d_1 is replaced by the distance from the pile to the center of gravity of the group.

The total reaction on any pile, found by adding the load shown in Fig. 20.5*c* to that in *b*, may be expressed in the form of eq. 20.10.

$$P = \frac{\Sigma V}{n} \pm \frac{\Sigma M d}{\Sigma d^2} \qquad 20.10$$

where P = total pile reaction resulting from moment and direct load

ΣV = sum of vertical loads acting on the foundation

ΣM = sum of moments about the center of gravity of the group. ΣM is sometimes expressed as ΣVe.

n = number of piles in the group

d = distance from the center of gravity of the group to pile in question

Σd^2 = sum of the squares of the distances to each pile from the center of gravity of the group

Inspection of eq. 20.10 reveals that it is no more than a special form of the basic formula for stress on a section or for pressure beneath a soil-supported footing when either is subjected simultaneously to direct load and moment. The number of piles n is substituted for the area, and the term Σd^2 replaces the moment of inertia of the area. For this reason Σd^2 is sometimes called the moment of inertia of the group of piles. The analogy between the terms of the two equations is shown in Fig. 20.5.

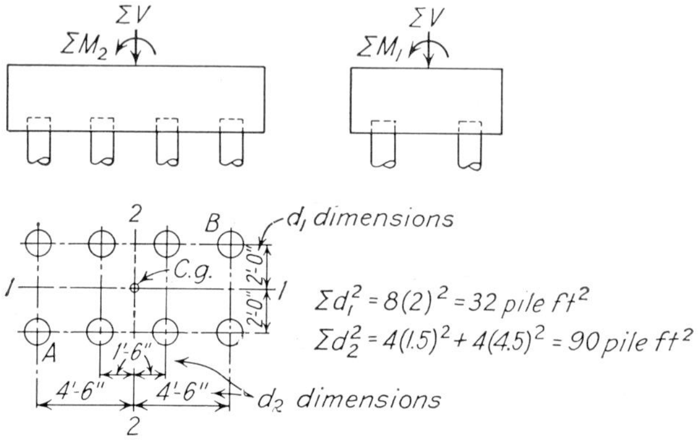

FIG. 20.6. Group of piles subjected to direct load and to moments about both axes.

Most groups of piles contain several rows. Furthermore, moment about both axes is not uncommon. Equation 20.11 applies to these conditions.

$$P = \frac{\Sigma V}{n} \pm \frac{\Sigma M_1 d_1}{\Sigma d_1{}^2} \pm \frac{\Sigma M_2 d_2}{\Sigma d_2{}^2} \qquad 20.11$$

The subscript of the moment denotes the centroidal axis about which the moment acts. The subscript of the distance indicates the centroidal axis to which the distance from the pile is measured. These symbols are shown in Fig. 20.6.

If the moments have the directions shown in Fig. 20.6, it is apparent that pile A carries the greatest load whereas pile B carries the least. Both ΣM_1 and ΣM_2 increase the reaction at A and decrease that at

B. Thus, it is possible to select by inspection the proper signs in the application of eq. 20.11 to any pile.

The determination of Σd^2 for large groups of piles may be considerably simplified by the use of eq. 20.12 which applies to a single row of piles with equal spacing.

$$\Sigma d^2 \text{ (one row)} = \frac{s^2}{12} n_1(n_1^2 - 1) \qquad 20.12$$

where s = spacing of piles in the row

n_1 = number of piles in the row

20.8. Piles Subjected to Tension

Ordinarily the piles beneath a footing are expected to act in compression, and only nominal provision is made to anchor them to the footing. This condition exists whenever all the pile reactions computed in accordance with eq. 20.10 and 20.11 are positive. If some of the reactions are negative but the piles are not anchored to the cap, the situation is analogous to that described in Arts. 20.3 and 20.5 which deal with footings having only part of their bases in compression. If the piles corresponding to the negative reactions cannot resist the tensile forces, the compression in the other piles is increased.

On the other hand, piles are often used specifically to resist tensile forces beneath several common types of structures such as towers, gas storage tanks, and tall stacks. Beneath such structures the tensile forces are usually temporary and are almost always caused by the moment due to wind. Under these conditions, if the piles are capable of withstanding tension and are adequately anchored to the cap, the loads in each pile may be computed by means of eq. 20.10 and 20.11.

20.9. Illustrative Design. DP D-4. Bridge Pier

The base of a bridge pier is a common example of a footing subjected to vertical loads together with moment about both axes. The vertical loads are due to the dead weight and live load of the superstructure and to the weight of the pier itself. Moments and shears on the foundation are produced by horizontal forces such as centrifugal force and those due to traction, nosing, wind, current, and ice. For the most unfavorable combination of these loadings, the allowable soil pressure or pile reaction beneath the base is commonly increased from 25 to 50 per cent above the value permitted under dead plus live load.

DP D-4
Bridge Pier Foundation
Sh. 1 of 3

Bridge Pier. Design of Foundation

General Data:

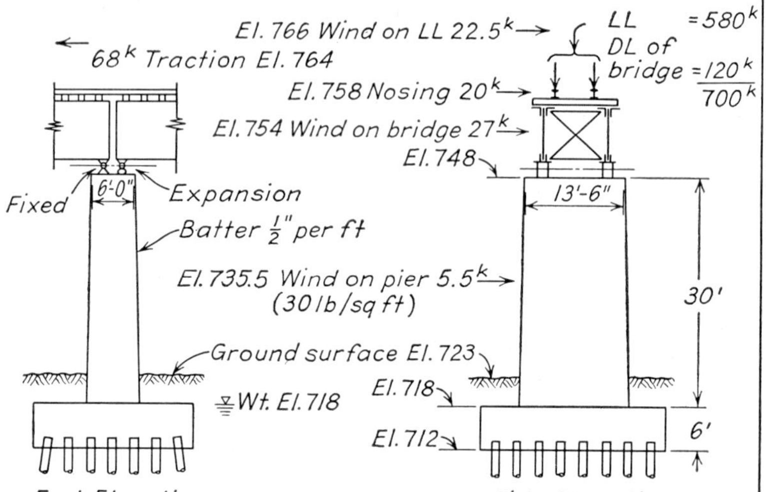

End Elevation

Side Elevation

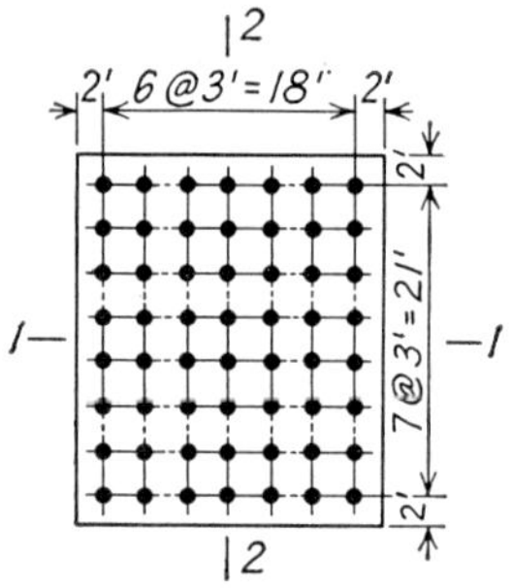

Plan of Base

56 Piles

$\Sigma d_1^2 = 7 \text{ rows} \times \frac{3^2}{12}(8)(8^2-1) = 2646 \text{ pile ft}^2$

$\Sigma d_2^2 = 8 \text{ rows} \times \frac{3^2}{12}(7)(7^2-1) = 2016 \text{ pile ft}^2$

Section Moduli:

Axis 1-1 = 2646 ÷ 10.5 = 252 pile ft

Axis 2-2 = 2016 ÷ 9.0 = 224 pile ft

DP D-4
Bridge Pier Foundation
Sh. 2 of 3

Bridge Pier. Design of Foundation

Pile Reactions Due to Vertical Loads:

Shaft: Top area = 13.5 × 6 = 81.00 sq ft
Mid " × 4 = 14.75 × 7.25 × 4 = 427.75
Bott. " = 16.0 × 8.5 = 136.00
644.75 sq ft
$\times \frac{30}{6}$ = 3224 cu ft

3224 cu ft × 0.15 = 484^k

Footing: 22 × 25 × 6 = 3300 cu ft
× 0.15 = 495

Earth: 22 × 25 = 550 sq ft
15.8 × 8.3 = 131
419 sq ft
× 5 = 2095 cu ft
× 0.12 = 251

DL (Superstructure): 120
1350^k

Buoyancy: 22 × 25 × 6 = 3300 cu ft
× 0.0625 = 206^k
1144^k = total DL

LL: 580^k
1724^k = total DL + LL
÷ 56 = 30.8^k/pile

Pile Reactions Due to Moment:

Traction: $68^k \times 52' = 3536'^k$
÷ 224 = 15.8^k/pile

Transverse wind:
On bridge $27^k \times 42' = 1134'^k$
÷ 252 = 4.5^k/pile
On LL $22.5^k \times 54' = 1212'^k$
÷ 252 = 4.8
On end of shaft $5.5^k \times 23.5' = 129'^k$
÷ 252 = 0.5
} 9.8^k/pile

Nosing: $20^k \times 46' = 920'^k$
÷ 252 = 3.7^k/pile

} 29.3^k/pile

Maximum pile reaction: 60.1^k/pile

Minimum pile reaction: $30.8^k - 29.3^k$ = 1.5^k/pile

DP D-4
Bridge Pier Foundation
Sh. 3 of 3

Bridge Pier. Design of Foundation

Soil Pressure if Piles Are Omitted:

Base 22' x 25' A = 550 sq ft

Section moduli: Axis 1-1 $= \frac{1}{6}(22)(25)^2 = 2290\ ft^3$

Axis 2-2 $= \frac{1}{6}(25)(22)^2 = 2020\ ft^3$

Vertical load on base $= 1724^k$

Moments on Base:

1134'k
1212
129
920

$M_{1-1} = 3395'^k$ $M_{2-2} = 3536'^k$

Maximum soil pressure: $\frac{1724}{550} + \frac{3395}{2290} + \frac{3536}{2020} = 6.4^k/sq\ ft$

Minimum soil pressure: $3.1 - 1.5 - 1.8 = -0.2^k/sq\ ft$

A typical pier for a single track E-60 railroad bridge is shown in Sh. 1, DP D-4. The transverse spacing of the girders and the dimensions of the base plates for the bridge shoes determine the size of the top of the pier shaft. The forces that must be considered in the analysis and design of the structure are given; a discussion of the basis for determining the magnitude of each force is beyond the scope of this text.

The object of the design plate is to demonstrate the application of the principles discussed in Arts. 20.5 and 20.7. Hence, the computations are limited mainly to those for determining the critical pile reactions and for evaluating the pressures against the base if piles are omitted. The number of piles and dimensions of the base are usually determined by trial and revision. The computations on Sh. 2 show that the pile loads resulting from traction, nosing, and wind amount to almost 50 per cent of the total reaction on the most highly stressed pile. Therefore, the total number of piles required is considerably greater than that determined by dividing the total vertical load by the allowable capacity per pile. The vertical loads in the piles are found by means of eq. 20.11; it should be noted that the computation of the moment of inertia of the group of piles is simplified considerably by the use of eq. 20.12. In Sh. 3, the distribution of pressure beneath the base if no piles are used is found by means of eq. 20.7.

As indicated on Sh. 1, most pile foundations for bridge piers contain some batter piles to provide greater stability against horizontal forces. However, computations concerned with this phase of the design have been omitted because the analysis of foundations containing batter piles is explained in Art. 22.6 as part of the discussion of pile-supported retaining walls.

A complete structural design of the base is not included because the choice of critical sections and the subsequent computations would involve the duplication of much that is shown in previous design plates. The computations for the shaft are also omitted because this phase of the design is exclusively a problem in reinforced-concrete analysis and design.

Combined Footings

Construction photograph showing two trapezoidal footings, each of which is to support an interior column and an exterior wall column. The reinforcement for the footing in the foreground has been placed but the concrete has not been cast. The footing in the rear has been completed.

PLATE 21.

CHAPTER 21

Combined Footings and Rafts

21.1. Purpose of Combined Footings

The need for combined footings arises most commonly either when two columns are spaced so closely that individual footings are not practicable or when a wall column is so close to the property line that it is impossible to center an individual footing under the column.

A combined footing is so proportioned that the centroid of the area in contact with the soil lies on the line of action of the resultant of the loads applied to the footing; thus, the distribution of soil pressure is fairly uniform. In addition, the dimensions of the footing are chosen such that the allowable soil pressure is not exceeded. When these two criteria are satisfied, the footing should neither settle nor rotate excessively.

In the design of a combined footing that rests on piles, the total number of piles is found by dividing the sum of the vertical loads on the footing, including the weight of the footing, by the safe load per pile. The piles are then arranged in a pattern such that the centroid of the group is close to the line of action of the resultant vertical load. As a consequence, all piles under the footing are subjected to approximately equal loads.

21.2. Combined Footings of Rectangular and Trapezoidal Shapes

A combined footing is usually made rectangular in plan, Fig. 21.1*a*, if the rectangle can extend beyond each column whatever distance is necessary to make the centroid of the rectangle coincide with the point at which the resultant R of the column loads intersects the base. If the footing is to support an exterior column at the property line where the projection is limited, a rectangular shape can still be used, provided the interior column carries the greater load. The length L of the combined footing is established by adjusting the projection d of the footing beyond the interior column. The width is then calculated by dividing the sum of the vertical loads by the product of the length and the allowable soil pressure.

If for some reason the footing cannot project the required distance beyond one or both columns, a trapezoidal combined footing, Fig. 21.1*b*, is commonly used. The location of the resultant of the column loads establishes the position of the centroid of the trapezoid. The length L is usually limited by the property line at one end and adjacent construction at the other. The dimensions B_1 and B_2 can be determined from the solution of two simultaneous equations: one equation expresses the location of the centroid; and the other equates

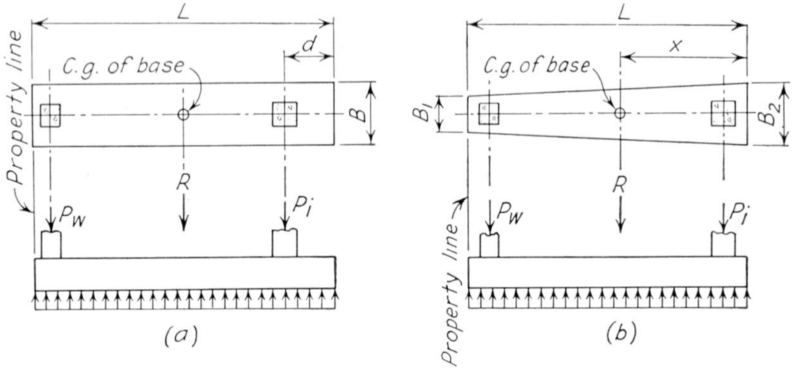

FIG. 21.1. Diagram illustrating forces acting on combined footings of (*a*) rectangular and (*b*) trapezoidal shape.

the sum of the loads to the product of the allowable soil pressure and the area of the footing. In this manner the two criteria discussed in Art. 21.1 are satisfied.

The solution of the two equations mentioned in the preceding paragraph leads to the following expressions for the widths of a trapezoidal footing.

$$B_1 = \frac{2A}{L}\left(\frac{3x}{L} - 1\right) \qquad 21.1$$

and

$$B_2 = (2A/L) - B_1 \qquad 21.2$$

where A is the area determined by dividing the sum of the loads by the allowable soil pressure; other symbols are shown in Fig. 21.1*b*. Inspection of eq. 21.1 shows that the width B_1 is zero when x is equal to one third the length of the footing. If this condition exists, a triangular footing is needed to satisfy the requirement of uniform soil pressure. In tentative designs, whenever the distance x approaches or is less than $L/3$, the length L should be increased by increasing the projection at the wide end.

21.3. Cantilever Footings

A third relatively common type of combined construction is the cantilever footing, Fig. 21.2. It is designed to support a wall column near its edge without causing nonuniform soil pressure. The cantilever principle, largely concealed in actual footings of this type, is illustrated in Fig. 21.2*b*. The combined footing may be regarded as two individual footings connected by a strap beam. It may be inferred from Fig. 21.2*b* that the proportioning of the areas of the

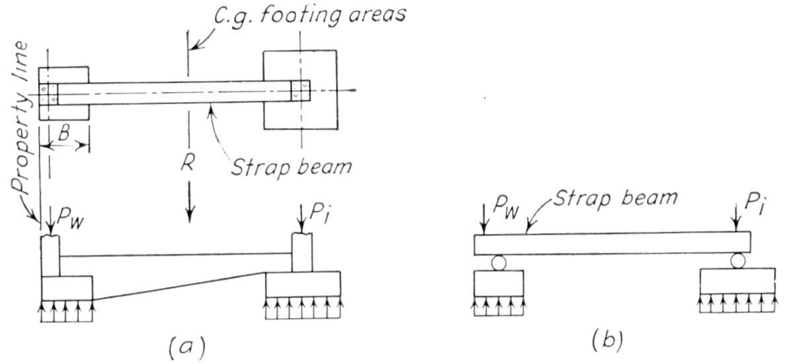

Fig. 21.2. Diagram illustrating (*a*) forces acting on cantilever footing; (*b*) principle of cantilever footing.

two individual footings is a problem in statics if the allowable soil pressure is known and if the dimension B of the wall footing is either fixed or assumed. Furthermore, the centroid of the two areas must lie on the line of action of the resultant load. This requirement may not be obvious because the two areas are usually found more or less independently from reactions determined by statics.

21.4. Choice of Column Loads

Except for the choice of the allowable soil pressure, the most difficult step in proportioning a combined footing is the selection of the appropriate column loads. Generally, it is considered good practice to proportion the footing for a uniform soil pressure under the dead load plus only the amount of live load that is likely to govern the settlement. If the footing is to rest on clay, the appropriate live load is the average likely to act on the footing over a period of years; if on sand, it is the maximum probable value. In either instance, the centroid of the footing is made to lie on the line of action of the resultant of column loads that consist of dead load plus only a fraction of the maximum live load specified in the building code for design of the columns.

Even though a combined footing may be so proportioned that uniform soil pressure is produced by dead load plus a reduced live load, Fig. 21.3*a*, sound engineering practice requires that the structural design of the footing be based on the loads for which the columns are designed. Moreover, the factor of safety against a bearing capacity failure under this loading should not be less than 2. Inasmuch as the resultant of the maximum loads does not ordinarily coincide with the resultant of the loads used in proportioning the footing, the allowable soil pressures appropriate for the maximum loads must be selected with proper consideration for the nonuniform distribution of the soil pressure, Fig. 21.3*b*. In this connection, the procedures given in Arts. 14.2 and 15.2 for determining the allowable toe pressure beneath footings acted upon by a triangular distribution of soil pressure may serve as a guide. The shears and moments corresponding to the nonuniform soil pressure are the basis for the structural design of the footing slab, Art. 21.5.

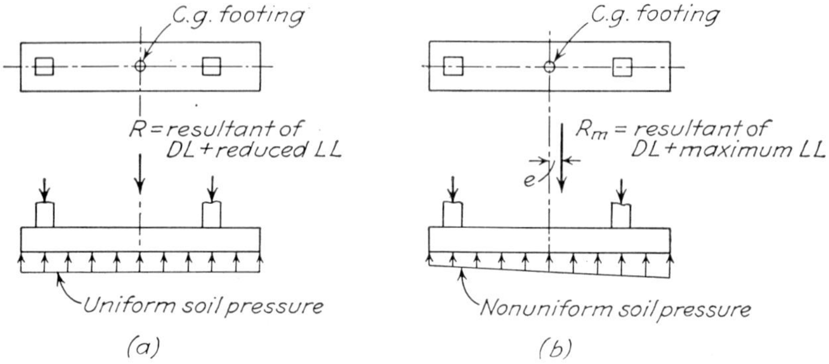

FIG. 21.3. Forces to be considered (*a*) in determining dimensions of combined rectangular footing and (*b*) in making structural design of the footing.

21.5. Structural Design of Combined Footings

The rigidity of ordinary combined footings is in general somewhat less than that of most individual spread footings. In spite of this fact, the design of a combined footing is commonly based on the assumption that the soil pressure beneath the footing has a planar distribution. This assumption is usually satisfactory and conservative in view of the other uncertainties associated with the problem.

The principal reinforcement in a combined footing is placed in the longitudinal direction. The amount of steel is usually determined by assuming that the footing acts as a one-way slab. However, transverse reinforcement is also provided at the bottom of the footing near the columns. The critical section for transverse bending is taken at the faces of the columns or pedestals. The transverse reinforcement is divided into groups proportionate in sectional area to the column loads, and the reinforcement at each column is placed uniformly within a band having an arbitrarily chosen width often taken as the width of the footing at the column. Finally, shear reinforcement is almost always necessary to resist the high diagonal tension adjacent to the columns.

The procedure for the design of a combined footing such as that shown in Fig. 21.3 may be summarized as follows:

1. Determine the column loads appropriate for considerations of settlement. These consist of dead load plus only a portion of the live load specified for design of the columns.
2. Using the resultant of the loads in step 1, select the plan dimensions of the footing to obtain a uniform soil pressure that does not exceed the pressure appropriate for this condition of loading.
3. Determine the maximum column loads. These are usually the loads for which the columns are designed.
4. Find the magnitude and variation of the soil pressure for the loads in step 3. If the maximum soil pressure exceeds the value considered appropriate for this condition of loading, the width of the footing must be increased whereas the position of the centroid must remain unchanged.
5. Draw the shear and moment diagrams for the footing when it is subjected to the maximum conditions of steps 3 and 4.
6. Using step 5 as the basis for design, determine the depth of the footing and the necessary amount of reinforcing steel at appropriate sections.

21.6. Illustrative Design. DP D-5 *a, b, c.* Combined Footings

The preceding articles of this chapter have discussed in some detail the various types of combined footings. The purpose of these design plates is to present typical computations that demonstrate the application of some of the principles associated with the proportioning and structural design of three common types of combined footings.

The computations given in DP D-5*a* deal with a rectangular combined footing. The footing is first proportioned for uniform soil pressure under dead load plus a reduced live load. Second, the

DP D-5a
Combined Footing
Sh. 1 of 4

Design of Rectangular Combined Footing

General Data:

Allowable soil pressure:

3.0^k/sq ft for DL + reduced LL

4.5^k/sq ft for DL + maximum LL

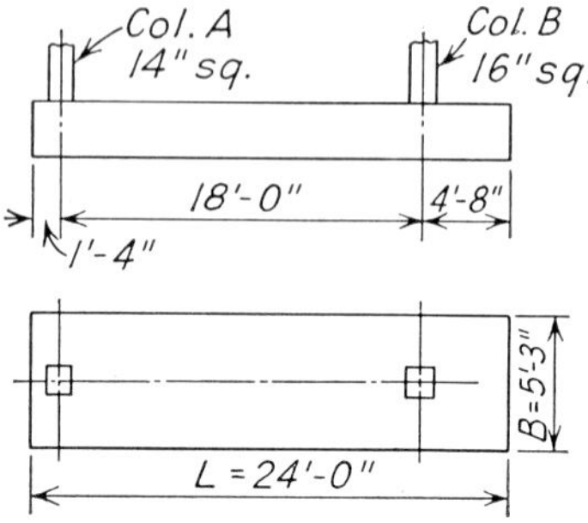

Column Loads:

	Col. A	Col. B
DL	111.2^k	139.5^k
Reduced LL	40.5	81.0
Maximum LL	81.0	162.0

Resultant Column Loads:

DL + red. LL = 151.7 + 220.5 = 372.2^k

DL + max. LL = 192.2 + 301.5 = 493.7^k

Proportion Footing for Uniform Pressure under DL + Red. LL:

Locate resultant col. load.

$$\frac{220.5 \times 18.0}{372.2} = 10.66' \text{ from col. A}$$

Length $L = 2(10.66 + 1.33) = 23.99'$ Use 24'-0".

Width $B = \dfrac{372.2}{24 \times 3.0} = 5.17'$ Use 5'-3"

Soil Pressure under DL + Max. LL

Locate resultant col. load.

$$\frac{301.5 \times 18.0}{493.7} = 10.99' \text{ from col. A}$$

10.67 from col. A to c.g. footing

$e = 0.32'$

$$q_{max} = \frac{493.7}{5.25 \times 24}\left(1 + \frac{6 \times 0.32}{24}\right) = 3.92\,(1 + 0.08) = 4.24^k/\text{sq ft} < 4.5 \text{ ok}$$

$$q_{min} = 3.92\,(1 - 0.08) = 3.60^k/\text{sq ft}$$

DP D-5a
Combined Footing
Sh. 2 of 4

Design of Rectangular Combined Footing

Structural Design of Footing:

Allowable stresses: $f_s = 20$ ksi, $v = 75$ psi (without stirrups)
$f_c = 1200$ psi, $v = 360$ psi (with stirrups)
$n = 10$, $u = 210$ psi (top bars)
$j = 0.87$, $u = 240$ psi (bottom bars)
$R = 197$ psi

Shear and Moment Diagrams:

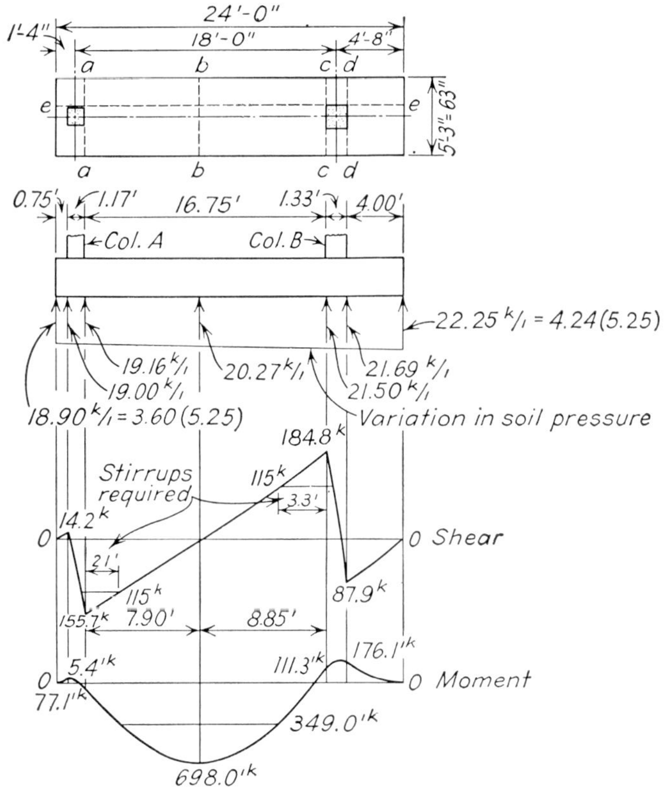

DP D-5a
Combined Footing
Sh. 3 of 4

Design of Rectangular Combined Footing

Structural Design of Footing: Cont. from Sh. 2

Depth of Footing:

Mom. at bb $= 698.0'^{k}$ $\quad d = \sqrt{\dfrac{698(12)}{0.197(63)}} = 26.0''$ $\quad$ Use $d = 28''$

Longitudinal Steel:

Top bars at bb: $A_s = \dfrac{698(12)}{20(0.87)(28)} = 17.2$ sq in.

Use 24-#8. $A_s = 18.96$ $\quad \Sigma_o = 75.3$

Top layer: 12-#8 x 20'-0" $\quad \Sigma_o = 37.6$
Bott. layer: 12-#8 x 13'-0"

Shear at point of inflection near col. B = say 175^{k}

$u = \dfrac{175}{37.6(0.87)(28)} = 191$ psi < 210 ok

Bott. bars at dd: $A_s = \dfrac{176.1(12)}{20(0.87)(28)} = 4.3$ sq in.

Use 16-#5 x 7'-0". $A_s = 4.96$ $\quad \Sigma_o = 31.4$

Shear at cc $= 184.8^{k}$

$u = \dfrac{184.8}{31.4(0.87)(28)} = 242$ psi > 240 say ok

Transverse Steel:

Bott. bars at ee: Shear at ee $= \dfrac{3.60 + 4.24}{2}(24)(1.96) = 184^{k}$

Mom. at ee $= 184 \times \dfrac{1.96}{2} = 180'^{k}$

$A_s = \dfrac{180(12)}{20(0.87)(28)} = 4.4$ sq in.

$\times \dfrac{192.2}{493.7} = 1.7$ sq in. at col. A

$\times \dfrac{301.5}{493.7} - 2.7$ sq in. at col. B

Use 7-#5 x 4'-9" @ 10" ctrs. at col. A. $A_s = 2.17$ $\quad \Sigma_o = 13.7$

Use 10-#5 x 4'-9" @ 7" ctrs. at col. B. $A_s = 3.10$ $\quad \Sigma_o = 19.6$

$\qquad 5.27 \qquad 33.3$

$u = \dfrac{184}{33.3(0.87)(28)} = 227$ psi < 240 ok

DP D-5a
Combined Footing
Sh. 4 of 4

Design of Rectangular Combined Footing

Structural Design of Footing: Cont. from Sh. 3

Stirrups:

Use #5  $A_v = 4 \times 0.31 = 1.24$ sq in.

Without stirrups $V_c = 0.075(63)(0.87)(28) = 115^k$

At col. A, $155.7 - 115.0 = 40.7^k$ } Shears to be resisted by vertical stirrups

At col. B, $184.8 - 115.0 = 69.8^k$ }

At col. A, $s = \dfrac{1.24(20)(0.87)(28)}{40.7} = 14.9''$ Use 2 spaces @ 14"

At col. B, $s = \dfrac{1.24(20)(0.87)(28)}{69.8} = 8.6''$ Use 2 spaces @ 8"
2 " " 14"

Sketch of Footing:

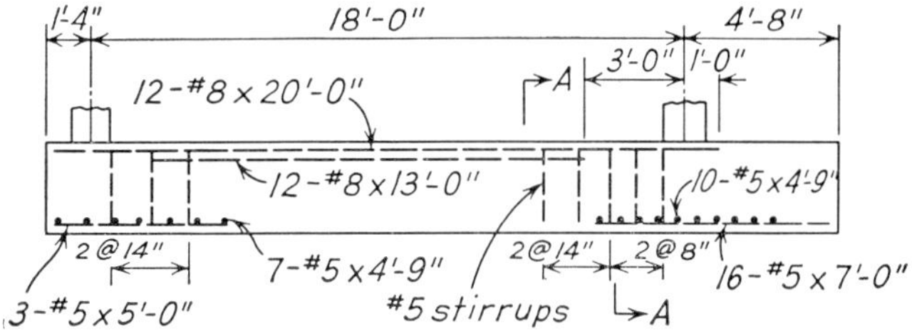

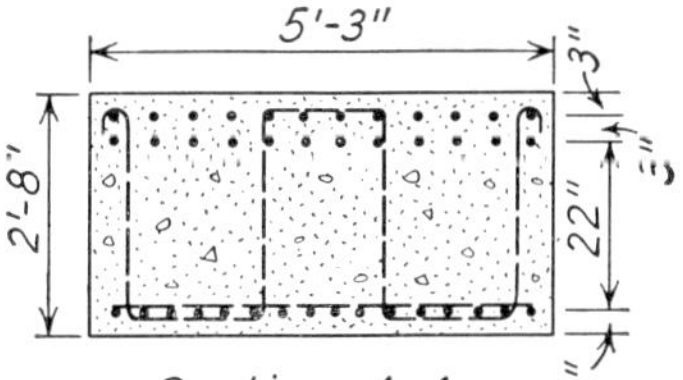

Section A-A

DP D-5b
Combined Footing
Sh. 1 of 1

Design of Trapezoidal Combined Footing
Required Area

General Data:

Refer to Sh. 1, DP D-5a, for allowable soil pressure, column loads, and the location of their resultants.

Proportion Footing for Uniform Pressure under DL + Reduced LL:

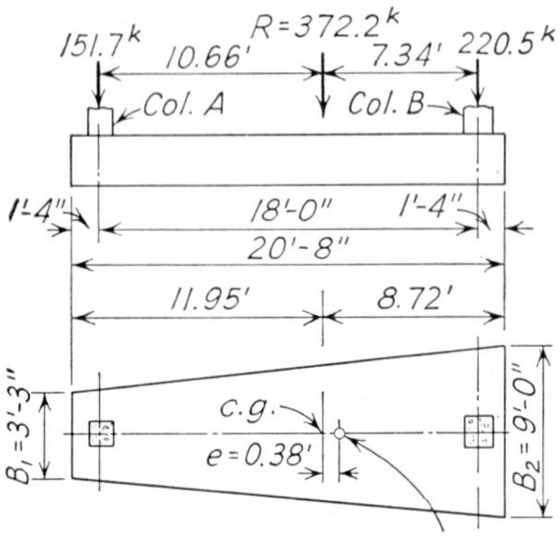

Resultant DL + max. LL = 493.7k

Total Area Required:

$$A = \frac{151.7 + 220.5}{3.0} = \frac{372.2}{3.0} = 124.1 \text{ sq ft}$$

$$B_1 = \frac{2 \times 124.1}{20.67}\left(\frac{3 \times 8.67}{20.67} - 1\right)$$

$$= 12.0\,(1.26 - 1) = 3.12'$$

Use $B_1 = 3'\text{-}3''$

$$B_2 = \frac{2 \times 124.1}{20.67} - 3.12 = 12.0 - 3.12 = 8.88'$$

Use $B_2 = 9'\text{-}0''$

Total area furnished $= \left(\dfrac{9.0 + 3.25}{2}\right) 20.67 = 126.5$ sq ft > 124.1 ok

Soil Pressure under DL + Max. LL:

Location of c.g $= \dfrac{20.67}{3}\left(\dfrac{2 \times 3.25 + 9.0}{12.25}\right) = 8.72'$ from B_2

Location resultant DL + max LL $= 8.34$ " "

$e = 0.38'$

$\times 493.7 = 188.0^{'k} = M$

Moment of Inertia:

$\frac{1}{3}(3.25)(20.67)^3 = 9580 \text{ ft}^4$

$\frac{1}{12}(5.75)(20.67)^3 = 4240$

$13820 \text{ ft}^4 = I$ about B_2

$126.5\,(8.72)^2 = 9620$

$4200 \text{ ft}^4 = I$ about c.g

$$q_{max} = \frac{493.7}{126.5} + \frac{188.0 \times 8.72}{4200} = 3.90 + 0.39 = 4.29^k/\text{sq ft} < 4.5 \text{ ok}$$

$$q_{min} = \frac{493.7}{126.5} - \frac{188.0 \times 11.95}{4200} = 3.90 - 0.54 = 3.36^k/\text{sq ft}$$

DP D-5c
Combined Footing
Sh. 1 of 1

Design of Cantilever Combined Footing

Required Areas

General Data:

Refer to Sh. 1, DP D-5a, for allowable soil pressure and column loads.

Proportion Footing for Uniform Pressure under DL + Red. LL:

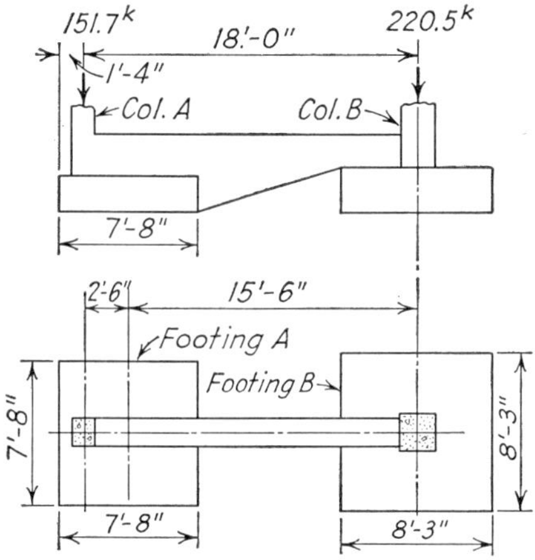

Footing A:

Assume length = 7'-8"

$151.7 \times \frac{18.0}{15.5} = 176^k$

$\div 3.0 = 58.7$ sq ft

Width $= \frac{58.7}{7.67} = 7.66'$

Use 7'-8" x 7'-8". A = 58.8 sq ft

Footing B:

Col. B $= 220.5^k$

$151.7 \times \frac{2.5}{15.5} = 24.4$

196.1^k

$\div 3.0 = 65.3$ sq ft

Use 8'-3" x 8'-3". A = 68.0 sq ft

Total area required $= \frac{151.7 + 220.5}{3.0} = \frac{372.2}{3.0} = 124.1$ sq ft $< 58.8 + 68.0$ ok

Soil Pressures under DL + Max. LL:

Footing A:

$192.2 \times \frac{18.0}{15.5} = 223.0^k$

$\div 58.8 = 3.8^k$/sq ft

Footing B:

Col. B $= 301.5^k$

$192.2 \times \frac{2.5}{15.5} = 31.0$

270.5^k

$\div 68.0 = 4.0^k$/sq ft < 4.5 ok

variation in soil pressure is determined for the maximum conditions of loading by methods discussed in Art. 20.2. Both of these steps are given on Sh. 1. Next, the shear and moment diagrams are drawn, Sh. 2. These form a basis for the design computations, Shs. 3 and 4. The computations on Shs. 3 and 4 should be familiar to anyone having a knowledge of the principles of reinforced-concrete design. However, the reader is referred to Art. 21.5 for a discussion of the distribution of the transverse steel.

The proportions of a trapezoidal combined footing are determined in DP D-5*b*. Equations 21.1 and 21.2 are used to obtain appropriate end widths for uniform soil pressure under dead load plus reduced live load. In order to determine the soil pressures under the maximum conditions of loading, the location of the center of gravity of the footing is obtained by means of the equation

$$\bar{x} = \frac{L}{3}\left(\frac{2B_1 + B_2}{B_1 + B_2}\right)$$

The moment of inertia of the footing is conveniently determined by the parallel-axis relationship

$$I_{cg} = I_{B_2} - A\bar{x}^2$$

Finally the soil pressures are obtained by means of the equation

$$q = \frac{\Sigma V}{A} \pm \frac{Mc}{I}$$

The computations given in DP D-5c demonstrate the method of proportioning cantilever footings. The reader is referred to Fig. 21.2*b* for illustration of the cantilever principle. Otherwise, the computations are self-explanatory.

21.7. Design of Raft Foundations

The conditions under which a raft foundation is suitable have been discussed in Part B, and the methods for selecting the allowable soil pressure and for estimating the settlement in Part C. This article deals with the structural design of the raft itself.

In its simplest common form a raft consists of a reinforced-concrete slab which supports the columns and walls of a structure and which distributes the load therefrom to the underlying soils. Such a slab is usually regarded and designed as an inverted continuous flat-slab floor supported without upward deflection at the columns and walls. The soil pressure acting against the slab is commonly assumed to be uniformly distributed and equal to the total weight of the building divided by the area of the raft. The moments and shears in the slab are determined by the use of appropriate coefficients listed in specifications for the design of flat-slab floors.

This procedure is usually followed if the columns are more or less equally loaded and equally spaced. However, if the downward loads on some areas are on the average much heavier than on others, the slab is often considered to be subdivided into such areas and the average soil pressure on each area determined by dividing the total load on the subdivision by the corresponding area. Continuity of the slab between areas is commonly provided, although vertical construction joints through the slab and superstructure may be advisable between areas of extreme differences in pressure, to allow for differential settlement without structural or architectural damage.

On account of the erratic variations in compressibility of almost every soil deposit, there are likely to be correspondingly erratic deviations of the soil pressure from the average value. Since the moments and shears are determined on the basis of the average pressure, it is considered good practice to provide the slab with more than the theoretical amount of reinforcement and to use the same percentage of steel at top and bottom.

The flat-slab analogy has been widely used, often with complete success. On the other hand, it has frequently led to structural failure not only of the slab but also of the superstructure. Therefore, its limitations must be clearly understood.

The analogy is valid only if the differential settlement between columns will be small and, furthermore, if the pattern of the differential settlement will be erratic rather than systematic. These limitations are necessary because the design of a flat-slab floor is based on the tacit assumption that there will be insignificant differential settlements among the points of support at columns or walls.

Adequate structural design of the slab of a raft foundation on this assumption is, unfortunately, no guarantee that the deflections of the raft will actually be unimportant. Indeed, if the structure covers a fairly large area and significantly increases the stresses in an underlying deposit of compressible clay or silt, it is likely to experience large systematic differential settlements. These cannot be avoided merely by providing great strength in the slab; it is also necessary to provide stiffness. However, a stiff foundation is likely to be subjected to bending moments far in excess of those corresponding to the flat-slab analysis. These moments may be so great as to require deep beams, trusses, or even utilization of the superstructure to provide the necessary strength. The raft foundation must now consist of two almost independent elements: the base slab, which may still be designed by the flat-slab analogy; and the stiffening members, which have the function of preventing most of the differential settlement of the points of support for the base slab.

The design of the stiffening members is a difficult structural problem for which no straightforward procedure as yet exists. Experience and mature judgment are essential. Increasing the stiffness of a raft foundation above a compressible soil leads to a redistribution of the soil pressure against the base slab; the raft should not deflect excessively when acted upon by the redistributed, nonuniform pressure. Unfortunately, no reliable rational estimate of the real distribution of pressure can be made because of the extremely complex relationships among stress, strain, and time, not only for soils but also for building frames.

According to theory, if a rigid structure rested on an elastic subgrade, the pressure on the base of the structure would vary from a minimum near the middle of the base to a maximum at the edges. This fact has occasionally been used as a basis for estimating the soil pressure for design of a stiff raft foundation above a subsoil of compressible clay or silt. It appears unlikely that the edge pressure could exceed twice the average pressure. Consequently, it is considered conservative to design the stiffening elements and the slab of such a raft foundation for two conditions: a soil pressure uniform over the entire slab; and a pressure varying in some arbitrary fashion from a minimum at the middle to twice the average at the edges. The average pressure is, of course, the same for both conditions. In any part of the foundation the strength is made adequate for whichever distribution leads to the more severe conditions, and for the nonuniform distribution the distortion of the foundations must not be excessive.

This is a logical basis for design, but it may often be too conservative and hence uneconomical. The choice of the most suitable edge pressure often taxes the ability of the most experienced foundation engineer.

These paragraphs suggest that the design of stiff rafts above sand, stiff clay, or other relatively incompressible materials is a somewhat complex but otherwise routine structural problem. On the other hand, if the subsoil contains highly compressible layers, the problem is by no means routine.

As an alternative to the relatively great expense of a stiff raft of large size above a compressible deposit, substantial economy can be realized by designing a flexible raft and superstructure that can deform without structural damage into the shape corresponding to the compression of the subsoil. Obviously this alternative cannot be chosen if architectural or functional considerations demand a relatively unyielding structure. On the other hand, many types of structures such as large steel tanks and one- or two-story steel-frame industrial buildings with corrugated metal or asbestos siding can experience large deformations without any detrimental consequences. It may often prove preferable to accept the deformations if the cost of a stiff foundation can be avoided.

The design of a flexible raft foundation cannot readily be based upon the calculation of stresses in the slab. Instead, it is necessary to estimate, on the basis of a settlement forecast, the maximum curvature to which the raft may be subjected, and to select the thickness of the slab and the amount of reinforcement such that the slab will not develop cracks large enough to be unsightly or to permit serious leakage of groundwater even if it is deformed in accordance with the estimated curvature. As a rough guide, the quantity of steel may be taken as 1 per cent in each of two directions at right angles to each other, equally divided between the top and bottom of the slab. The thickness of the slab should not be generally greater than about 0.01 times the radius of curvature, but local increases in thickness near columns and walls may be required to prevent shear failures. In any particular project, many factors enter into the final choice of the criteria for design, and expert judgment and experience are required. Hence, the design of a flexible raft above compressible deposits, like that of a stiffened raft, is not a routine problem.

Charles Augustin Coulomb (1736–1806)

French army engineer. After impairing his health by nine years of service in the tropics he devoted himself to purely scientific research and became immortal by his fundamental contributions to our knowledge of friction, electricity, and magnetism. However, at the beginning of his career as a scientist, in 1773, he published a paper on the earth pressure against retaining walls. The theory contained in this paper is distinguished by its clarity and simplicity, and by its realistic assumptions derived from field observations regarding the shape of the surface of rupture. Today it is still the basis for the design of retaining walls.

PLATE 22.

CHAPTER 22

Retaining Walls and Abutments

22.1. Introduction

A properly designed retaining wall or abutment must satisfy two almost independent requirements. In the first place, to make the structure safe against failure by overturning and excessive settlement, the pressure beneath the base must not exceed the allowable soil pressure; furthermore, the structure as a whole must have an adequate factor of safety with respect to sliding along its base or along some weak stratum below its base. The evaluation of the ability of the structure to satisfy these requirements is known as the *stability analysis*. The second requirement is that the entire structure as well as each of its parts must possess adequate strength. The evaluation of the strength is known as the *structural analysis*. In practice the stability and structural analyses are made more or less simultaneously. Therefore, they are considered together in this chapter.

Retaining walls and abutments of the gravity and semigravity types are widely used today, especially where a high degree of permanence under unfavorable climatic conditions is desired. The design of such structures, however, is relatively simple in comparison with that of cantilever walls and abutments. Therefore, this chapter is concerned primarily with the latter type. The discussion of cantilever retaining walls is contained in Arts. 22.2 to 22.5, inclusive, and that of cantilever abutments in Art. 22.8.

In general, the procedure for the design of retaining walls and abutments, like that for many other structures, is essentially one of trial and correction. Tentative dimensions must be assumed before either a stability analysis or a structural analysis can be made. After analysis, certain dimensions may have to be revised before a satisfactory design is obtained.

22.2 Proportions of a Cantilever Retaining Wall

Base. The base of the ordinary cantilever retaining wall should be as narrow as possible for economy, but at the same time it must be wide enough to provide adequate stability against overturning and sliding, and to reduce the soil pressure to a tolerable value. The ratio of the width of the base to the over-all height of the wall commonly varies from 0.40 to 0.65. The smaller ratio is appropriate if the base is supported by firm soil and if the backfill has a horizontal surface and consists of clean sand or gravel. On the other hand, as the strength of the subsoil or backfill decreases, and as the slope of the backfill increases, the ratio may approach or even exceed 0.65. Furthermore, the width of the base is influenced by additional loads on the fill behind the wall, such as those due to a railroad, highway, or structure.

The thickness of the base is a function of the shears and moments at sections located at the front and back faces of the stem. Therefore, the thickness is significantly influenced by the position of the stem on the base. If the stem is located so that the projection of the toe from the front face of the wall is approximately 1/3 the width of the base, the thickness of the base commonly lies in the range of 1/12 to 1/8 the height of the wall.

The depth of the base below the ground surface in front of the wall should be sufficient to avoid the movements associated with freezing and thawing of the soil. The necessary depth varies from a few inches in the southernmost states to as much as 8 ft in some of the northern regions of the United States. Even if the depth of frost is small, the base should preferably be placed below the zone of seasonal volume change caused by the variation in moisture content of the subsoil. This is particularly important if swelling clays are encountered. In many instances it becomes necessary to establish the base at a depth much greater than these minimum values to reach a stratum of soil adequate to withstand the pressures imposed upon it.

Stem. The thickness of the stem must be sufficient to resist safely the shears and moments due to the earth pressure against the back of the wall. For this reason, the strength and slope of the backfill have considerable influence on the thickness of the stem. The thickness at the top of the wall should be great enough to permit easy placement of the concrete. The critical section for shear and moment is at the junction of the stem with the base. In order to provide adequate strength at this section it is customary to increase the thickness of the stem with depth by ¼ to ¾ in. per ft.

22.3. Forces Acting on Retaining Walls

Summary of Forces. For the stability and structural analyses, a cantilever retaining wall, Fig. 22.1*a*, is regarded as three separate cantilever beams, Fig. 22.1*b*. The principal forces commonly considered are the earth pressure P_A which acts against the vertical section *ab* through the heel, the earth pressure P_P which acts against the vertical section *cd* through the toe, the soil pressure ΣV which acts vertically on the base *db*, the shear along the base *db*, and the weights of the various parts of the wall and of the masses of earth above the base.

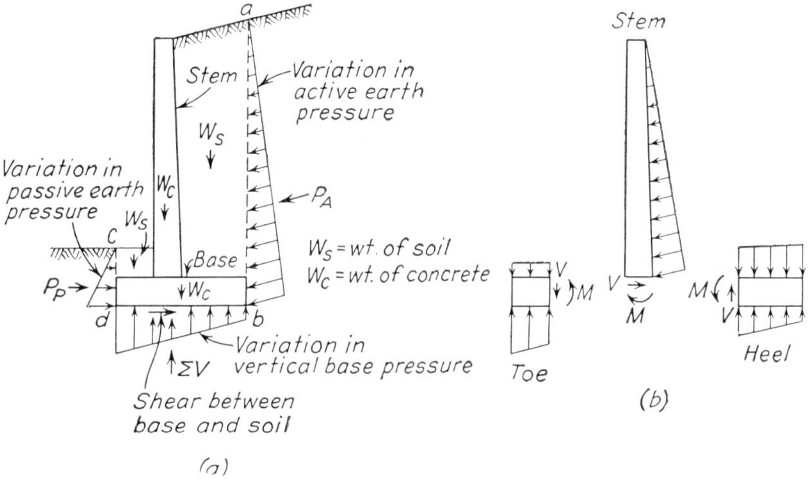

FIG. 22.1. Cantilever retaining wall. (*a*) Forces considered in stability analysis; (*b*) forces acting upon the principal structural elements of the wall.

Earth Pressure. The pressure exerted by an earth backfill against a retaining wall can be computed with reasonable accuracy on the basis of theory only for conditions rarely encountered in practice. In the first place, the designer must know what materials are to be used for the backfill and in what state they will be placed. This implies careful supervision of the backfilling operations. Moreover, the backfill, once it is placed, must be protected in such a way that its physical properties remain constant. These conditions cannot be satisfied economically on any but the largest and most important projects; ordinarily the designer can hope to learn in advance of construction little more than the general type of backfill material. Hence, instead of expending time and effort on theoretical earth-pressure computations it is preferable to estimate the earth pressure on the basis of charts or rules having a partly theoretical, partly empirical basis. Such charts are shown in Fig. 22.2. They can be used if the type of backfill can be classified into one of the four categories listed in the diagram.

Nevertheless, intelligent use of the charts requires a knowledge of the basic principles that govern the magnitude of the earth pressure. These are summarized in the following paragraphs.

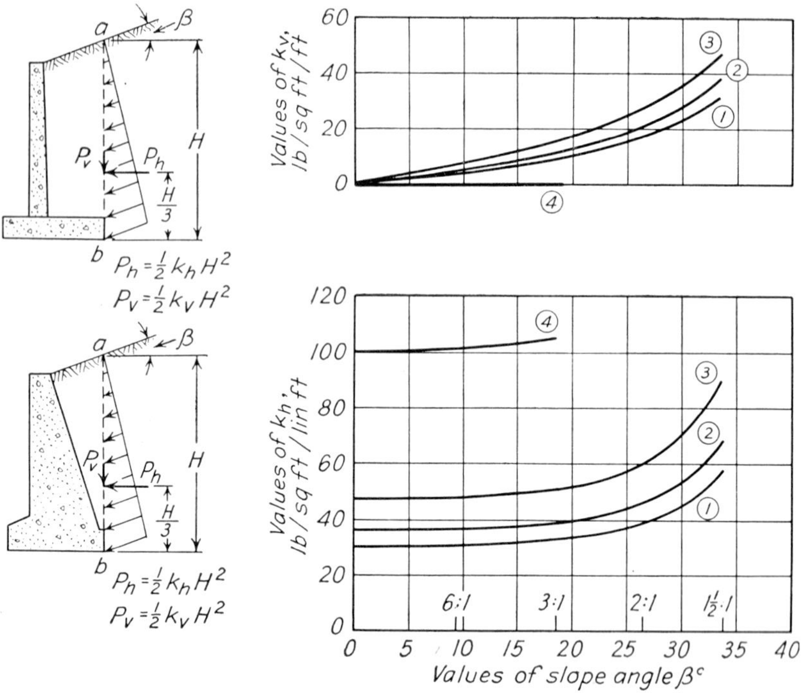

FIG. 22.2. Chart for estimating pressure of backfill against retaining walls supporting backfills with plane surface. Use of chart limited to walls not over about 20 ft high.

(1) Backfill of coarse-grained soil without admixture of fine particles, very permeable, as clean sand or gravel.

(2) Backfill of coarse-grained soil of low permeability due to admixture of particles of silt size.

(3) Backfill of fine silty sand, granular materials with conspicuous clay content, and residual soil with stones.

(4) Backfill of very soft or soft clay, organic silt, or silty clay.

If backfill is placed behind a perfectly unyielding retaining wall, the pressure exerted against the wall has a definite magnitude that depends on the kind of backfill and on the method of placement. This pressure is known as *earth pressure at rest*. Its intensity increases directly with the depth below the ground surface.

If the wall moves away from the backfill, the earth pressure decreases in a continuous manner until, at a specific amount of wall movement, a slip occurs in the backfill along an inclined, slightly curved surface of sliding, as shown in Fig. 22.3*a*. At this stage the shearing strength *s* of the soil is fully mobilized along the surface of sliding, and the direction of the shearing forces is such as to reduce the earth pressure. Hence, at this stage, the earth pressure reaches its minimum value, known as the *active earth pressure*. Further outward movement does not significantly alter the earth pressure.

Laboratory tests and experience have demonstrated that the movement necessary to reduce the earth pressure to the active state is on

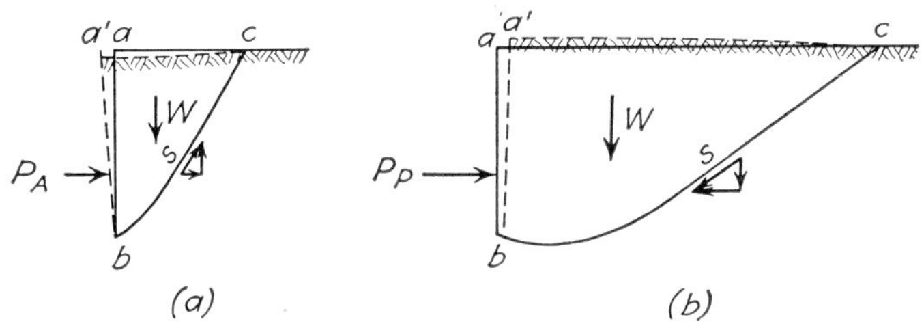

FIG. 22.3. Surface of sliding in backfill of vertical wall if wall is (*a*) allowed to yield away from fill, (*b*) forced inward toward the fill.

the order of 0.1 per cent of the height of the wall. Moreover, it has been found that most retaining walls can and do deflect or displace this far without any undesirable consequences. Therefore, it is reasonable to base the design of ordinary retaining walls on the active earth pressure. The charts Fig. 22.2 give both the vertical and horizontal components of the active pressure.

In contrast, if a wall is forced to move toward the backfill, Fig. 22.3*b*, the earth pressure increases in a continuous manner until, at a particular amount of movement, a slip occurs in the backfill along an inclined somewhat curved surface of sliding which, for theoretical reasons, has a shape different from that of the surface corresponding to the active state. When the slip occurs, the shearing resistance is again fully mobilized along the surface of sliding but in such a direction as to increase the pressure against the wall. At this stage, the earth pressure reaches a maximum value known as the *passive earth pressure*. Further displacement of the wall does not significantly change the pressure. The movement required to develop the passive pressure is somewhat greater than that associated with the active state. The magnitude of the passive pressure may be estimated roughly by means of the charts, Fig. 22.2, and the expression

$$k_p = \tfrac{2}{3}\gamma\,(\gamma/k_h) \qquad 22.1$$

where γ is the unit weight of the soil and k_h, taken from the charts, is the coefficient of the horizontal component of the active pressure. The total passive pressure is

$$P_P = \tfrac{1}{2}k_pH_1^2 \qquad 22.2$$

where H_1 is the height of the face against which the pressure acts. Only the horizontal component of the passive pressure is ordinarily considered in connection with retaining walls.

For the materials included in the categories described in Fig. 22.2, the intensity of both the active and passive pressures can be assumed to increase directly with the depth below the ground surface.

The passive pressure P_P is often disregarded in the stability analyses of retaining walls because of the possibility that the soil in front of the wall may not yet be in position when the backfill is being placed or may at some time be removed by excavation or scour. Furthermore, the strength of the soil may be lowered as a result of frost action, infiltration of water, or cracking due to shrinkage.

Vertical Pressure against Base. Experience has shown that most failures of retaining walls have been the result of misjudgment of the foundation conditions. Therefore, a careful evaluation of the strength and compressibility of the soil beneath both the base and the backfill is the most important single step in the design of a retaining wall. All the factors discussed in Part C in connection with the selection of suitable types of foundations and the choice of appropriate soil pressures or pile loads are as worthy of consideration with respect to retaining walls as to any other type of structure.

The pressure against the base of a retaining wall is commonly assumed to vary in planar fashion as shown in Fig. 22.1*a*. The position of the resultant of all the forces acting on the supporting material at the level of the base can be found by eq. 20.1. For ordinary walls supported by soil the resultant is commonly required to intersect the base within the middle third; hence, the entire area beneath the base is theoretically subjected to compression. If this condition exists, the magnitude of the soil pressure against the base can be determined by eq. 20.4.

Two useful criteria for the design of soil-supported walls may be established from the preceding discussion: (1) The eccentricity of the resultant force, measured from the center line of the base, should not exceed one-sixth the width of the base, and (2) the maximum pressure should not exceed the allowable soil pressure. One or the other of these two criteria commonly controls the width of the base.

On the other hand, retaining walls may be supported by rock, in which case the first criterion given above is commonly changed to permit larger eccentricities. However, in order to provide adequate safety against overturning, most designers prefer to limit the eccentricity to one-fourth the width of the base. That is, the resultant force must intersect the plane of the bottom of the base within the middle half, even though the pressure at the toe may be considerably less than the allowable pressure for the rock. When the resultant lies outside the middle third, the maximum pressure at the toe must be computed by eqs. 20.5 and 20.6, because compression does not exist over the entire area of the base.

Forces Resisting Sliding. According to Fig. 22.1*a*, the horizontal component of the earth pressure P_A must be resisted by the shear between the soil and the base and by the passive earth pressure of the soil in contact with the front of the structure. The ratio between the resisting forces and the horizontal component of P_A is known as the *factor of safety against sliding.* This ratio should be not less than 1.5. Moreover, the passive earth pressure should be disregarded in computing the factor of safety unless local conditions permit reliable evaluation of its lower limiting value and unless the existence of the pressure is assured during the placing of the fill behind the wall.

The shearing resistance between the base and the soil is greatly influenced by the character of the soil. If the surface of contact between the concrete and soil is rough, the maximum shearing strength of the soil can be counted upon. Procedures for determining the shearing strength of soils of different types have been discussed in Arts. 2.10, 2.11, and 2.12. However, in the absence of tests, the total shearing resistance between the base and a soil that derives most of its strength from internal friction may be taken as the normal force ΣV times a coefficient of friction selected from the following values. For coarse-grained soil without silt, the coefficient of friction may be taken as 0.55; for coarse-grained soil with silt, 0.45; and for silt, 0.35. If the base of the retaining wall rests on clay, the shearing resistance against sliding should be based on the cohesion of the clay, which can be conservatively estimated as one-half the unconfined compressive strength. If the clay is stiff or hard, its surface should be roughened before the concrete base is placed.

If the factor of safety against sliding is less than 1.5, the design should be revised. The resistance to sliding may be increased by the use of a key that projects into the soil below the base, as shown in Fig. 22.4, or the base may be widened to increase the surface of sliding. For the same volume of concrete, a key is ordinarily considered to be somewhat more effective than an increase in base width, but, on the other hand, the width of the base can usually be extended at less cost.

The effectiveness of short keys is often overestimated. Consideration of the equilibrium of the block of soil *bcde,* Fig. 22.4, leads to the

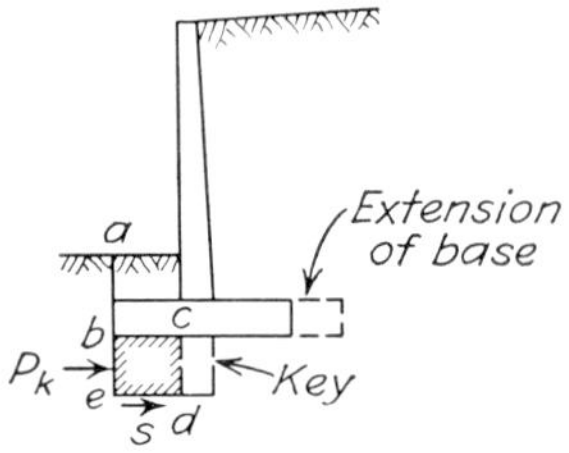

FIG. 22.4. Horizontal forces resisting movement of key beneath retaining wall.

conclusion that the total horizontal force acting on the key can be no larger than the sum of the force P_K and the shearing force S developed on the surface *de*. However, S is not likely to be much greater than the shearing resistance available along *bc* if no key were employed. For this reason, any additional horizontal force gained by the use of a key can be no larger than P_K. Hence, the additional resistance to sliding offered by the projection of a key below the base can be determined only by an evaluation of the lower limiting passive resistance of the soil against the surface *be* during the time the backfill is being placed and throughout the subsequent life of the retaining wall. Any estimate of the passive resistance against the plane *abe,* or on any portion of this plane is, at best, only a crude approximation. Moreover, the excavation for a key is likely to disturb the subsoil during construction and in some instances may conceivably do more harm than good.

On the other hand, if the base of a retaining wall is supported by rock or by a very stiff cohesive soil, a key may provide an effective means of creating additional resistance to sliding. If one or the other of these subsoil conditions exists, a larger, more reliable passive resistance P_K, Fig. 22.4, acts against the plane *be*. Moreover, the shearing force S developed on the surface *de* is likely to exceed the shearing resistance available along *bc* if no key were employed.

22.4. Summary of Procedure for Design of Cantilever Retaining Wall

The preceding articles have dealt with the proportions of a cantilever retaining wall and the principal forces acting upon the structure. The analysis and design of a retaining wall can be summarized by the following steps:

1. Choose tentative proportions for the structure, including dimensions for the stem and the base as well as the position of the stem on the base.
2. Estimate the magnitude of all the forces acting above the bottom of the base, as indicated in Fig. 22.1.
3. Determine the point of intersection of the resultant of the forces found in step 2 with the plane of the bottom of the base. The location of this point constitutes a check on the stability of the wall with respect to overturning.
4. Determine the magnitude of the foundation pressure against the base.
5. Check the factor of safety against sliding.
6. Make the structural design of the stem and of the toe and heel of the base.

The computations involved in steps 3 to 6 almost always indicate necessary revisions in the tentative dimensions of step 1.

22.5. Illustrative Design. DP D-6. Cantilever Retaining Wall

The application of the principles discussed thus far in Chapter 22 is demonstrated in this design plate. It is noted that the computations may be divided into two separate steps: (1) the stability analysis in which the soil pressures beneath the base are determined and the factor of safety against sliding along the base is estimated, and (2) the structural design, in which the strengths of the cantilever elements, Fig. 22.1*b*, are made adequate to resist the applied forces.

It may be noted that the maximum soil pressure at the toe computed by methods discussed in Art. 20.2 is somewhat less than the allowable value. This would suggest that the width of the base might be reduced. However, further investigation shows that very little, if any, reduction in width can be made without resulting in an inadequate factor of safety against sliding.

The computations for required depths of section, areas of steel, bond stresses, and resisting moments are typical for reinforced-concrete design.

DP D-6
Retaining Wall
Sh. 1 of 5

Design of Cantilever Retaining Wall

General Data:

Backfill-group 2 soil (see Fig. 22.2). Assume $\gamma = 125\ lb/ft^3$
Allowable soil pressure $= 3.00^k/sq\ ft$
Friction between soil and base $= 0.55$

$f_s = 20\ ksi$ $R = 197\ psi$
$f_c = 1200\ psi$ $v = 75\ psi$
$n = 10$ $u = 240\ psi$
$j = 0.87$

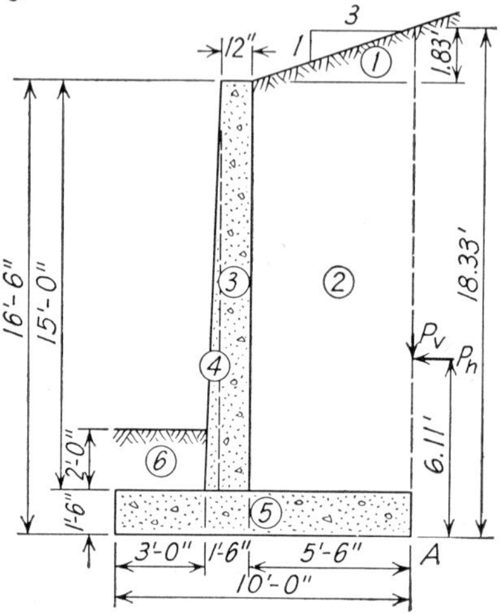

Note: All loads, shears, and moments are per lin ft of retaining wall.

See Fig. 22.2.

$\underline{P_v} = \frac{1}{2}(12)(18.33)^2 = \underline{2.0^k}$

$\underline{P_h} = \frac{1}{2}(39)(18.33)^2 = \underline{6.6^k}$

Stability Computations:

Moments about A:

Area			Force	Arm	Mom.
①	$\frac{1}{2} \times 5.5 \times 1.83$	$= 5.03 \times 0.125 =$	0.63^k	1.83'	$1.2'^k$
②	5.5×15.0	$= 82.5 \times 0.125 =$	10.31	2.75	28.4
③	1.0×15.0	$= 15.0 \times 0.150 =$	2.25	6.00	13.5
④	$\frac{1}{2} \times 0.5 \times 15.0$	$= 3.75 \times 0.150 =$	0.56	6.67	3.7
⑤	10.0×1.5	$= 15.0 \times 0.150 =$	2.25	5.00	11.3
⑥	3.0×2.0	$= 6.0 \times 0.125 =$	0.75	8.50	6.4
P_v			2.00		
P_h		$\Sigma H = 6.60^k$		6.11	40.4
			$\Sigma V = 18.75^k$		$\Sigma M_A = 104.9'^k$

DP D-6
Retaining Wall
Sh. 2 of 5

Design of Cantilever Retaining Wall

Stability Computations:

Location of Resultant:

From point A, $\frac{104.9}{18.75} = 5.6'$ then $e = 5.6 - \frac{10}{2} = 0.6' < \frac{10'}{6}$ ok

Soil Pressure at Base:

At toe $q_{max.} = \frac{18.75}{10}\left(1 + \frac{6 \times 0.6}{10}\right) = 1.875(1+0.36) = 2.55^k/sq\,ft$

ok allow $3.00^k/sq\,ft$

At heel $q_{min.} = 1.875(1-0.36) = 1.20^k/sq\,ft$

Sliding:

Shear available along base $= 18.75^k \times 0.55 = 10.3^k$

Passive force at toe: $k_p = \frac{2}{3}\frac{(0.125)^2}{0.036} = 0.29$

$P_p = \frac{1}{2}(0.29)(3.5)^2 = 1.8^k$

Min. F.S. $= \frac{10.3}{6.60} = 1.6$

Max. F.S. $= \frac{10.3 + 1.8}{6.60} = \frac{12.10}{6.60} = 1.8$ — ok without key

Design of Stem:

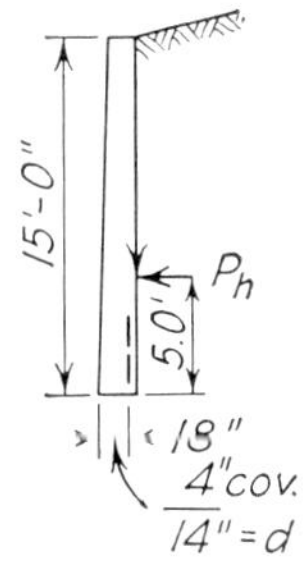

P_h = shear $= \frac{1}{2}(39)(15)^2 = 4.4^k$

$\times 5.0 = 22.0'^k = 264''^k$ = mom.

Required Depths:

For shear, $d = \frac{4.4}{0.075 \times 0.87 \times 12} = 5.6''$

For moment, $d = \sqrt{\frac{264}{0.197 \times 12}} = 10.6''$

ok 14" available

Steel at Bottom of Wall:

$A_s = \frac{264}{20 \times 0.87 \times 14} = 1.08$ sq in./ft

Use #6 @ $4\frac{1}{2}''$ ctrs. $A_s = 1.17$ $\Sigma_o = 6.3$

$u = \frac{4.4}{6.3 \times 0.87 \times 14} = 57$ psi ok allow 240 psi

DP D-6
Retaining Wall
Sh. 3 of 5

Design of Cantilever Retaining Wall

Design of Stem : Cont. from Sh. 2.

Moment Computations for Steel Cutoffs.:

5 ft down, $P_h = \frac{1}{2}(39)(5)^2 = 0.49^k$

$\times \frac{5}{3} \times 12 = 9.7''^k = M_5$

10 ft down, $P_h = \frac{1}{2}(39)(10)^2 = 1.95^k$

$\times \frac{10}{3} \times 12 = 78.0''^k = M_{10}$

15 ft down, bottom, (see Sh. 2) $264.0''^k = M_{15}$

Resisting Moments:

#6 @ $4\frac{1}{2}''$ ctrs., d = 14", $M_r = 1.17(20)(0.87)(14) = 285''^k$

#6 @ $4\frac{1}{2}''$ ctrs., d = 8", $M_r = 1.17(20)(0.87)(8) = 163''^k$

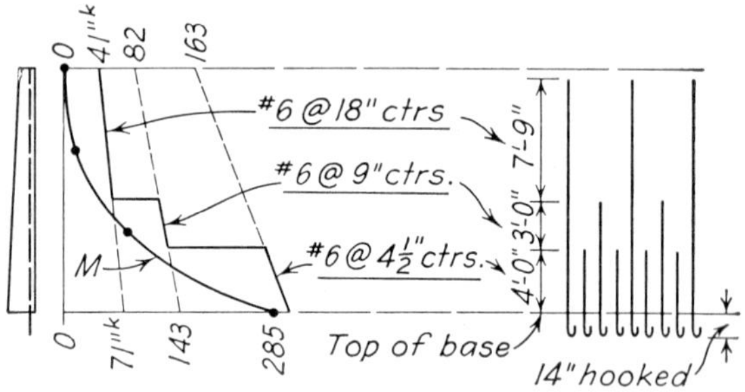

Design of Base

Toe:

$0.23^k/\text{sq ft} = 1.5 \times 0.15 = \text{wt. footing}$

$0.25 \quad = 2.0 \times 0.125 = \text{wt. soil}$

$0.48^k/\text{sq ft}$

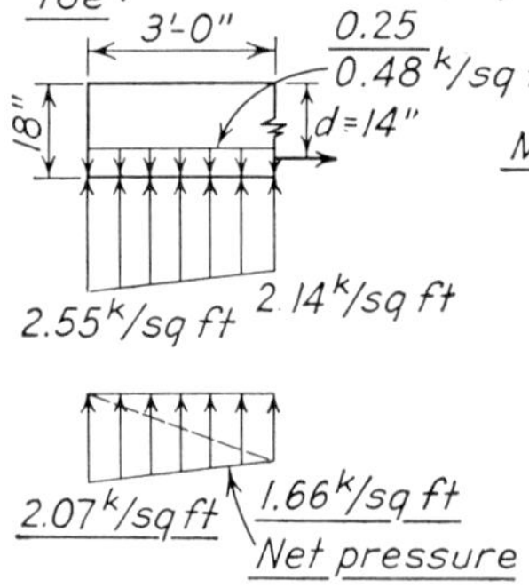

Mom. and Shear:

$\frac{1}{2}(2.07)(3.0) = 3.11^k$

$\times 2.0 = 6.22'^k$

$\frac{1}{2}(1.66)(3.0) = 2.49^k$

$\times 1.0 = 2.49'^k$

$V = 5.60^k \quad M = 8.71'^k = 105''^k$

d = 14" ok by inspection after computations for stem, Sh. 2.

DP D-6
Retaining Wall
Sh. 4 of 5

Design of Cantilever Retaining Wall

Design of Base: Cont. from Sh. 3.

Steel in Toe:

$$A_s = \frac{105}{20 \times 0.87 \times 14} = 0.43 \text{ sq in./ft} \quad \text{Use \#6 @ 12'' ctrs.} \quad A_s = 0.44, \ \Sigma_o = 2.4$$

$$u = \frac{5.60}{2.4 \times 0.87 \times 14} = 190 \text{ psi ok allow 240 psi}$$

$$\text{Extend bars } \frac{20 \times 0.75}{4 \times 0.24} = 16'' \text{ back of front face of wall to provide anchorage}$$

Heel:

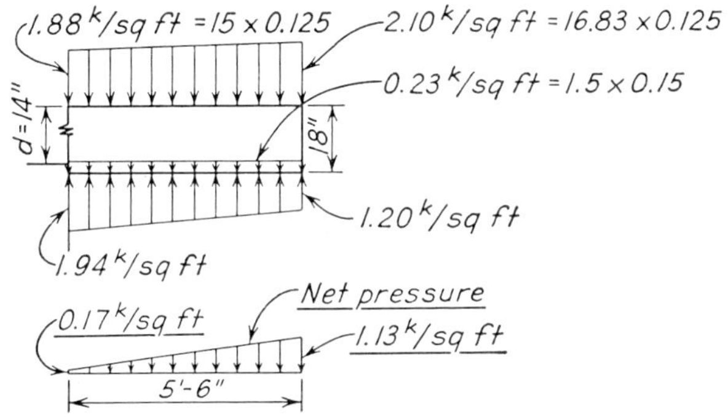

Mom. and Shear:

$$\tfrac{1}{2}(1.13)(5.5) = 3.11^k \quad \times \tfrac{2}{3}(5.5) = 11.4'^k$$

$$\tfrac{1}{2}(0.17)(5.5) = 0.47^k \quad \times \tfrac{1}{3}(5.5) = 0.9'^k$$

$$V = 3.58^k \qquad M = 12.3'^k = 148''^k$$

$d = 14''$ ok

Steel in Heel:

$$A_s = \frac{148}{20 \times 0.87 \times 14} = 0.61 \text{ sq in./ft} \quad \text{Use \#6 @ } 8\tfrac{1}{2}'' \text{ ctrs.} \quad A_s = 0.62, \ \Sigma_o = 3.3$$

$$u = \frac{3.58}{3.3 \times 0.87 \times 14} = 89 \text{ psi ok allow 240 psi}$$

DP D-6
Retaining Wall
Sh. 5 of 5

Design of Cantilever Retaining Wall

Steel in Stem for Shrinkage and Temperature:

Use 0.5 sq in./ft of height.

Use #4 @ 6" ctrs. front face. $A_s = 0.40$ sq in./ft

Use #4 @ 18" ctrs. back face. $A_s = \underline{0.13}$

$\underline{0.53}$ sq in./ft

Sketch of Retaining Wall

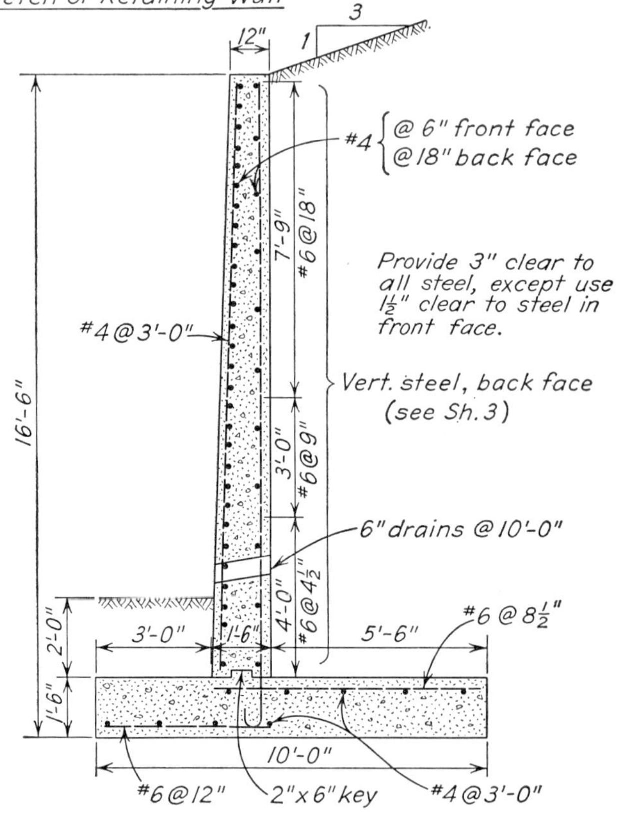

22.6. Pile-supported Retaining Walls

Introduction. A retaining wall is commonly established on a pile foundation when the soil for a considerable depth is too weak or compressible to provide adequate support for the structure. Furthermore, if excessive dimensions of the base are required for an adequate factor of safety with respect to sliding or to keep the vertical pressure within the allowable limit, it may be more economical to provide the retaining wall with a pile foundation as shown in Fig. 22.5.

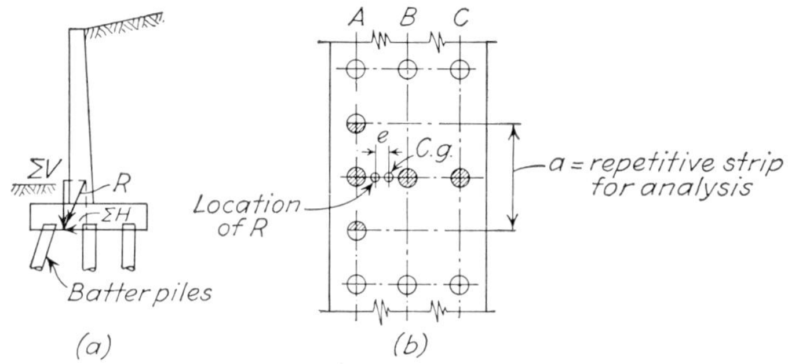

FIG. 22.5. Pile foundation beneath retaining wall. (*a*) Section; (*b*) foundation plan.

If the horizontal force against the retaining wall cannot otherwise be resisted, some of the piles are ordinarily driven on a batter. The arrangement of the piles indicated in Fig. 22.5 is typical for retaining walls approximately 15 to 20 ft in height, subjected to moderate horizontal loads. For higher walls and more extreme conditions of loading, it is sometimes necessary to use the same number of piles along all rows and to provide batter piles in more than one row.

The design of a batter-pile foundation, as described in the following paragraphs, is based on the assumption that the only horizontal forces to be resisted are those that act above the base level of the retaining wall. In many instances, possibly in most, the lateral forces against that part of the subsoil in which the piles are buried are very much greater than those acting on the wall itself. The batter piles may then be ineffective. The circumstances under which this may occur have been discussed in Art. 15.7.

Vertical Forces on Piles. If the piles are arranged in plan as shown in Fig. 22.5*b*, the centroid of the piles usually lies very near the line of action of the resultant *R*, and the loads on the piles are approximately equal. The loads on the vertical piles in rows *B* and *C*, as well as the vertical component of the load on the piles in row A, are commonly determined by eq. 20.10. It is evident from Fig. 22.5b that the strip a containing four piles of the foundation is repetitive. For this reason, the section properties of these four piles, along with the loads and moment acting upon strip a, can be used in eq. 20.10. Figure 22.5 indicates that the resultant force R lies to the left of the centroid of the piles. Hence, the vertical load per pile in row A is larger than that in rows B and C. On the other hand, if the eccentricity were to lie to the right of the centroid, the piles in rows B and C would be subjected to a greater load than those in row A. The

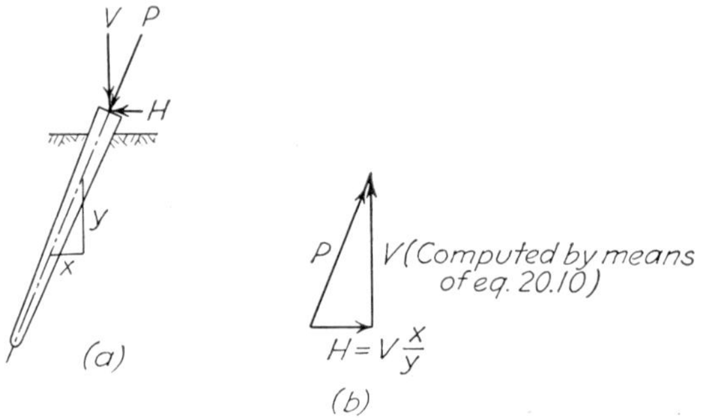

FIG. 22.6. (a) Batter pile; (b) force diagram for single batter pile.

magnitude of the moment that exists at the junction of the piles and the base is the product of the summation of the vertical loads ΣV and the eccentricity e. The eccentricity is measured from the centroid of the piles, not from the centerline of the base, unless the two coincide.

Horizontal Forces on Piles. The horizontal load ΣH at the base of a pile-supported retaining wall is transferred to the piles and thence to the soil in which the piles are embedded. Thus the vertical piles may not appreciably alter the factor of safety of the wall against sliding on its base or within some weak deposit beneath its base. If the strength of the soil in which the piles are located is inadequate to provide the necessary factor of safety, the available resistance may be supplemented by the use of batter piles. Usually the resistance of the batter piles and of the soil beneath the wall are considered to act simultaneously; this presupposes that sufficient lateral movement of the wall can take place before failure to develop the ultimate strength of both elements.

It is commonly assumed that the vertical load carried by a batter pile is the same as it would be if the pile were vertical. Thus, the vertical component of load in a batter pile may be computed by means of eq. 20.10. If the line of action of the resultant load P, Fig. 22.6a, is made to coincide with the axis of the pile, the horizontal component of this resultant can be determined readily if the vertical component and the slope of the pile are known. The force diagram for a single batter pile is shown in Fig. 22.6b.

Steps in Design. In the design of a batter-pile foundation, two criteria must be satisfied: (1) The axial load in any pile must not

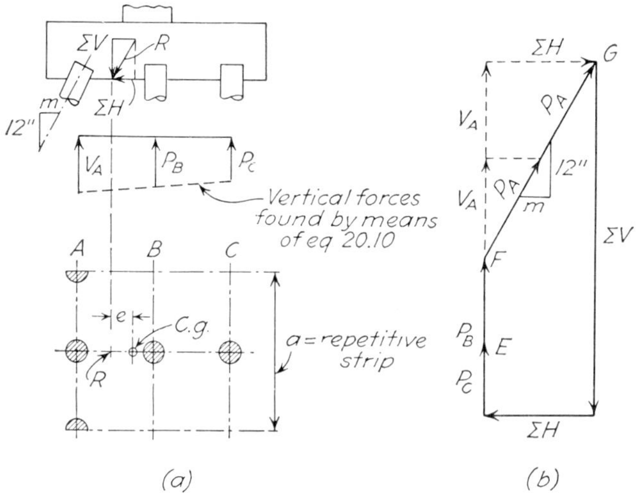

FIG. 22.7. Diagrams illustrating method of computation of forces acting in piles beneath retaining wall. (a) Arrangement of piles; (b) force polygon.

exceed the safe load per pile, and (2) the sum of the horizontal components of the forces in the batter piles must equal the applied force ΣH. Since in a batter pile the relationship between the axial load and its horizontal component depends upon the slope or batter, which is not initially known, the design is most conveniently made by a method of trial and correction. The steps in this method for a typical problem (Fig. 22.7) are as follows:

1. Make a tentative layout of the foundation piles for the section at the base of the wall.

2. Compute the vertical forces V_A, P_B, and P_C in each pile by means of eq. 20.10.

3. Compare the vertical forces in the batter piles with the safe axial load for the piles. The maximum vertical force in a batter pile should be limited tentatively to a value about 8 per cent less than the safe axial load to allow for the influence of the batter.

4. Construct the force polygon, Fig. 22.7*b*. According to this diagram, if the piles in row A are to resist the entire horizontal force ΣH, they must have the inclination of line FG. Thus the required batter is

$$m = 12 \frac{\Sigma H}{\Sigma V'} \qquad 22.3$$

where m = batter of the piles expressed in inches (horizontal) per foot (vertical)

ΣH = horizontal force to be resisted by batter piles

$\Sigma V'$ = summation of the vertical forces acting upon the batter piles. For the conditions shown in Fig. 22.7, $\Sigma V' = 2V_A$.

5. Check whether the required batter is reasonable (see Art. 9.1). If it is not, a new layout is required or else batter piles must be used in more than one row as explained in the following paragraph.

6. Compute the axial load in the batter piles, and compare with the safe axial load per pile. If the safe load is exceeded, a new layout is required.

If the horizontal force ΣH is large, it may be necessary to batter more than one row of piles. If batter piles are used in row B as well as row A, and if the same slope is desired in both rows, a line connecting E and G, Fig. 22.7*b*, gives the appropriate slope. Hence, it is apparent that eq. 22.3 remains applicable, provided the vertical summation $\Sigma V'$ is adjusted to include P_B. In some designs the slope of the batter piles differs from row to row. In many instances, either engineering judgment or tests of the soil into which the piles are driven may lead the designer to the decision that only a fraction of the total horizontal force need be resisted by batter piles. In any event a force diagram similar to that shown in Fig. 22.7*b* is helpful in establishing the proper slopes for the different rows.

The value m as given by eq. 22.3 is commonly expressed to the nearest inch. More refinement cannot be justified in view of the many assumptions and uncertain field conditions involved in this type of problem.

22.7. Illustrative Design. DP D-7. Batter-pile Foundation

The purpose of this design plate is to illustrate a common method of design for pile foundations in which some of the piles are driven on a batter to increase the stability of the structure against horizontal loads. A cantilever retaining wall has been chosen for the example

DP D-7
Retaining Wall
Sh. 1 of 2

Foundation Design for Pile-supported Retaining Wall

General Data:

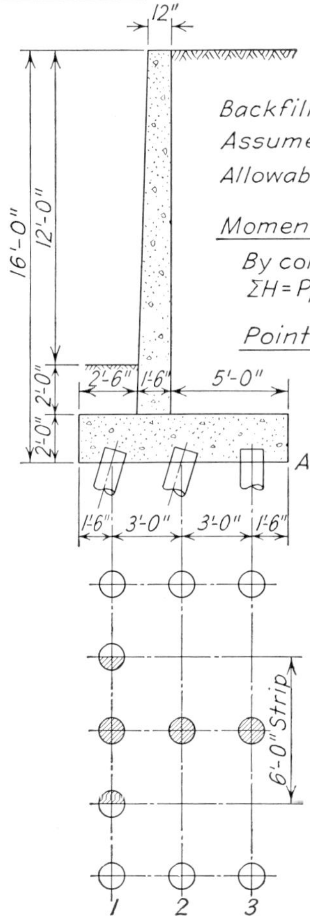

Backfill-group 2 soil (see Fig. 22.2)
Assume $\gamma = 125$ lb/cu ft
Allowable load/pile $= 25^k$

Moments about A:

By computations similar to DP D-6, Sh. 1,
$\Sigma H = P_h = 4.6^k$, $\Sigma V = 14.7^k$, $\Sigma M_A = 78.4'^k$

Point A to resultant $= \dfrac{78.4}{14.7} = 5.33'$
or 0.83' left of row 2

C.G. of Piles and Eccentricity:

$2 \times 3 = 6$
$1 \times 3 = 3$
$3 \div 4 = 0.75'$ left of row 2
Then $e = 0.83 - 0.75 = 0.08'$

Σd^2 and Section Moduli:

$2(2.25)^2 = 10.12$ pile ft^2
$1(0.75)^2 = 0.56$
$1(3.75)^2 = 14.06$
$\Sigma d^2 = 24.74$ pile ft^2
Row 1 $\div 2.25 = 11.0$ pile ft
Row 2 $\div 0.75 = 33.0$ pile ft
Row 3 $\div 3.75 = 6.6$ pile ft

Loads and Moment on 6' Strip:

$\Sigma H = 4.6 \times 6 = 27.6^k$
$\Sigma V = 14.7 \times 6 = 88.2^k$
$\Sigma M = \Sigma Ve = 88.2 \times 0.08 = 7.1'^k$

DP D-7
Retaining Wall
Sh.2 of 2

Foundation Design for Pile-supported Retaining Wall

Vertical Components of Pile Reactions:

Row 1: $\frac{88.2}{4} + \frac{7.1}{11.0} = 22.1 + 0.6 = \underline{22.7^k}$

Row 2: $\frac{88.2}{4} - \frac{7.1}{33.0} = 22.1 - 0.2 = \underline{21.9^k}$

Row 3: $\frac{88.2}{4} - \frac{7.1}{6.6} = 22.1 - 1.1 = \underline{21.0^k}$

Horizontal Loads on Piles:

Assume 2^k/pile resisted by moment in piles combined with passive resistance of soil.

$\times 4 = 8.0^k$ So batter piles for $27.6 - 8.0 = 19.6^k$

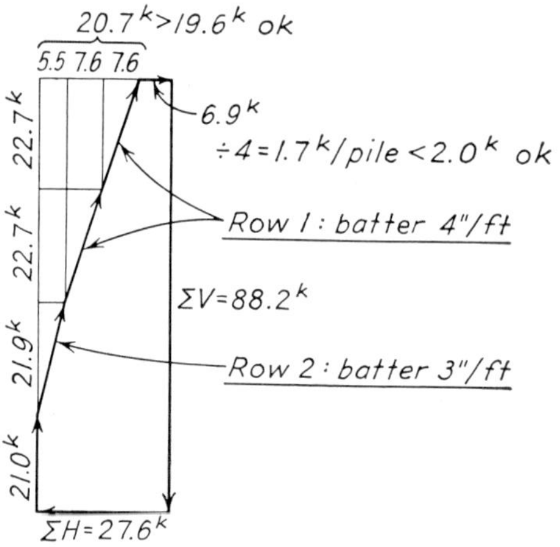

Maximum Pile Reaction:

$\sqrt{22.7^2 + 7.6^2} = \sqrt{575} = \underline{24.0^k/\text{pile} < 25^k \text{ ok}}$

because a similar soil-supported structure has been designed in DP D-6.

First, the vertical components of the pile reactions are found in accordance with the discussion of pile foundations subjected to moment, Art. 20.7. Second, a value of 2 kips is estimated as the horizontal load that may be assigned to each pile as a consequence of the strength of the soil into which the piles are driven, combined with the flexural stiffness of the piles. Next, the values of the batter for rows 1 and 2 are determined conveniently by a force polygon similar to Fig. 22.7*b*. Finally, the resultant load per pile in row 1 is compared with the allowable value.

22.8. Abutments

General. The principles governing the analysis of the stability and strength of cantilever retaining walls are for the most part applicable to cantilever abutments and wing walls. In this article the discussion of cantilever retaining walls is extended to include some

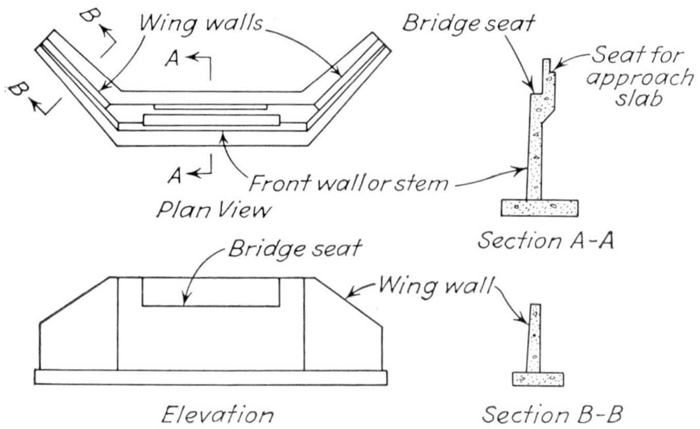

FIG. 22.8. Typical cantilever abutment for highway bridge.

of the problems associated with the design of cantilever abutments. A typical cantilever abutment with wing walls is shown in Fig. 22.8. Other types of abutments have been described in Art. 11.3.

The forces that are commonly considered in the design of the individual parts of an abutment are dependent to a considerable degree on the tendency of these parts to act together as one unit. For example, the wing walls are generally assumed to behave as ordinary retaining walls and they are designed accordingly. Yet, it is apparent that the force P_A, Fig. 22.9*a*, must cause appreciable shear, moment, and direct tension on vertical sections near the corner *A*. Therefore, if continuity is maintained at this junction, the action of the walls in this region differs considerably from that of pure vertical cantilevers. Moreover, as explained in Art. 22.3, some tilt is necessary to establish the condition of active earth pressure in the soil behind a retaining wall; otherwise, the total earth pressure may be appreciably larger than the assumed active value.

Abutments Built Monolithically. Where no joint has been provided between the wing walls and the front of abutments, field

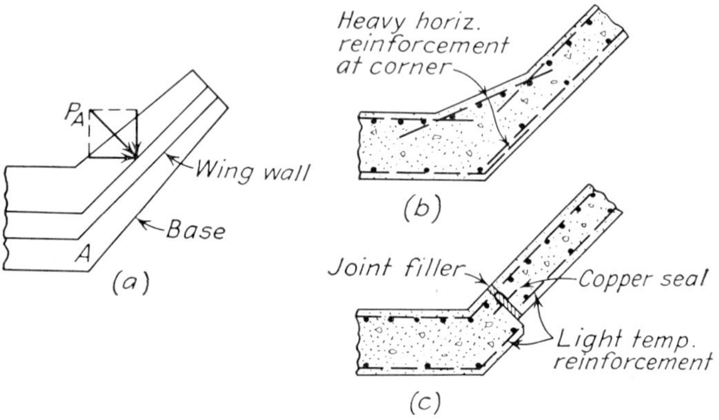

FIG. 22.9. (*a*) Detail at junction of front wall and wing wall of abutment; (*b*) reinforcement required for monolithic construction; (*c*) joint required for independent action.

evidence has indicated a tendency to crack near the corner *A*, Fig. 22.9*a*. However, most of the failures can be attributed to inadequate structural strength in this region. A detail of the corner for monolithic construction is shown in Fig. 22.9*b*. The proper amount of horizontal reinforcement and the appropriate thickening of the concrete at the corner can be determined only by judgment. Both should increase as the angle between the wing walls and the front of the abutment increases. If a detail similar to that shown in Fig. 22.9*b* is used, and if the structure is otherwise properly designed, there is no reason to expect unsatisfactory performance of the abutment at the corner.

Abutments with Joints at Wing Walls. Some engineers consider it better practice as well as more economical design to provide a joint at each corner as shown in Fig. 22.9*c*. Such a joint has no structural strength and permits the wing walls as well as the front wall of the abutment to act as true vertical cantilevers. It also accommodates the movements associated with temperature changes. The offset in the concrete at the front of the joint permits considerable relative movement without detracting from the appearance of the structure.

Action of Base Slab. The base of an abutment such as that shown in Fig. 22.8 is commonly cast as a continuous unit extending beneath the wing walls and the front wall. Nevertheless, in the structural analysis, regardless of the presence or absence of joints at the corners, a fictitious joint is commonly assumed to extend across the base at the junction of each wing wall with the front of the abutment. Thus, the abutment is assumed to consist of three entirely independent units.

The assumptions for design, as outlined in the preceding paragraph, are generally considered conservative from the standpoint of the stability of the structure as a whole because the resisting moment of the entire area of the base acting as a unit is greater than that of the assumed three components acting separately, and the resistance increases rapidly with increasing angle between the wing walls and the front of the abutment. On the other hand, continuity of the base beneath the entire abutment and continuity between the wings and the front face undoubtedly reduce the tilt of the walls. The influence of this reduction in tilt upon the earth pressure acting against the walls is usually ignored. This involves a relatively small error, but one that is on the unsafe side.

Loads in Addition to Earth Pressure. The preceding discussion has indicated that wing walls are commonly designed as simple cantilever retaining walls; therefore, the design procedure is identical to that given in design plates DP D-6 and DP D-7. On the other hand, the front section of the abutment must resist certain forces in addition to the normal active earth pressure against the back of the wall. Two important additional forces are the reaction of the superstructure of the bridge and the increase in earth pressure due to wheel loads on the backfill adjacent to the wall. The remainder of this article is devoted to a discussion of the influence of these additional loads.

The reactions from the superstructure may be transmitted to the bridge seat of an abutment in several ways. Two details, typical for small bridges, are shown in Fig. 22.10. The roller shown in Fig. 22.10*a* provides for expansion of the bridge; consequently, if rolling friction is ignored, the force transmitted to the abutment is normal to the lower bearing plate. On the other hand, since the detail shown in Fig. 22.10*b* fixes the end of the bridge, both horizontal and vertical forces are transmitted to the abutment wherever this type of bearing is used. The number and spacing of the beams, girders, or trusses that comprise the superstructure determine the number and location of the concentrated reactions that must be resisted by the abutment.

Although the vertical and horizontal reactions from the superstructure represent more or less concentrated loads, they are com-

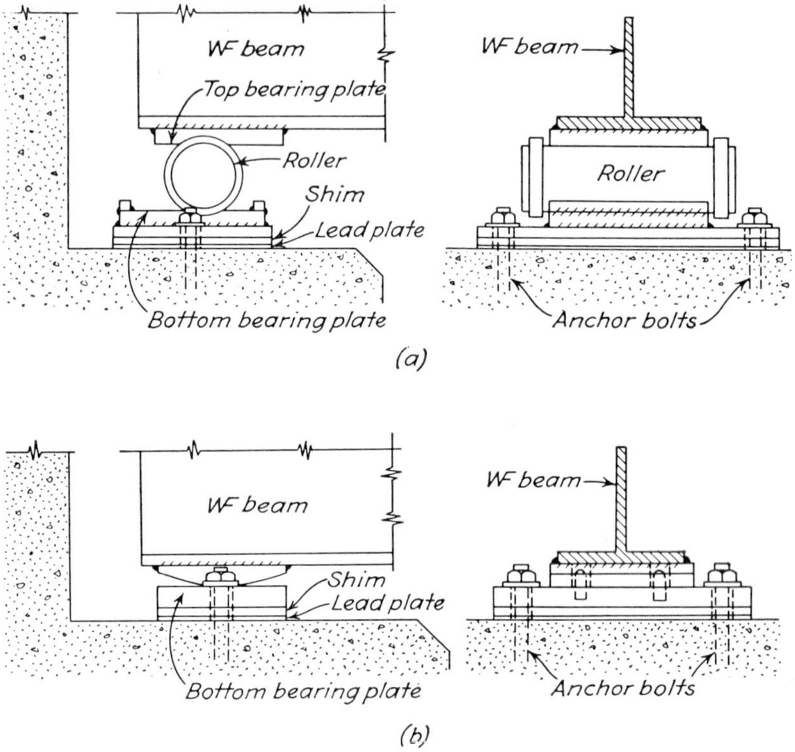

FIG. 22.10. Typical methods for transferring reactions from superstructure to bridge seat. (*a*) Roller; (*b*) rocker.

monly assumed to be distributed over the entire length of the front wall of the abutment. That is, the sum of the reactions, either horizontal or vertical, is divided by the length of the wall to obtain a load per foot to be used in both the stability analysis and the structural design. This procedure is probably sufficiently exact for most design purposes. However, in the design of low abutments where the reactions from the superstructure are widely spaced, considerable judgment must be exercised in the establishment of a reasonable width over which each reaction is distributed.

The earth pressure against the back of an abutment is increased whenever wheel loads are transmitted to the fill immediately behind the wall. The magnitude of this increase depends upon the type of soil and the position of the wheels relative to the abutment, as well as on the wheel loads themselves. An allowance for this increase in earth pressure must be included in the analysis and design of the abutment. Usually the wheel loads are assumed to be equivalent to

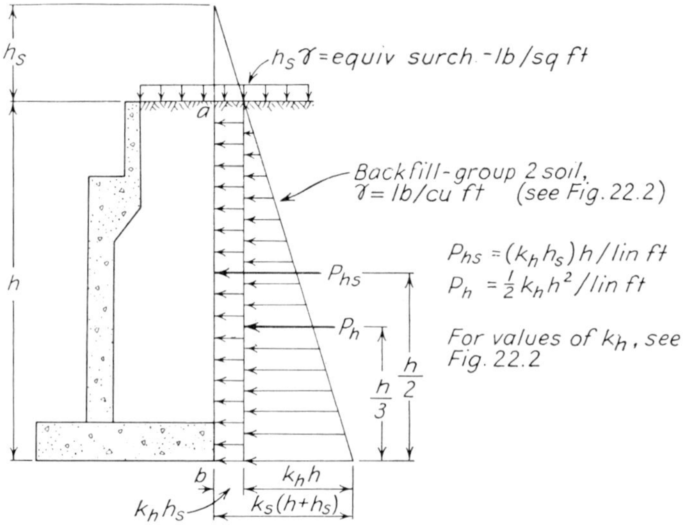

FIG. 22.11. Diagram illustrating horizontal forces acting against cantilever abutment.

a uniformly distributed load often taken as 240 lb/sq ft for H-10 highway loading and as 140 lb/sq ft for Cooper's E-10 railway loading. The uniform surcharge is commonly considered as an additional layer of backfill with a height h_s, Fig. 22.11. The corresponding additional horizontal pressure is assumed to be uniformly distributed and equal to $k_h h_s$, where k_h is a coefficient depending upon the type of backfill. Values of k_h may be determined by means of Fig. 22.2. Since the additional pressure $k_h h_s$ is uniformly distributed, the resultant force P_{hs} is assumed to act at mid-height of the vertical section *ab*.

In the preceding discussion it has been assumed that the wheel loads can bear directly upon the fill behind the abutment. However, on highways with concrete pavement, the use of approach slabs adjacent to abutments is common practice. These are heavily reinforced concrete slabs, seated on the back of the abutment wall, and extending some 15 or 20 ft back of the abutment. The principal purpose of such construction is to provide a slab with enough structural strength to bridge the depression caused by any differential settlement between the fill behind the abutment and the abutment itself. A typical

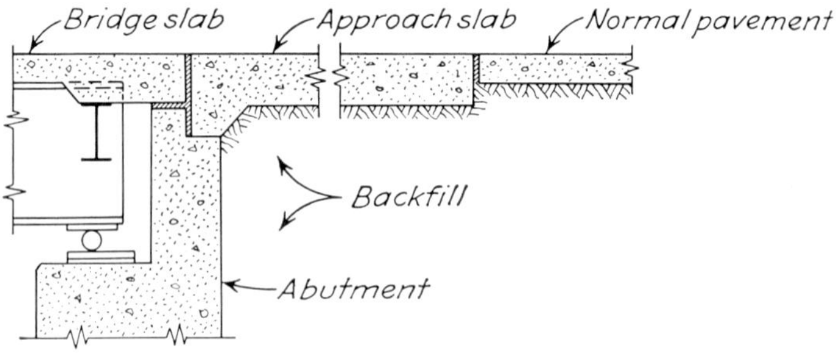

FIG. 22.12. Approach slab for highway bridge abutment.

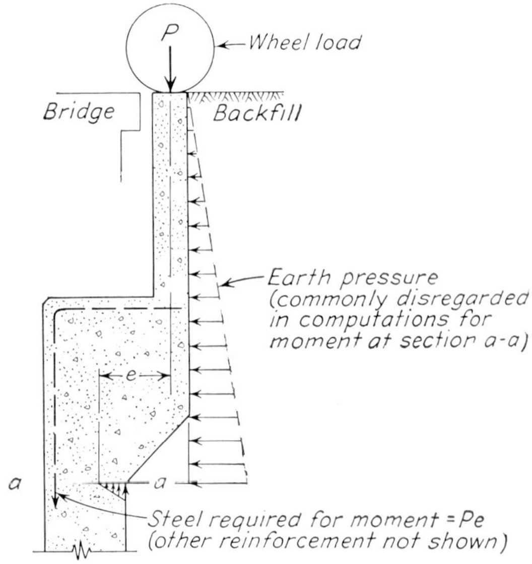

FIG. 22.13. Diagram illustrating cause of moment near top of cantilever abutment due to wheel load directly over parapet wall.

approach slab is shown in Fig. 22.12. For this type of construction, it should be noted that most of the reaction due to the wheel loads on the approach slab is transmitted to the abutment at the seat which provides the support for the slab. Therefore, for design purposes, the force P_{hs}, Fig. 22.11, may usually be replaced by the vertical reaction of the approach slab upon the abutment.

Regardless of whether or not an approach slab is utilized, the design of abutments for highway bridges should consider the possibility that wheel loads may be located directly over the parapet wall as shown in Fig. 22.13. It is apparent that wheels in this position will cause moment at section *a-a*. Therefore, in some cases, considerable reinforcement may be required at the front of the wall to provide adequate strength against a backward movement of the portion of the abutment above this section.

22.9. Illustrative Design. DP D-8. Cantilever Bridge Abutment

This design plate is concerned only with the stability of the front section of a cantilever bridge abutment. It is noted that many of the computations correspond to those given in DP D-6. However, the reactions at the bridge seat as well as the surcharge resulting from loads on the fill behind the abutment are additional factors peculiar to this type of structure. Therefore, the principal purpose of this design plate is to demonstrate the effects of these additional loads as discussed in the preceding section.

DP D-8
Bridge Abutment
Sh.1 of 2

Design of Bridge Abutment Stability
Analysis of Front Wall

General Data:

Backfill - group I soil (see Fig. 22.2). Assume $\gamma = 110\ lb/cu\ ft$
Equivalent surcharge = 3 ft.
Allowable soil pressure = $3.5^k/sq\ ft$
Allowable shear between soil and base = $0.45\ \Sigma V$

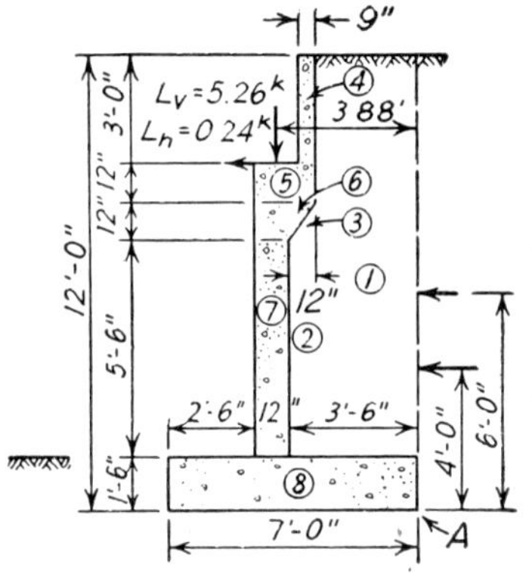

Note:
All loads and moments are per lin ft of abutment.

$P_{hs} = 30(3.0)(12) = 1.08^k$

$P_h = \frac{1}{2}(30)(12)^2 = 2.16^k$

Stability Computations:

Moments about A

Area		Force	Arm	Mom.
①	2.5 × 10.5 × 0.110	= 2.89^k	1.25'	$3.61'^k$
②	1.0 × 5.5 × 0.110	= 0.61	3.00	1.83
③	½ × 1.0 × 1.0 × 0.110	= 0.06	2.83	0.15
④	0.75 × 3.0 × 0.150	= 0.34	2.88	0.98
⑤	2.0 × 1.0 × 0.150	= 0.30	3.50	1.05
⑥	½ × 1.0 × 1.0 × 0.150	= 0.08	3.17	0.25
⑦	1.0 × 6.5 × 0.150	= 0.97	4.00	3.88
⑧	7.0 × 1.5 × 0.150	= 1.58	3.50	5.53
L_v		5.26	3.88	20.40
L_h	0.24^k		9.00	2.16
P_{hs}	1.08		6.00	6.48
P_h	2.16		4.00	8.64
	$\Sigma H = 3.48^k$	$\Sigma V = 12.09^k$		$\Sigma M_A = 54.96'^k$

DP D-8
Bridge Abutment
Sh. 2 of 2

Design of Bridge Abutment. Stability Analysis of Front Wall

Stability Computations:

Location of Resultant

From point A, $\frac{54.96}{12.09} = 4.55'$ then $e = 4.55 - \frac{7.0}{2} = 1.05' < \frac{7}{6}$ ok

Soil Pressure at Base:

At toe $q_{max.} = \frac{12.09}{7}\left(1 + \frac{6 \times 1.05}{7}\right)$

$q_{max.} = 1.73\,(1 + 0.9) = 3.3^k/\text{sq ft} < 3.5$ ok

At heel $q_{min.} = 1.73\,(1 - 0.9) = 0.2^k/\text{sq ft}$

Sliding:

Shear available along base $= 12.09^k \times 0.45 = 5.43^k$

Factor of safety $= \frac{5.43}{3.48} = 1.6 > 1.5$ ok

Problems

Art. 1.5

1. Determine the effective size and uniformity coefficient for materials 4–6 inclusive, Fig. 1.1.

Art. 1.6

1. A sample of saturated clay has a water content of 23 per cent and a unit weight of 128 lb/cu ft. Compute the void ratio and porosity.

Ans. 0.625; 38.4%

2. A sample of saturated glacial clay has a water content of 47 per cent. On the assumption that $s_s = 2.70$, compute the void ratio, porosity, and weight per cubic foot.

3. The weight of an undried specimen of clay was 34.62 g. The oven-dry weight of the same specimen was 20.36 g. Before drying, the specimen was immersed in mercury and its volume found to be 24.26 cu cm. If the specific gravity of the solids was 2.70, what were the water content, void ratio, and degree of saturation of the material?

Ans. 70.4%; 2.22; 85.4%

4. The density of a rolled earth fill was checked in the field by excavating a hole in the fill and carefully saving all the excavated material. The volume of the hole was determined by measuring the weight of loosely deposited standard Ottawa silica sand required to fill it. The unit weight of similarly deposited sand was found by calibration tests to be 92.0 lb/cu ft.

The weight of sand required to fill the hole was 13.62 lb. The weight of the excavated soil at its field moisture content was 18.63 lb, its oven-dry weight was 14.96 lb, and the specific gravity of its solid constituents was 2.74. What were the dry density, water content, and degree of saturation of the fill?

5. A sample of quartz sand was found to have a water content of 25 per cent and a unit weight of 120 lb/cu ft. Laboratory tests on the same material indicated that the void ratios in the loosest and densest possible states were 0.90 and 0.50, respectively. Compute the relative density and degree of saturation of the sample.

6. The following data were obtained in a standard Proctor test on a silty sand:

Trial	Weight Moist Soil in Mold, lb	Moist Weight of Water Content Sample, g	Dry Weight of Water Content Sample, g
1	4.39	100.9	96.4
2	4.53	108.1	101.6
3	4.67	129.2	119.1
4	4.64	124.9	113.9
5	4.54	137.7	123.4

Plot the moisture-density curve, and determine the values of γ_d (lb/cu ft) and optimum water content. On the assumption that $s_s = 2.64$, compute the void ratio and degree of saturation for each trial.

7. Specifications for a proposed earth fill require that the soil be compacted to 95 per cent standard Proctor compaction. Tests on a sandy borrow material adjacent to the project indicate a dry density of 115 lb/cu ft at 100 per cent compaction. The borrow material in its natural condition has a void ratio of 0.65. The specific gravity of the solid matter is 2.65. How many cubic feet of borrow will be required to make 1 cu ft of fill?

8. The natural water content of a sample of clay from Mexico City is 326 per cent. The unit weight of the solids is 2.35 g/cu cm. On the assumption that the clay is saturated, what is its unit weight in pounds per cubic foot?

9. The values of e_{min} and e_{max} for a pure silica sand were found to be 0.46 and 0.66, respectively. What is the corresponding range in the saturated unit weight in pounds per cubic foot?

10. The dry density of a compacted sand is 114 lb/cu ft, and the solids have a specific gravity of 2.67. What is the water content of the material when saturated?

ART. 1.7

1. Laboratory tests were made on a sample from a clay stratum with the following results:

$$w = 24.1\%$$

$$L_w = 42.5\%$$

$$P_w = 18.3\%$$

Compute the plasticity index and the liquidity index.

Ans. 24.2%; 0.24.

2. Tests on a sample of silty clay indicated the following index properties: liquid limit, 49.1 per cent; plastic limit, 26.5 per cent; shrinkage limit, 21.0 per cent; standard Proctor optimum moisture content, 24.5 per cent; maximum dry density, 90.5 lb/cu ft; natural water content, 45.6 per cent; diameter of 60 per cent size, 0.0060 mm; diameter of 10 per cent size, 0.0005 mm; unconfined compressive strength of undisturbed sample, 2.0 tons/sq ft; unconfined compressive strength of remolded sample, 0.9 ton/sq ft.

Compute the uniformity coefficient, the plasticity index, the liquidity index, and the sensitivity of the clay.

3. A soil with a liquidity index of 1.21 has a liquid limit of 72.8 per cent and a water content of 81.3 per cent. What are its plastic limit and plasticity index? In what region would the soil be located on the plasticity chart, Fig. 1.7*a*?

4. A soil with a liquidity index of −0.15 has a liquid limit of 36.0 per cent and plasticity index of 12.5 per cent. What is its natural water content?

5. An extrasensitive clay has a remolded unconfined compressive strength of 0.20 ton/sq ft. What is the minimum probable value of the undisturbed strength?

ART. 1.8

1 With the aid of the grain-size curves in Fig. 1.1, classify soils 1, 3, 5, and 7 on the basis of the original triangular chart proposed by the U.S. Bureau of Soils and Chemistry.

2. Given the following additional data, classify each of the soils 1, 3, 5, and 7, Fig. 1.1, according to the Airfield Classification system.

Description	L_w	P_w
1. Pea gravel		N. P.
3. Glacial till	22.8	13.6
5. Glacial rock flour	24.0	N. P.
7. Silty clay	41.6	24.7

N. P. indicates nonplastic.

3. A sample of soil was analyzed in a set of U.S. Standard sieves (see Fig. 1.1) containing the following sizes: nos. 4, 10, 20, 40, 100, 200, and 270. The weights of soil passing through each, expressed as percentages of the whole, were, respectively, 95.0, 90.2, 74.6, 58.1, 40.2, 3.6, and 1.2. What percentages were gravel, sand, and silt, according to the MIT Classification?

Art. 2.2

1. A sand with rounded grains has an effective size of 0.080 mm and a uniformity coefficient of 3.5. Estimate its coefficient of permeability.

2. During a constant-head permeability test on a sample of sand, 150 cu cm of water were collected in 2 min. The sample had a length of 10 cm and a diameter of 2 in. The head was maintained at 20 cm. Compute the coefficient of permeability.

3. A falling-head permeability test was performed on a sample of uniform clean sand. The standpipe consisted of a graduated burette, and it was observed that 1 min was required for the water level to fall from the 0 cu cm to the 50 cu cm graduation. The initial hydraulic head was 90 cm and the final head 40 cm. The sample was 20 cm long and had a diameter of 4.0 cm. Calculate the coefficient of permeability.

4. A pumping test was made in pervious gravels and sands extending to a depth of 50 ft where a bed of clay was encountered. The normal groundwater level was at the ground surface. Observation wells were located at distances of 10 and 25 ft from the pumping well. At a discharge of 48 gpm from the pumping well a steady state was attained in about 24 hours. The drawdown at 10 ft was 5.5 ft, and at 25 ft was 1.2 ft. Compute the coefficient of permeability.

Ans. 0.22×10^{-2} cm/sec

5. A pumping test was made in a medium sand extending to a depth of 50 ft where a bed of clay was encountered. The normal groundwater level was at the ground surface. Observation wells were established at distances of 10 and 20 ft from the pumping well. A steady state was reached at a rate of pumping equal to 1.2 gpm. The drawdown at the outer well was 0.5 ft, and at the inner well 1.5 ft. What was the coefficient of permeability of the sand?

Art. 2.4

1. A clay layer 12 ft thick rests beneath a deposit of submerged sand 26 ft thick. The top of the sand is located 10 ft below the surface of a lake. The saturated unit weight of the sand is 125 lb/cu ft and of the clay is 117 lb/cu ft. Compute the total vertical pressure, the porewater pressure, and the intergranular vertical pressure at mid-height of the clay layer.

2. The surface of a saturated clay deposit is located permanently below a body of water. Laboratory tests have indicated that the average natural water content of the clay is 47 per cent and that the specific gravity of the solid matter is 2.74. What is the vertical intergranular pressure at a depth of 37 ft?

3. If the water level, Prob. 2, remains unchanged and an excavation is made by dredging, how many feet of clay must be removed to reduce the intergranular pressure at 37 ft by 1000 lb/sq ft?

4. The water table is lowered from a depth of 10 ft to a depth of 20 ft in a deposit of silt. All the silt is saturated, even after the water table is lowered. Its water content is 26 per cent. Estimate the increase in effective pressure at a depth of 34 ft on account of lowering the water table.

5. Compute the critical hydraulic gradient for the following materials:

(*a*) Coarse gravel, $k = 10$ cm/sec; $s_s = 2.67$; $e = 0.65$
(*b*) Sandy silt, $k = 10^{-6}$ cm/sec; $s_s = 2.67$; $e = 0.80$

Art. 2.8

1. The water content of a soft clay is 54.2 per cent and the liquid limit 57.3 per cent. Estimate the compression index.

2. A consolidation test was made on a sample of saturated glacial clay in a circular ring with a height of 1.5 in. and an area of 90.18 sq cm. The sample weighed 645.0 g at the start of the test and 477.8 g in the oven-dry state after the test. An independent test indicated a specific gravity of 2.74 for the solid matter. Primary consolidation under each addition of load was complete at 1000 min. The dial readings corresponding to this time are recorded in the following table:

Pressure, tons/sq ft	Dial Reading, in. $\times 10^{-4}$
0.0000	0
0.0665	70
0.133	110
0.266	216
0.5325	386
1.064	734
2.13	1333
4.26	1995
8.52	2630

Compute the void ratio at the start of the test and after each increment of load. Plot the e-log p curve.

3. Using the graphical construction proposed by A. Casagrande, estimate the maximum overburden pressure that has ever acted upon the soil represented by the sample in Prob. 2. Determine the compression index C_c obtained by means of eq. 2.24, and compare with that derived from the consolidation test.

4. From the data in Prob. 2, compute a_v and m_v for the increment of pressure from 0.5325 to 1.064 tons/sq ft.

5. The dial readings observed during the consolidation of the sample, Prob. 2, under the pressure of 1.064 tons/sq ft are tabulated on the next page.

Time, min	Dial Reading, in. × 10^{-4}	Time, min	Dial Reading, in. × 10^{-4}
0.00	396.0	15	549.0
0.10	427.0	30	594.7
0.25	436.0	60	636.8
0.50	443.8	135	670.7
1.0	453.5	240	689.9
2.0	466.0	1180	729.0
4.0	485.0	1600	737.0
8.0	511.8	2625	749.2

Plot the dial reading–time curve, and determine the readings corresponding to 0 and 100 per cent consolidation.

6. Compute the value of the coefficient of consolidation c_v and the coefficient of permeability k for the time-deformation curve plotted in Prob. 5.

7. The sample, Prob. 2, was taken from a layer of clay 12 ft thick, drained top and bottom. After how many days would this layer reach 30 per cent consolidation under a suddenly applied load of about 1 ton/sq ft?

8. If the layer of clay, Prob. 7, contained a thin freely draining parting of sand at a depth of 3 ft below its upper surface, how many days would be required for it to reach an average degree of consolidation equal to 30 per cent?

9. The time to reach 60 per cent consolidation is 32.5 sec for a sample 0.5 in. thick tested in the laboratory under conditions of double drainage. How long will the corresponding layer in nature require to reach the same degree of consolidation if it is 15 ft thick and drained on one side only?

10. A layer of clay beneath a building has consolidated and caused a settlement of 9.2 in. in 200 days since the weight of the building was added. According to the results of laboratory consolidation tests this corresponds to 30 per cent consolidation of the layer. Plot the time-settlement curve for the structure for the first 5 years of its existence.

Art. 2.9

1. It is believed that the shearing strength of a soil under certain conditions in the field will be governed by eq. 2.31, wherein $c = 1.6$ tons/sq ft and $\phi = 22°$. What minimum lateral pressure would be required to prevent failure of the soil at a given point if the vertical pressure were 4.5 tons/sq ft?

Art. 2.10

1. A sample of dry sand was tested in a direct shear device under a vertical pressure of 20 lb/sq in. on the horizontal plane of shear. Failure occurred at a shearing stress of 14 lb/sq in. Compute the angle of internal friction of the sand.

2. The sand in a deep natural deposit has an angle of internal friction of 40° in the dry state, a dry density of 110 lb/cu ft, and a saturated unit weight of 131 lb/cu ft. If the water table is at a depth of 20 ft, what is the shearing resistance of the material to sliding along a horizontal plane at a depth of 10 ft?

3. Compute the shearing resistance under the conditions specified in Prob. 2 if the water table is at the ground surface.

4. In a drained triaxial test on dense sand the all-around pressure was 1.5 tons/sq ft and the added vertical pressure to cause failure was 5.43 tons/sq ft.

Compute the angle of internal friction ϕ and the angle of inclination α of the failure planes, on the assumption that equation 2.31 is valid.

5. In a triaxial test on a saturated sample of dense sand, the sample was consolidated under an all-around pressure of 1.5 tons/sq ft. Further drainage was then prevented. During the addition of vertical load the porewater pressure in the sample was measured. At the instant of failure it amounted to 1.20 tons/sq ft. The added vertical pressure at this time was 1.45 tons/sq ft. What was the value of ϕ for the sand?

ART. 2.12

1. In an undrained triaxial test on a sample of saturated clay the all-around pressure is maintained at 1.0 ton/sq ft. The unconfined compressive strength is 3.62 tons/sq ft. At what vertical load in addition to the all-around pressure should the sample fail?

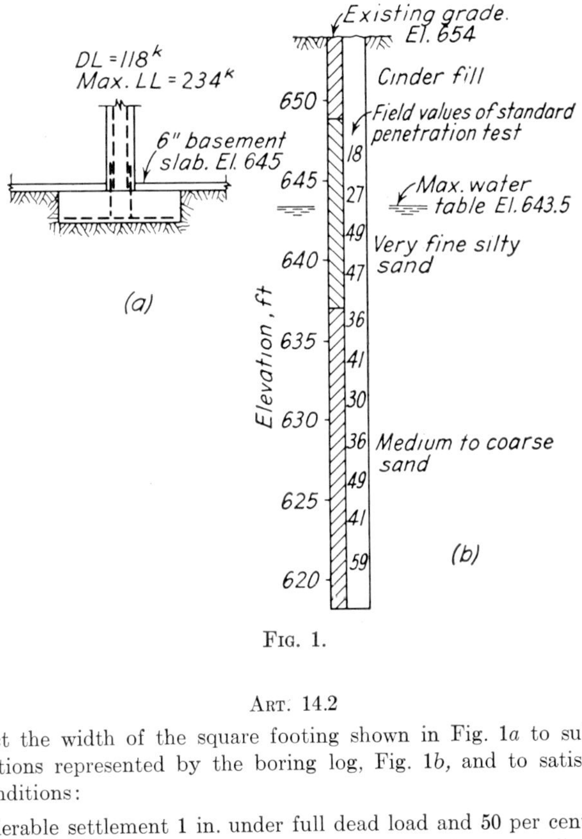

FIG. 1.

ART. 14.2

1. Select the width of the square footing shown in Fig. 1*a* to suit the subsoil conditions represented by the boring log, Fig. 1*b*, and to satisfy the following conditions:

(*a*) Tolerable settlement 1 in. under full dead load and 50 per cent live load.
(*b*) Factor of safety equal to 3 for dead load plus 50 per cent live load.
(*c*) Factor of safety equal to 2 for dead load plus maximum live load.

2. Estimate the settlement of the footing shown in Fig. 2. Compute the factor of safety against a bearing-capacity failure.

3. (*a*) If the footing shown in Fig. 3 is not to settle more than 1 in., what is the highest level to which the water table should rise with respect to the base of the footing? (*b*) What will be the factor of safety with respect to a bearing-capacity failure if the water table rises to the surface of the ground?

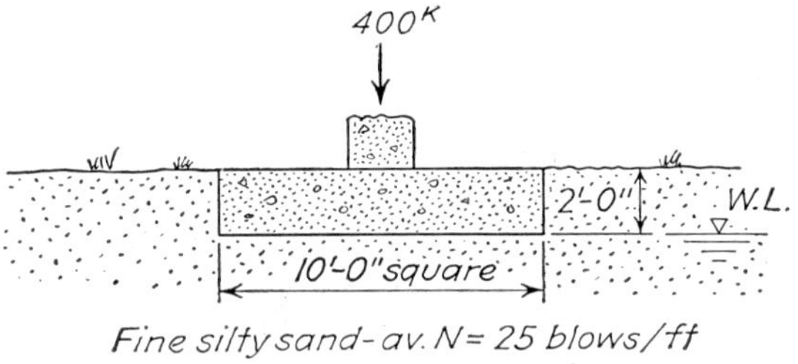

FIG. 2.

4. A footing 5 ft square and 1.5 ft thick is to be located on sand. The surface of the ground is at the top of the footing and the water table is 10 ft below the base. What should be the minimum N value of the sand if the footing must support a column load of 200 kips without (*a*) experiencing a settlement of more than 1 in. or (*b*) possessing a factor of safety of less than 3 against a bearing-capacity failure?

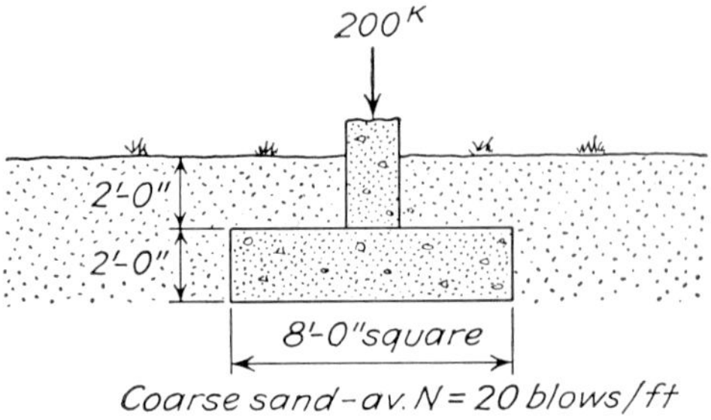

FIG. 3.

5. A footing 12 ft square and 2 ft thick is supported by sand with an average N value of 30 blows per foot. The surface of the ground is 3 ft above the top of the footing, and the water table is 4 ft below the base. Compute the maximum load that the footing can support if the settlement must not exceed 0.5 in.

ART. 14.3

1. A raft 30 ft wide and 50 ft long is to be supported 10 ft below the surface of the surrounding ground. The soil has been explored by means of the standard penetration test in four borings, each of which encountered rock at 40 ft. The borings indicate that the subsoil is a fairly uniform very fine sand. The minimum average penetration resistance in any boring between depths of 10 and 40 ft is 19 blows per foot. The water table is located 10 to 12 ft below the ground surface; hence, it can be assumed that the sand below the

base of the raft is completely submerged. If the structure is to have a basement, and if the soil to be excavated has a unit weight γ of 100 lb/cu ft, what is the maximum soil pressure that should be allowed at the base of the raft?

Art. 14.5

1. A caisson for a bridge pier is 20 ft wide and 56 ft long. It extends through soft clay and is supported on sand that has an N value of 50 blows per foot. The water table is considerably higher than the base of the pier. If the settlement is limited to 0.5 in., what total load can be supported by the foundation? Neglect the difference between the weight of the caisson and the weight of the clay that is removed during construction.

Art. 15.2

1. A continuous wall footing will rest upon a saturated clay that possesses an unconfined compressive strength of 1.1 tons/sq ft. At a load of 4.3 tons/foot of wall a factor of safety of 3 is required; in addition, the factor of safety should not be less than 2 if the footing should be subjected to a load of 5.8 tons/ft. Ignore the difference in the unit weights of concrete and clay, and determine the width of the footing.

2. Proportion a square footing to carry a column load of 160 tons at a factor of safety of 2.5. The clay beneath the footing has an unconfined compressive strength of 1.4 tons/sq ft.

3. A footing 10 ft square rests on clay that has an unconfined compressive strength of 1.2 tons/sq ft. If the factor of safety is not to be less than 2.5, what is the maximum column load that can be supported by the footing?

Art. 15.3

1. A building is to be supported on a reinforced-concrete raft covering an area of 46 x 69 ft. The subsoil is clay with an unconfined compressive strength of 0.8 ton/sq ft. The pressure on the soil, due to the weight of the building and the loads it will carry, will be 1.4 tons/sq ft at the base of the raft. If the unit weight of the excavated soil is 120 lb/cu ft, at what depth should the bottom of the raft be placed to provide a factor of safety of 3?

2. A raft 60 x 72 ft in plan has its base 10 ft below the surface of a deposit of clay with a unit weight of 120 lb/cu ft. The unconfined compressive strength of the clay is 0.75 ton/sq ft. The factor of safety against a bearing-capacity failure must be 3. What total weight of building plus foundation can safely be supported by the raft?

Art. 15.4

1. A group of friction piles is shown in Fig. 4. If a factor of safety of 3 is required, find the maximum load P as determined by

(*a*) The piles acting as individuals.
(*b*) The piles acting as a group.

2. A test pile with a diameter of 12 in. penetrates 30 ft of clay to a stratum of dense sand. The clay has an average unconfined compressive strength of 0.4 ton/sq ft. Estimate how much of the ultimate test load may be the result of skin friction in the clay deposit.

ART. 15.6

1. Estimate the settlement of the center of the tank, Fig. 5, due to consolidation of the 20-ft layer of clay.

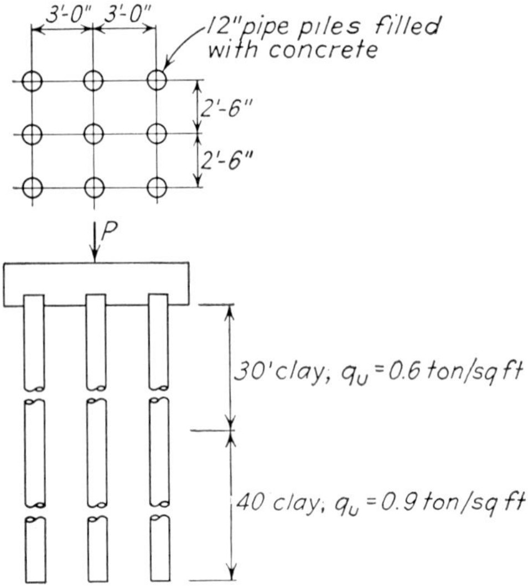

FIG. 4.

2. Estimate the settlement at the center and at a corner of the raft in DP C-5 if the thickness of the layer of sand is 46 ft instead of 36 ft. All other data are the same as those given on Sh. 1 of DP C-5.

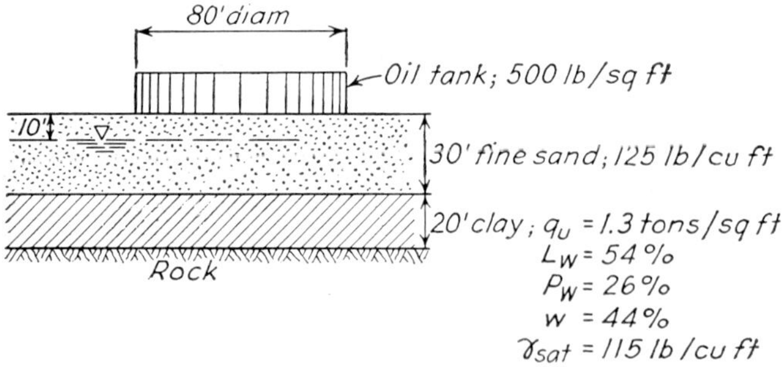

FIG. 5.

ART. 15.7

1. What is the approximate factor of safety against failure for a 30-ft fill, weighing 125 lb/cu ft, if it rests on clay that possesses an unconfined compressive strength of 1.5 tons/sq ft?

ART. 19.6

1. Design the interior footing, DP C-1a, on the basis of the data given in DP D-1 for the concrete and steel.

2. Design the rectangular footing shown in Fig. 6 on the basis of the following data for the concrete and steel: $f_c = 1125$ lb/sq in., $f_s = 18$ kips/sq in., $v = 75$ lb/sq in., $u = 200$ lb/sq in.

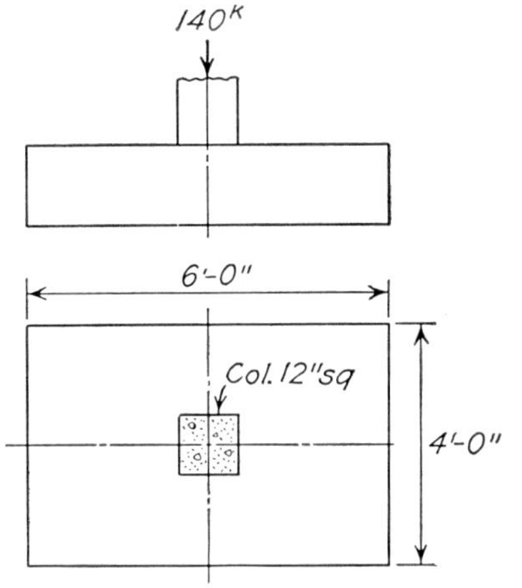

FIG. 6.

ART. 19.8

1. Redesign the pile footing, DP D-2, for a column load of 350 kips. The allowable load per pile is 40 kips. All other data are the same as those given in DP D-2.

2. Design the pile footing shown in Fig. 7 on the basis of the following data for the concrete and steel: $f_c = 1125$ lb/sq in., $f_s = 20$ kips/sq in., $v = 75$ lb/sq in., $u = 200$ lb/sq in.

ART. 20.4

1. If the vertical load on the footing shown in DP D-3 remains 242 kips, compute the maximum moment that can be resisted on the assumption that the area between the soil and the footing is entirely in compression. What is the magnitude of the maximum soil pressure for this condition of loading?

2. Compute the maximum soil pressure acting against the base of the footing shown in DP D-3 if the vertical column reaction is 180 kips and the moment is 360 ft-kips.

3. A vertical load of 135 kips is applied at the center of a footing 5×9 ft in plan.

(*a*) How much moment in the direction of the long axis can the footing resist before the soil pressure exceeds 5 kips/sq ft;

(*b*) How much moment in the same direction can the footing resist before the soil pressure exceeds 8 kips/sq ft?

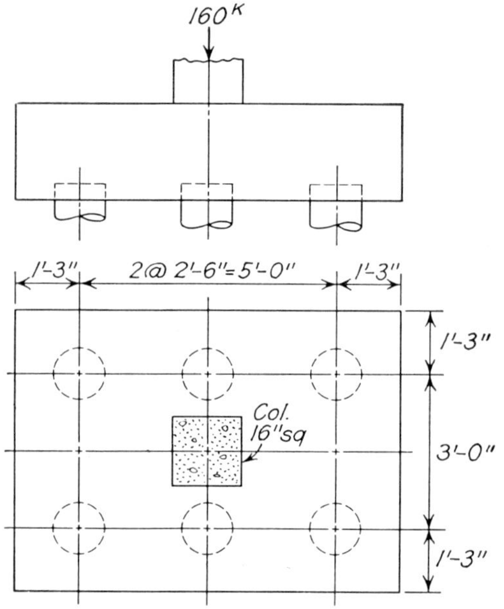

FIG. 7.

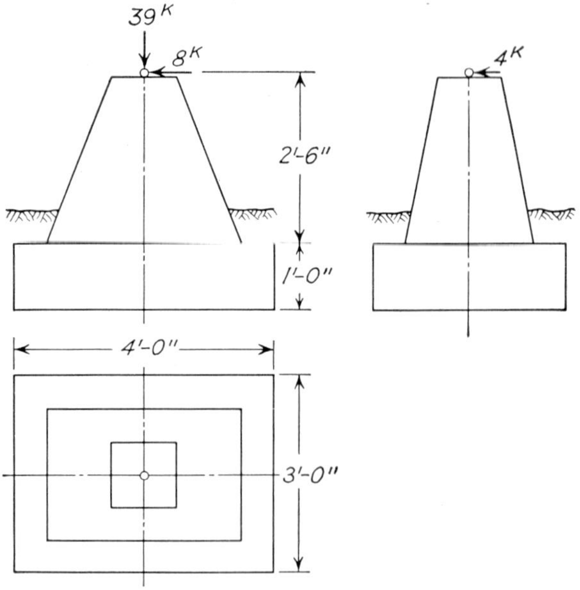

FIG. 8.

ART. 20.9

1. Compute the maximum and minimum soil pressures at the base of the footing shown in Fig. 8. The weight of the footing itself is 3.4 kips. Compute the magnitude of the pressures at the other two corners.

2. Compute each of the pile reactions beneath the footing shown in Fig. 20.6 if $\Sigma V = 162$ kips, $M_1 = 90$ ft-kips, and $M_2 = 236$ ft-kips. Neglect the weight of the footing itself. Check sum of pile reactions against ΣV.

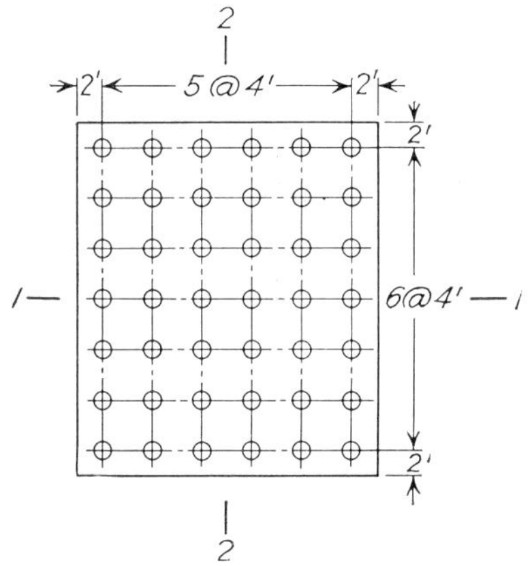

FIG. 9.

3. Substitute the pile foundation, Fig. 9, for the foundation of the bridge pier as shown in DP D-4, and compute the maximum and minimum pile reactions. Compute the maximum and minimum soil pressure beneath the base if the piles are omitted.

ART. 21.6

1. Given the following column loads:

	Col. *A*	Col. *B*
DL	94.4 kips	116.6 kips
Reduced LL	46.2	74.8
Maximum LL	98.8	160.4

Use the allowable soil pressures given in DP D-5a and

(*a*) Proportion a rectangular combined footing to support these column loads.

(*b*) Proportion a trapezoidal combined footing to support these column loads.

(*c*) Proportion a cantilever combined footing to support these column loads.

2. Complete the structural design of the footing proportioned in part *a* of the preceding problem. Use the general data given in DP D-5a.

ART. 22.5

1. Redesign the retaining wall, DP D-6, using group 1 backfill (see Fig. 22.2). Try a reduction in heel width from 5′ 6″ to 1′ 6″. All other data are the same as those given on Sh. 1 of DP D-6.

ART. 22.7

1. Compute the batter required in row 1 of the pile-supported retaining wall DP D-7, if the piles in row 2 are vertical.

2. Refer to the pile-supported retaining wall, DP D-7. Compute the pile reactions and the amount of batter required if the stem is shifted into such a position that the toe dimension is increased to 3′–0″ and the heel dimension is decreased to 4′–6″.

ART. 22.9

1. Check the stability of the abutment shown in DP D-8 if the backfill is group 2 soil (see Fig. 22.2). All other data are the same as those given in the design plate. If the resultant force on the base falls outside the middle third, make the necessary changes in the width of the base and in the position of the stem to maintain compression everywhere between the soil and the base.

Index